1st Online National Conference on Sustainable Material and Practices for Infrastructure Development (SMPID-2020)

Held online due to COVID-19

IOP Conference Series: Materials Science and Engineering
Volume 970

Odisha, India
17 - 18 September 2020

Editors:

Dilip Kumar Bagal
Jajati Keshari Naik
Gopal Charan Behera
Pradyut Kumar Muduli
Kishor Chandra Panda

ISBN: 978-1-7138-3295-9
ISSN: 1757-8981

Printed from e-media with permission by:

Curran Associates, Inc.
57 Morehouse Lane
Red Hook, NY 12571

Some format issues inherent in the e-media version may also appear in this print version.

Institute of Physics
Dirac House, Temple Back
Bristol BS1 6BE UK

Phone: 44 1 17 929 7481
Fax: 44 1 17 920 0979

techtracking@iop.org

Additional copies of this publication are available from:

Curran Associates, Inc.
57 Morehouse Lane
Red Hook, NY 12571 USA
Phone: 845-758-0400
Fax: 845-758-2633
Email: curran@proceedings.com
Web: www.proceedings.com

TABLE OF CONTENTS

STRUCTURAL ENGINEERING

TRANSPORTATION ENGINEERING/SUSTAINABLE PAVEMENT MATERIALS/DESIGN

ENVIRONMENTAL ENGINEERING/URBAN WATER TREATMENT AND SEWAGE SYSTEMS

PREFACE

With immense pleasure, I introduce the proceedings of the 1st Online National Conference on Sustainable Materials and Practices for Infrastructure Development (SMPID-2020) which was organized by Department of Civil Engineering, Government College of Engineering, Kalahandi, Bhawanipatna, Odisha, India in collaboration with Government College of Engineering, Jalgaon, Maharashtra, India on 17th & 18th September, 2020.

SMPID-2020 focused on emerging research and developments in infrastructure with emphasis on eco-friendly materials and practices. The infrastructures, which are being constructed today, will shape our lives tomorrow. The complex and diverse nature of the impacts due to climate change / weather extremes on civil infrastructures can be seen in our roadways, railways, bridges, flyovers, harbours, buildings etc. Extreme summer temperatures, flash floods, and rising numbers of freeze-thaw cycles pose challenges for civil infrastructure and can endanger public safety. Thus, civil infrastructures require constant attention, maintenance, and upgradation. Use of sustainable resources and design practices in various infrastructural sectors is the key to respond the impending crisis. Structural, geotechnical, environmental aspects along with transportation networks and water resources management are critical areas to be considered. SMPID-2020 provided a forum for the dissemination of latest technical information, publication of the high-quality research results and debate on the new trends in the area of infrastructure with lower ecological footprint.

The conference program witnessed keynote lectures and oral presentations delivered by renowned scholars working in the areas of civil and its associative field from all over India. To enrich the quality of program, High quality and cutting edge research papers have been selected among the received and presented work. Keeping the COVID-19 pandemic situation in mind, the conference was swiftly held in online platform with the help of Cisco Webex Web conferencing and Videoconferencing platform by taking all the safety measures.

Our conference committee and the authors were committed to support and made the conference successful. All accepted papers were subject to peer-review by conference committee and invited reviewers. The papers were selected based on their quality and their relevance to the conference. Off course, Plagiarism platforms were employed for monitoring the authenticity and quality of the presented research articles.

I would like to express special thanks to all the members of the conference committee. I would also like to thank the reviewers for sharing their valuable time and advice which helped in improving the quality of the papers selected for presentation and publication in the proceedings. Finally, I want to thank all the authors and other conference participants for their unconditional support of SMPID-2020.

Prof. (Dr.) Dulu Patnaik
Patron, SMPID-2020
Principal,
Government College of Engineering, Kalahandi,
Bhawanipatna, Odisha, India

IOP Conf. Series: Materials Science and Engineering **970** (2020) 011002 doi:10.1088/1757-899X/970/1/011002

COMMITTEES OF SMPID-2020

Chief Patron

Prof. (Dr.) C. R. Tripathy
Vice Chancellor,
Biju Patnaik University of Technology, Odisha, India

Patron

Prof. (Dr.) D. Patnaik
Principal,
Government College of Engineering, Kalahandi, Bhawanipatna, Odisha, India

Prof. (Dr.) R. D. Kokate
Principal,
Government College of Engineering, Jalgaon, Maharashtra, India

Chairman

Prof. (Dr.) K. C. Panda
Professor & Head,
Department of Civil Engineering,
Government College of Engineering, Kalahandi, Bhawanipatna, Odisha, India

Convenor

Dr. G. C. Behera* **Dr. P. K. Muduli***
*Associate Professor,
Department of Civil Engineering,
Government College of Engineering, Kalahandi, Bhawanipatna, Odisha, India

Co-Convenor

Dr. K. P. Sethy* **Prof. J. K. Naik*** **Prof. S. Panda*** **Prof. S. K. Sahu***
*Assistant Professor,
Department of Civil Engineering,
Government College of Engineering, Kalahandi, Bhawanipatna, Odisha, India

Publishing Committee

Prof. D. K. Bagal
Assistant Professor,
Department of Mechanical Engineering,
Government College of Engineering, Kalahandi, Bhawanipatna, Odisha, India

Prof. J. K. Naik
Assistant Professor,
Department of Civil Engineering,
Government College of Engineering, Kalahandi, Bhawanipatna, Odisha, India

Finance Committee

Prof. J. K. Naik* **Prof. S. K. Sahu*** **Prof. P. Harish*** **Prof. N. Gupta***
*Assistant Professor,
Department of Civil Engineering,
Government College of Engineering, Kalahandi, Bhawanipatna, Odisha, India

Advisory Committee

Prof. (Dr.) S. K. Barai
Director, BITS Pilani, Rajasthan, India

Prof. (Dr.) A. Mishra
State Project Administrator, SPIU, Odisha, India

Prof. (Dr.) S. K. Dash
Professor, Department of Civil Engineering, IIT Kharagpur, West Bengal, India

Prof. (Dr.) S. K. Das
Professor & Head, Department of Civil Engineering, IIT(ISM) Dhanbad, Jharkhand, India

IOP Conf. Series: Materials Science and Engineering **970** (2020) 011002 doi:10.1088/1757-899X/970/1/011002

Prof. (Dr.) S. B. Singh
Professor, Department of Civil Engineering, BITS Pilani, Rajasthan, India
Prof. (Dr.) C. B. K. Rao
Professor, Department of Civil Engineering, NIT Warangal, Telangana, India
Prof. (Dr.) P. K. Parhi
Professor, Department of Civil Engineering, CET Bhubaneswar, Odisha, India
Prof. (Dr.) S. S. Pusadkar
Professor & Head, Department of Civil Engineering, GCE, Jalgaon, Maharashtra, India
Prof. (Dr.) T. Sahu
Principal, GCE, Keonjhar, Odisha, India
Prof. (Dr.) T. K. Nath
Professor & Head, Department of Civil Engineering, IGIT Sarang, Odisha, India
Prof. (Dr.) S. S. Das
Professor, Department of Civil Engineering, VSSUT Burla, Odisha, India
Prof. (Dr.) B. Panda* **Prof. (Dr.) P. Pani***
Professor, Department of Civil Engineering, IGIT Sarang, Odisha, India
Prof. (Dr.) C. R. Mohanty
Professor, Department of Civil Engineering, PMEC Berhampur, Odisha, India
Dr. R. P. Nanda
Associate Professor, Department of Civil Engineering, NIT Durgapur, West Bengal, India
Dr. B. B. Das
Associate Professor, Department of Civil Engineering, NIT Surathkal, Karnataka, India
Dr. S. R. Dash
Assistant. Professor, Department of Civil Engineering, IIT Bhubaneswar, Odisha, India

Technical Committee

Prof. (Dr.) T. D. G. Rao
Professor, Department of Civil Engineering, NIT, Warangal, Telangana, India
Prof. (Dr.) S. K. Patra
Professor, Department of Civil Engineering, VSSUT, Burla, Odisha, India
Dr. S. Sahoo
Associate Professor, Department of Civil Engineering, NIT, Meghalaya, India
Dr. G. K. Pothal
Associate Professor, Department of Civil Engineering, IGIT, Sarang, Odisha, India
Dr. R. K. Panigrahi
Associate Professor, Department of Civil Engineering, VSSUT, Burla, Odisha, India
Dr. D. Giri
Associate Professor, Department of Civil Engineering, VSSUT, Burla, Odisha, India
Dr. R. N. Behera
Assistant. Professor, Department of Civil Engineering, NIT, Rourkela, Odisha, India
Dr. T. Jena
Assistant. Professor, Department of Civil Engineering, ITER, Bhubaneswar, Odisha, India

Organizing Members

Dr. C. R. Dash
Head of Department,
Department of Basic Science, GCE, Kalahandi, Bhawanipatna, Odisha, India
Dr. M. J. Sable
TEQIP-III Coordinator, Government College of Engineering, Jalgaon, Maharashtra, India
Dr. D. P. Dash
Head of Department,
Department of Electrical Engineering, GCE, Kalahandi, Bhawanipatna, Odisha, India
Dr. M. Rout
Head of Department,
Department of Mechanical Engineering, GCE, Kalahandi, Bhawanipatna, Odisha, India
Prof. D. R. Nayak
Head of Department,
Department of Computer Science and Engineering, GCE, Kalahandi, Bhawanipatna, Odisha, India

Dr. Priya S.
Head of Department,
Department of Humanities, GCE, Kalahandi, Bhawanipatna, Odisha, India
Dr. R. K. Swain* **Dr. A. K. Baliarsingh***
*Associate Professor,
Department of Electrical Engineering, GCE, Kalahandi, Bhawanipatna, Odisha, India
Prof. M. K. Nath
*Assistant Professor,
Department of Electrical Engineering, GCE, Kalahandi, Bhawanipatna, Odisha, India
Dr. L. P. Panda* **Dr. S. Patnaik***
Assistant Professor,
Department of Humanities, GCE, Kalahandi, Bhawanipatna, Odisha, India
Prof. A. K. Patnaik
Assistant Professor,
Department of Mechanical Engineering, GCE, Kalahandi, Bhawanipatna, Odisha, India
Prof. S. Sahoo* **Prof. M. Priyadarshini*** **Prof. M. Patnaik*** **Prof. V. M. Reddy***
*Assistant Professor,
Department of Civil Engineering, GCE, Kalahandi, Bhawanipatna, Odisha, India

Technical Session Chair Committee

Dr. B. Basa
Assistant Professor, Department of Civil Engineering, ITER, SOA University, Bhubaneswar, Odisha India
Dr. R. P. Nanda
Associate Professor, Department of Civil Engineering, NIT Durgapur, West Bengal, India
Prof. (Dr.) S. S. Das
Professor, Department of Civil Engineering, VSSUT, Burla, Odisha, India
Dr. D. Giri
Associate Professor, Department of Civil Engineering, VSSUT, Burla, Odisha, India

Guest of Honour

Dr. A. K. Nayak
Joint Secretary to Government, Skill Development & Technical Education, Govt. of Odisha, India
Prof. (Dr.) A. Mishra
State Project Administrator, SPIU, Odisha, India
Prof. (Dr.) S. K. Barai
Director, BITS Pilani, Rajasthan, India
Mr. S. A. Noor
Commercial & Contracts Manager, United Mechanical Electromechanical LLC, Dubai, UAE
Mr. A. P. Mishra
Vice-President & Head, Commercial Contracts & Regulatory, GMR Group, Energy Sector, India

Keynote Speaker

Prof. (Dr.) S. K. Barai
Director, BITS Pilani, Rajasthan, India
Prof. (Dr.) C. B. K. Rao
Professor, Department of Civil Engineering, NIT Warangal, Telangana, India
Prof. (Dr.) S. K. Das
Professor & Head, Department of Civil Engineering, IIT(ISM) Dhanbad, Jharkhand, India
Dr. R. P. Nanda
Associate Professor, Department of Civil Engineering, NIT Durgapur, West Bengal, India
Dr. B. B. Das
Associate Professor, Department of Civil Engineering, NIT Surathkal, Karnataka, India

IOP Conf. Series: Materials Science and Engineering **970** (2020) 011003 doi:10.1088/1757-899X/970/1/011003

EDITORS OF SMPID-2020

Prof. DILIP KUMAR BAGAL
Assistant Professor,
Department of Mechanical Engineering,
Government College of Engineering, Kalahandi, Bhawanipatna, Odisha, India
Email: dilipbagal90@gmail.com; dkbagal@gcekbpatna.ac.in

Prof. JAJATI KESHARI NAIK
Assistant Professor,
Department of Civil Engineering,
Government College of Engineering, Kalahandi, Bhawanipatna, Odisha, India
Email: jajatinaik007@gmail.com

Dr. GOPAL CHARAN BEHERA
Associate Professor,
Department of Civil Engineering,
Government College of Engineering, Kalahandi, Bhawanipatna, Odisha, India
Email: beheragb@gmail.com

Dr. PRADYUT KUMAR MUDULI
Associate Professor,
Department of Civil Engineering,
Government College of Engineering, Kalahandi, Bhawanipatna, Odisha, India
Email: pmuduli@gcekbpatna.ac.in

Prof. (Dr.) KISHOR CHANDRA PANDA
Professor and Head,
Department of Civil Engineering,
Government College of Engineering, Kalahandi, Bhawanipatna, Odisha, India
Email: kishoriit@gmail.com

IOP Conf. Series: Materials Science and Engineering **970** (2020) 011004 doi:10.1088/1757-899X/970/1/011004

1st Online National Conference on Sustainable Materials and Practices for Infrastructure Development (SMPID-2020)

Organized by Department of Civil Engineering, Government College of Engineering, Kalahandi, Bhawanipatna, Odisha, India in collaboration with Government College of Engineering, Jalgaon, Maharashtra, India on 17th & 18th September, 2020

Sponsored by:

TEQIP-III

IOP Conf. Series: Materials Science and Engineering **970** (2020) 011005 doi:10.1088/1757-899X/970/1/011005

Peer Review Declaration

All papers published in this volume of ***IOP Conference Series: Materials Science and Engineering*** have been peer reviewed through processes administered by the Editors. Reviews of the professional and scientific standards required by the proceedings journal published by IOP Publishing have been carried out by expert referees.

- **Type of peer review:** Double-blind
- **Conference submission management system:** Via mail to smpid2020@gmail.com
- **Number of submissions received:** 51
- **Number of submissions sent for review:** 45
- **Number of submissions accepted:** 35
- **Acceptance Rate (Number of Submissions Accepted / Number of Submissions Received X 100):** 68.6%
- **Average number of reviews per paper:** 3
- **Total number of reviewers involved:** 17
- **Any additional info on review process:**

Originality, novelty, technological soundness, applicability, and lucidity of presentation and relevance prior to their selection for publication in the proceedings will be re-examined in the papers adopted in the entire study.

Contact person for queries:

Prof. Dilip Kumar Bagal
Email: dkbagal@gcekbpatna.ac.in, dilipbagal90@gmail.com

IOP Conf. Series: Materials Science and Engineering **970** (2020) 012001 doi:10.1088/1757-899X/970/1/012001

Evaluation of Strength Deterioration Factor on Blended Concrete Exposed to Magnesium Sulphate Solution

Trilochan Jena[1], Kishor Chandra Panda[2], Chittaranjan Panda[1]

[1]Department of Civil Engineering, ITER, SOA (Deemed to be University), Bhubaneswar, Odisha, India

[2]Department of Civil Engineering, Government College of Engineering, Kalahandi, Bhawanipatna, Odisha, India

E-mail: trilochanjena@soa.ac.in

Abstract. This paper reports the combined effects of industrial waste as fly ash (FA) and agro-waste such as silpozz on the compressive strength and strength deterioration factor (SDF) of blended cement concrete exposed to 5% and 10% MgSO4 solution and normal water. Silpozzis manufactured by burning of agro-waste rice husk in designed furnace in between 600° to 700°C which is one of the main agriculture.al residues obtained from the outer covering of rice grains during the milling process. There are four mix series taken with control mix. The control mix made 0% replacement of FA and silpozz with Ordinary Portland Cement (OPC). The first mix series made 0% FA and 10-30% replacement of silpozz with OPC. The second mix series made with 10% FA and 10-40% replacement of silpozz with OPC. The third mix series made 20% FA and 10-30% replacement of silpozz with OPC and the fourth mix series made 30% FA and 10-20% silpozz replaced with OPC. The samples (cubes) are prepared and cured in normal water and 5% and 10% MgSO4 solution for 7, 28 and 90 days. The studied parameters are compressive strength and strength deterioration factor (SDF) for 7, 28 and 90 days. The regression equations and value of correlation coefficient R2 between compressive strength and partial replacement of silpozz is obtained and revealed from these results that value of correlation coefficient is varied from 0.9602 to 1.

1. Introduction

Durability of concrete is completely dependable on amount of moisture and its rate of diffusion and also other dangerous chemicals which may enter into the concrete from the outer environment. The corrosion formation is completely because of diffusion of sulphate ions from the cover or outer layer of concrete to its inner implanted reinforcing bars is normally considered as one of the direct process of deterioration in reinforcement of concrete structures. Normally, there are various phases of damages in concrete structures with reinforcement due to corrosion but among them only three phases are considered. The first phase is influenced by the perforation of sulphide, during which sulphide ions are said to be dispersed throughout the concrete and reaches toward the bars of reinforcement. When the sulphate concentration increases and its critical value are reached inside the concrete then the steel begins to corrode and the first phase hereby ends. In the second stage, there is almost formation of rust between reinforcement steel bars and concrete due to which expansion is caused [1-10].

Some of the researchers have reported the utilization of supplementary cementitious materials to improve concrete properties in marine climate. Jena and Panda [7] studied the development of compressive strength in blended concrete made with silpozzand fly ash which is used as a partial replacement of cement to improve the durability of marine structures. Jena and Panda [6] studied the mechanical properties on concrete blended with silpozz which is can be used as an effective material and substitute of silica fume (SF) improves the structural durability. Anwar and Roushdi [5] found the properties of concrete blended with PC, FA and SF for improvements in concrete to resist deterioration due to environmental effects. Andrade et al. [4] have conducted accelerated chloride migration test and done the modeling of chloride penetration into concrete to analyze chloride content in concrete exposed to saline climate. Yildirim et al. [10] analyzed on the effect of types of cement on chloride diffusion for the given compressive strength and the cement blended with blast-furnace slag have better resistance against chloride diffusion than OPC and sulfate resisting cement. Wegian [9] investigated the strength properties such as the compressive, tensile, flexural and bond strengths of concrete mixing and curing with sea water and the strength reduction factor increases with exposure duration. Aburawi and Swamy [3] reported the effects of salt weathering and exposure to curing on the properties of high durability concretes containing chlorides. This article focuses the study of SDF of pre-cast concrete samples with FA and silpozz in magnesium sulphate solution for 90 days curing period. Sunil [8] investigated the properties of precast plain and blended cement concrete made with FA were cured in seawater for a period of 1 year and the precast concrete samples performed better than the cast in situ samples.

2. Materials and Methods

The materials such as OPC 43 grade, FA, Silpozz, fine aggregate, coarse aggregate, normal water, magnesium sulphate solution and super plasticizer are used in this study. The experimental physical properties of OPC such as initial setting time 160 min, final setting time 360 min, standard consistency 32%, fineness 333 m^2/kg and specific gravity 3.15 are determined confirming to IS 8112:1989.

Table 1. Chemical and physical properties of cement.

Oxides (%)	Cement (OPC)	Silpozz	FA
SiO_2	20.99	88.18	58.13
Al_2O_3	6.05	1.61	31.00
Fe_2O_3	6.01	0.56	4.10
Carbon	-	2.67	-
CaO	62.74	1.59	0.60
MgO	1.33	1.63	0.10
K_2O	0.40	1.67	0.90
Na_2O	0.04	-	0.05
SO_3	1.82	-	0.12
TiO_2	.025	-	1.63
Others	-	2.09	0.011
Moisture content (%)	-	0.79	3.0
Loss on ignition (%)	1.14	0.04	0.29
Physical properties			
Bulk Density (gm/cc)	1.43	0.23	1.2
Specific gravity	3.15	2.3	2.12
Particle size (Micron)	35	25	34
Specific surface, m^2/g	0.33	17	33
Color	Gray	Gray black	Gray

The physical properties of coarse aggregates obtained as per IS: 383-1970 such as fineness modulus 7.0, specific gravity 2.86, water absorption 0.2% and crushing value 24% as well as the properties of fine aggregates such as fineness modulus 3.03 (Zone-III), water absorption 0.6% and specific gravity 2.68. CERA HYPERPLAST XR-W40 high end super plasticizers are used [11-13]. The chemical and physical properties of FA and silpozz given by the supplier are shown in Table 1.

2.1. Preparation of $MgSO_4$ solution

Magnesium sulfate is a complex chemical compound simply containing magnesium, sulphur and oxygen ($MgSO_4$). It is of two types: heptahydrate and monohydrate. Heptahydrate ($MgSO_4.7H_2O$) is commonly known as Epsom salt or sulfate mineral epos mite which usually comes in crystal form. The monohydrate ($MgSO_4.H_2O$) is normally found in mineral kieserite. It comes in powder form. In this work, heptahydrate is being used to form magnesium sulfate solution. We have taken 5 gm and 10 gm $MgSO_4$ in 100 ml of water to prepare 5% and 10% $MgSO_4$ solution respectively. The following equation represents how to prepare the $MgSO_4$ solution. The samples such as cubes, cylinders and prisms are cured in sulphate solution which is shown in Figure 1.

$$[MgSO_4\,(5gm) + H_2O\,(100\ ml) = 5\%\ MgSO_4\ solution] \quad (1)$$

$$[MgSO_4\,(10gm) + H_2O\,(100\ ml) = 10\%\ MgSO_4\ solution] \quad (2)$$

3. Results and Discussions

3.1 SDF (%) obtained from compressive strength in 5% and 10% $MgSO_4$ solution curing

The compressive strength test was performed on 150 mm cube specimens at the age of 7, 28 and 90 days for NWC, 5% and 10% $MgSO_4$ solution curing samples. The SDF obtained from compressive strength of 5% and 10% $MgSO_4$ solution curing samples are shown in Figure 1 - 2 respectively. It is observed from Figure 1 that the minimum SDF from 5% $MgSO_4$ solution is 0.18%, 0.35% and 0.74% at the age of 7, 28 and 90 days for the sample M1C80F10S10 respectively.

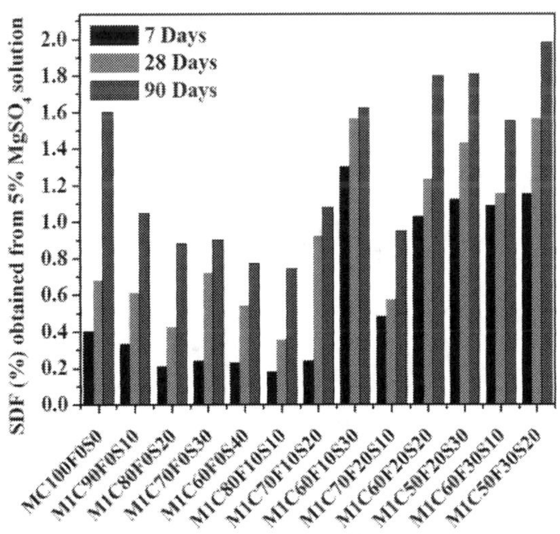

Figure 1. SDF obtained from 5% $MgSO_4$ solution verses Concrete Mix

The maximum SDF from 5% MgSO4 solution is 1.15%, 1.56% and 1.98% at the age of 7, 28 and 90 days for the sample M1C50F30S20 respectively. The maximum SDF from 10% $MgSO_4$ solution is 2.18%, 2.30% and 2.37% at the age of 7, 28 and 90 days for the sample M1C50F30S20 respectively. The 7, 28 and 90 days SDF in 5% and 10% $MgSO_4$ solution verses concrete mix is shown in Figure 3-

5 respectively. It observed from Figure 3 that the SDF obtained from 10% $MgSO_4$ solution-curing samples are more compared to 5% $MgSO_4$ solution curing samples including control mix. Similar trend is found from Figures 10 and 11 for 28 and 90 days respectively. The rate of SDF is studied more in between 7 to 28 days, but 28 to 90 days, the rate of SDF is slowly processed as compared to first 28 days. For each case, the silpozz-based specimens are giving low SDF as compared to FA based samples. The combined replacement of 10% FA and upto 20% silpozz with OPC showed optimal resistance against sulphate attack. Even without FA and upto 40% silpozz replaced with OPC also performed better in sulphate solution.

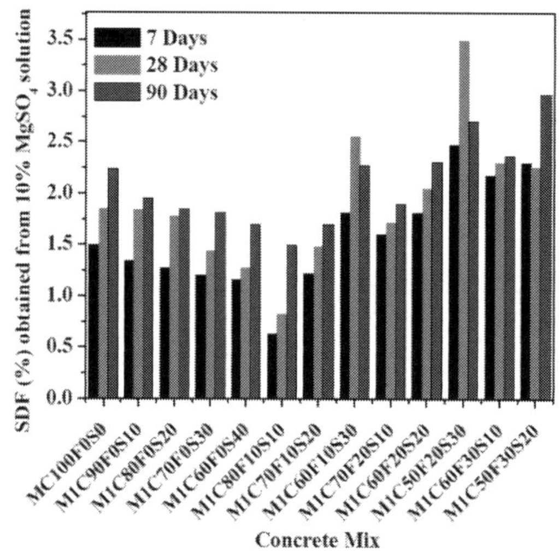

Figure 2. SDF obtained from 10% $MgSO_4$ solution verses Concrete Mix

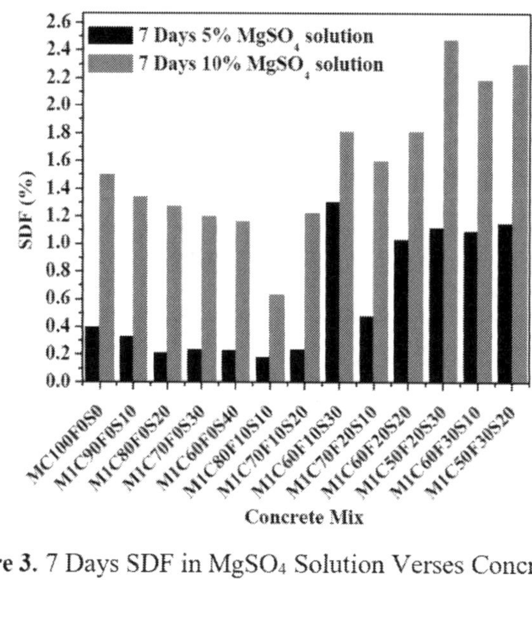

Figure 3. 7 Days SDF in $MgSO_4$ Solution Verses Concrete Mix

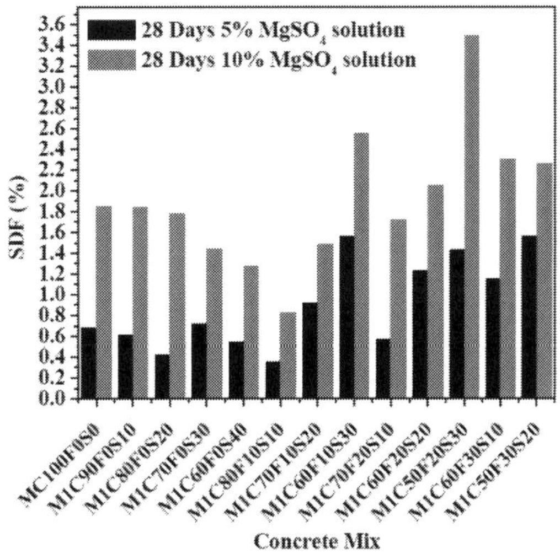

Figure 4. 28 Days SDF in $MgSO_4$ Solution Verses Concrete Mix

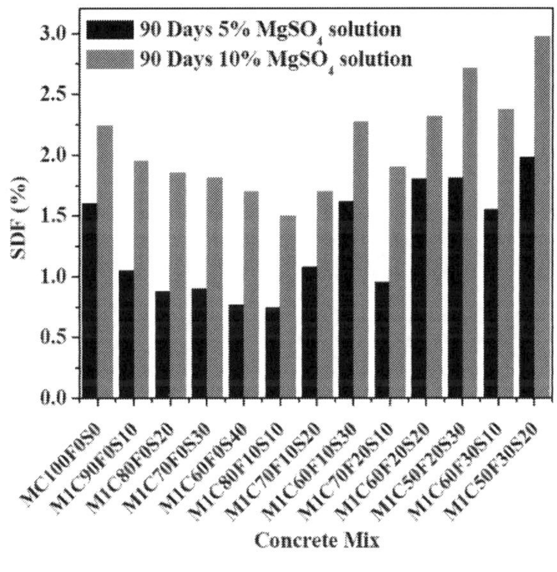

Figure 5. 90 Days SDF in $MgSO_4$ Solution Verses Concrete Mix

3.2 Compressive Strength Verses Partial Replacement of Silpozz

The regression equations and value of correlation coefficient R^2 between compressive strength and partial replacement of silpozz for NWC samples at the age of 28 and 90 are shown in Figure 6-7. As evident from these results that value of correlation coefficient is varied from 0.9602 to 1. The high values of correlation coefficient indicate that there is strong relationship for compressive strength and replacement of silpozz in percentage. From the study, it is observed that the limitation of SP varies from 0.2 to 0.6% for 0% FA with silpozz 10% to 40%. For 10% FA with 10% to 40% silpozz, the range of SP varies from 0.22 to 0.72% and for 20% FA with 10-30% silpozz, SP range is 0.25 to 0.56%. The correlation factor is very high and it means that, inclusion of silpozz in concrete gives better affect to its strength properties. It makes the matrix dense due to its particle fineness. Due to this, the quality of concrete in term of its density, homogeneity and lack of imperfections improved.

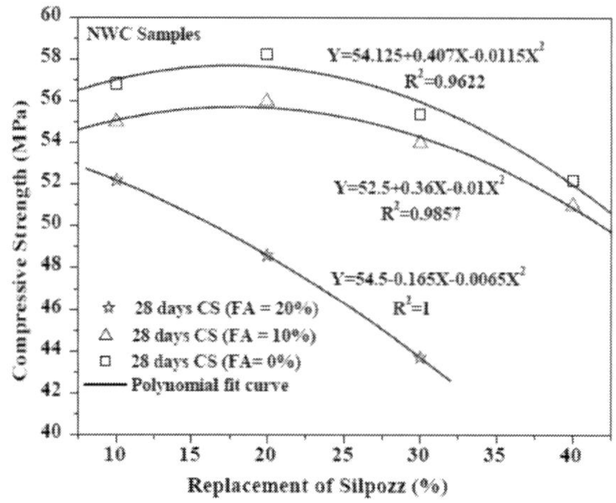

Figure 6. Replacement of Silpozz verses 28 Days Compressive Strength (FA 0-20 series)

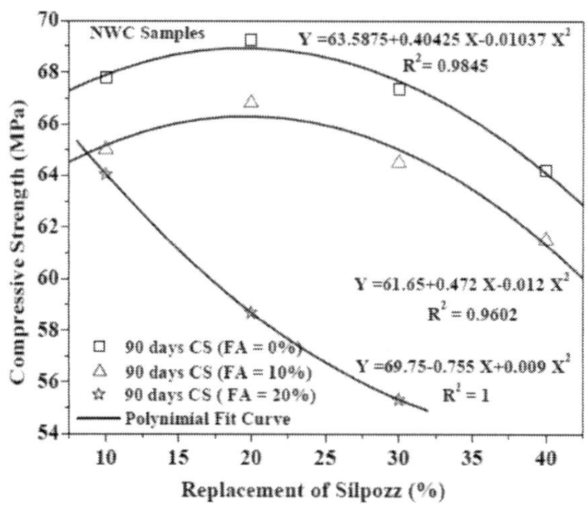

Figure 7. Replacement of Silpozz verses 90 Days Compressive Strength (FA 0-20 series)

4 Conclusion

It may be concluded from this study that the 10% FA and 10-20% silpozz replacement with OPC contributes higher compressive strength as compared to control mix cured in normal water, 5% and 10% $MgSO_4$ solution. When the percentage replacement of FA and silpozz increases more than 40% with OPC the strength properties decreases with respect to control sample for all types of curing specimens. It may be recommended that a suitable combination of silpozz and fly ash show satisfactory performance in both fresh and hardened concrete with proper doses of super plasticizer. It is predicted from regression analysis that good correlation obtained for strength properties of concrete and the partial replacements of silpozz varies from 10-40% with 0% FA and 10% FA series and 20% FA with 10-30% silpozz series. The addition of SCM such as FA and silpozz with super plasticizer enhances the homogeneity of cement paste and densify the microstructure of the cement considerably as well as improves the strength and durability properties of concrete in sulphate environment.

References

[1] Jena T and Panda K C 2015 Effect of fly ash and silpozz on strength and durability properties of concrete in sea water *Indian Journal of Science Technology* **8** 1-7

[2] Jena T and Panda K C 2015 Influence of sea water on strength and durability properties of concrete *Advances in structural engineering* **3** 1863-73
[3] Aburawi M F 2008 Effect of sea water for mixing and curing on structural concrete *The IES journal part A, Civil and Structural Engineering* **3** 235-43
[4] Andrade C, Prieto M, Tanner P, Tavares F and d'Andrea R 2013 Testing and modelling chloride penetration into concrete *Journal of Construction Building Materials* **39** 9-18
[5] Anwar M and Roushdi M 2013 Improved concrete properties to resist saline water using environmental by-product *Water Science Journal* **27** 30-8
[6] Jena T and Panda K 2018 Mechanical and Durability Properties of Marine Concrete using Fly Ash and Silpozz *Advancecs in Concrete Construction* **6** 47-68
[7] Jena T and Panda K C 2020 Effect of magnesium sulphate solution on compressive strength and sorptivity of blended concrete *Advances in concrete construction* **9** 267-78
[8] Kumar S 2000 Influence of water quality on the strength of plain and blended cement concretes in marine environments *J Cement Concrete Research* **30** 345-50
[9] Wegian F M 2010 Effect of seawater for mixing and curing on structural concrete *The IES Journal Part A: Civil Structural Engineering* **3** 235-43
[10] Yildirim H, Ilica T and Sengul O 2011 Effect of cement type on the resistance of concrete against chloride penetration *J Construction Building Materials* **25** 1282-8
[11] 1989 IS: 8112-1989, 43 grade OPC specifications (first revision), Bureau of Indian Standards, New Delhi, India
[12] 1970 IS: 383-1970, Indian Standard Specification for course and fine aggregates from natural sources for concrete, (second revision), Bureau of Indian Standards, New Delhi, India
[13] 2009 IS: 10262-2009, Guide lines for concrete mix design proportioning, Bureau of Indian Standards, New Delhi, India

Recent Developments and Future Prospects of Manufacturing of Broad Gauge Railway Sleepers Using Waste Materials in India

Dilip Kumar Bagal[1], Sangam Kumar Singh[2], Bibekananda Naik[3], Abhishek Barua[4], Siddharth Jeet[4], Biswajit Parida[4], Ajit Kumar Pattanaik[1]

[1] Department of Mechanical Engineering, Government College of Engineering, Kalahandi, Bhawanipatna, Odisha, India
[2] Department of Mechanical Engineering, Delhi Technical Campus, Bahadurgarh, Haryana, India
[3] Department of Civil Engineering, Bhubaneswar Institute of Technology, Bhubaneswar, Odisha, India
[4] Department of Mechanical Engineering, Centre for Advanced Post Graduate Studies, BPUT, Rourkela, Odisha, India

E-mail: dilipbagal90@gmail.com

Abstract. Railways form the backbone of all economies, transporting goods, and passengers alike. Sleepers play a pivotal role in track performance and safety in rail transport. Composite sleeper is becoming a suitable alternative for replacing the existing concrete, steel and particularly timber sleeper in both mainline and heavy haul rail network. Composite sleeper technologies are already available but they have gained limited acceptance by the railway industry. A number of composite railway sleeper technologies have been developed but their applications in rail tracks are still limited. This paper discusses in brief about the materials that have been used in making sleepers in the early stages of railways and rigorously reviews the recent developments on composite sleepers and identifies the critical barriers to their widespread acceptance and applications. It has been shown that majority of the sleepers do not last till their expected design life resulting in massive replacement and repair cost. Currently the composite sleeper technologies that are available ranges from sleepers made with recycle materials starting from plastic, rubber, construction waste etc. While these recycled material based sleepers are low cost, the major challenges of using this type of sleepers are their limited strength, stiffness and dynamic properties which in most cases, are incompatible with those of traditional concrete, steel or timber sleeper. This paper rigorously reviews the recent developments on composite sleepers and identifies the critical barriers to their widespread acceptance and applications.

1. Introduction

Conventional materials used in production of railroad sleepers are timber, concrete and steel with a span of service life 20, 50 and 50 years, individually [1–23]. During the most recent couple of many years, the railroad business has zeroed in on a concrete based cement as opposed to wood and steel sleepers. Mono-block prestressed solid sleepers are utilized in hefty take and rapid rail track developments all through the world. In Indian, pre-tensioned prestressed concrete sleepers are used in almost every region where the railway tracks have been laid. Though few sections of metre gauge tracks present in Rajasthan, Gujarat, Uttar Pradesh; narrow gauge railway track of Darjeeling Himalayan Railway, West Bengal; Kalka-Shimla Railway, Himachal Pradesh; Matheran Hill Railway, Maharashtra; Nilgiri Mountain Railway, Tamil Nadu; Kangra Valley Railway, Himachal Pradesh; steel girder bridges; diamond-crossings and complex layout still uses timber or steel railway sleepers. Figure 1 shows different types of railway sleepers.

Figure 1. Timber sleeper, steel sleeper and prestressed concrete sleeper

Besides, cement and steel sleepers require extraordinary latches and can't supplant wood ones out of a current track in view of their inconsistent conduct. All the previously mentioned issues have inspired scientists around the globe to create and research new and compelling elective sleeper advances for railroad industry. An ongoing report on the possible reasons for failures of railroad sleepers demonstrated that the conventional materials have not acceptably fulfilled the need necessities to oppose mechanical, organic and compound corruption. These days, the worldwide market for composites is quickly expanding a direct result of the numerous points of interest including high solidarity to-weight proportion, amazing obstruction against erosion, dampness and creepy crawlies, and warm and electrical non-conductivity. This paper gives a review of ongoing improvements of composite railroad sleepers and their constraints, and recommends an answer which beats the difficulties characteristic in their usage and acknowledgment.

2. Customary materials for sleepers with major failures

Numerous materials have been being used according to the overarching conditions in various pieces of the world. A portion of the materials generally utilized for railroad sleepers are recorded beneath.

2.1. Timber

Timber sleepers have been utilized in railroads for quite a while and have been solid and proficient. Utilization of wood likewise prompts deforestation and consequently the ecological effect is extremely high. Other than the utilization of synthetically impregnated lumber additionally present removal chances. These unfavourable impacts have brought about the utilization of different materials [4].

2.1.1 Failure occurs in timber sleeper

a) Fungal decay

Fungal deterioration is the dominating method of lumber sleeper failure as wood is powerless to bio-weakening from numerous miniature creatures since lumber is a natural material. Railroad sleepers can ingest dampness, which can spread starting with one sleeper then onto the next across non-wholesome surfaces that antagonistically influence a track's auxiliary honesty [3-7].

b) End splitting
The failure of a wood sleeper because of parting at its end emerges when the sleeper is exposed to enormous cross over shear stacking. Additionally, the rail is associated with every sleeper by an appropriate securing framework which incorporates a rail attaching cut, sleeper plate and screw-spike [3-5].

c) Termite attacks
At the point when a termite assaults wood, it disburses all the cellulose-containing constituents and bases lasting harm [3-7]. Figure 2 shows failure of timber sleeper due to various reasons.

Figure 2. Failure of wooden sleeper

2.2 Steel Sleeper
Steel sleepers functioned the railroads for a few decades. Aside from giving an unrivalled sidelong unbending nature, steel sleepers additionally have a high piece esteem. As a result of the better quality when looked at than cast iron sleepers, more consideration isn't needed subsequent to laying. Their protection from creep additionally gives them an additional edge. In any case, steel sleepers likewise represent a lot of detriments [3-7].

2.2.1. Steel sleeper's failures
Notwithstanding, a few scientists detailed that the steel's danger of erosion, high electrical conductivity, weariness splitting in the rail-seat locale and the trouble of pressing it with weight has made it a sub-par material for sleepers. Hence, an appropriate examination concerning the purposes behind its failure is fundamental [3-7].

a) Corrosion in steel
Steel sleepers experience the ill effects of erosion in the territories where the supporting soil or counterbalance is wealthy in pungent components. Different reasons, including a persistently damp condition, metallic slag-based balance and the presence of destructive materials, can improve erosion in a steel sleeper [3-7]. Figure 3 shows corrosion in steel sleeper.

Figure 3. Corrosion in steel sleeper

b) Fatigue cracking
A sleeper encounters both longitudinal and cross over burdens when a train is running over rails, as, an inclining pressure begins its rail-seat area which is for the most part on its top surface however can likewise happen the converse way relying upon the train's development and, after some time brings about weakness breaking [3-7].

2.3 Prestressed Concrete Sleepers
With a normal assistance life of about 50 years, solid sleepers are the strongest of the apparent multitude of sleepers referenced previously. By their heavyweight, they can give extraordinary parallel dependability to the tracks. Aside from being consumption safe when contrasted with different materials, solid sleepers show high proficiency in controlling wet blanket. Their capacity to oppose termite assault and their appropriateness with practically a wide range of soil make them broadly acknowledged. Despite the fact that the underlying expense of production of these sleepers is high, its recuperation in 50 years of its administration life more than makes up for the significant expense. The inflexible idea of the solid sleepers makes them hard to deal with. Their powerlessness to be as proficient to the shifting level of utilization in examination with wood sleeper, represent a few issues. When contrasted with wood, solid sleepers are unbending [3-7].

2.3.1 Failures of Concrete Sleepers.

a) Rail-seat deterioration
This failure is caused either by rail-seat scraped spot, hydro grating disintegration, water powered weight breaking, freeze defrost splitting or synthetic weakening of which rail-seat scraped spot is the most basic [4, 5].

b) Derailment
Imperfections in sleepers, which happen principally during the operational stage in light of labor blames and existing vague deformities in tracks. The essential drivers of wrecking failures are because of labour flaw and existing impermissible imperfections in track [5, 6].

c) Centre-bound damage
This failure is caused either where vertical splits are framed on sleeper because of an elastic crack in its upper focal portion which later engenders all through its focal fragment and framed a 'X' shape before unmistakably cracking [4, 5].

d) High-impact loading

The bowing breaks in a solid sleeper are regularly distinguished at its mid-range either by wheel or rail irregularities, for example, level haggles rails [5, 6]. Figure 4 shows failure of prestressed solid sleeper.

Figure 4. Failure of prestressed concrete sleeper

e) Ice forming in sleeper

Failure of sleeper is formed because of freezing of the water spilling into sleepers which makes an ice weight of 40 MPa that relates to 72–88 kN relying upon the zone of applied weight. This type of problems is mainly found in the railway tracks laid in Jammu and Kashmir valley; Kalka Shimla railway [1-7].

3. OVERCOMING THE CHALLANGES

Composite sleeper technologies have emerged as an effective alternative for railway track maintenance and renewal. However, there are barriers that still need to be overcome for their increased acceptance and use. This section discusses the emerging issues on composite sleepers and presents opportunities for overcoming these challenges [4].

3.1. Strength and Stiffness

The composite sleeper innovations have been created to supplant existing sleepers, therefore, the strength and stiffness of composite sleepers are expected to be similar to the existing sleeper. One approach to improve the strength and stiffness of this type of sleeper is to reinforce them either with long fibres or steel bars. The high shear strength of multidirectional fibre sleepers makes them ideal for bridge application.

3.2. Cost Issue

The cost of high fiber containing composite sleeper advancements (unidirectional and multidirectional fiber sleeper) is roughly 5 to multiple times more than timber sleeper. Nonetheless, the lower life cycle cost is foreseen to counterbalance its high starting cost which to pull in the consideration of the rail business, should be like, or inconsequential higher than, that of customary ones.

3.3. Design Guidelines

A complete design guideline for composite railway sleeper has been established in India by the Research Designs and Standards Organisation (RDSO) named "Provisional Specification for Composite Sleepers (2014)". From 2016, Ministry of Railway, Government of India has adopted the regular use of composite sleepers based on the performance under field trials and on recommendation of RDSO. According to the report, these sleepers will be used over steel girder bridges, special turnouts (where concrete sleeper design is not available), support for temporary girder, over temporary girder and for temporary restoration. From 2017-2018 onwards, these composite sleepers have been used over steel girder bridges for renewal of old sleepers as well as over all new steel girder bridges in projects of new lines, doubling and gauge conversion of railway tracks in Indian Railway [24].

4. Technical requirement of composite sleeper which will be used in Indian Railways

A composite material is a material shaped utilizing at least two discretely recognized materials (for example a polymer folio with fortification in polymer composites) to get explicit properties that are better than the individual material.

Built polymer composite sleeper consolidates a polymer network regularly post-customer reused high-thickness polyethylene (HDPE) as an essential segment, with fortifying strands and/or fillers to contribute improved properties. Poisonous additives will not be utilized to make the sleeper. The sleeper will oppose rot and bug assault. Water assimilation will not cause loss of solidarity requiring the sleeper to be supplanted. The sleeper will be non-unsafe and non-filtering. When all is said in done, composite sleeper is a material framework that fuses support (for example glass filaments) and/or other property modifiers in a polymer network [24]. In light of fortification and property modifiers, the Polymer composite sleepers might be named Fiber strengthened polymer composite, Particle fortified polymer composite and Hybrid composite.

5. Composite Sleepers Available All Over the World which can be used in Indian Railway

Several companies and research institutions have developed composite sleepers in different countries. These sleepers are made from a variety of materials: recycled plastic, FRP, natural rubber, sandwich composites and so on. Brief descriptions of the available composite sleepers are discussed in the following subsections.

a) FFU Synthetic Sleeper

In 1978, the Japanese company Sekisui Chemical Co. Ltd. developed a synthetic wood called ESLON Neo Lumber FFU (Fibre reinforced Foamed Urethane) for the manufacture of railway sleepers in which thermosetting rigid urethane resin foam is reinforced with long glass fibres and does not need to be impregnated with environmentally harmful chemicals. Due to good resistance to water absorption; light weight; easy drill ability; heat and corrosion; it has a life span more than 50 years. These sleepers are currently used in Japan, Germany, Austria, Australia and China [6, 7]. Figure 5 shows Fibre reinforced Foamed Urethane sleeper.

Figure 5. Fibre reinforced Foamed Urethane Sleeper [1-7]

b) Recycled waste plastic Sleeper

'Axion', a US green technology company manufactures composite sleepers made from 100% recycled consumer plastic (such as plastic coffee cups, plastic bags, laundry detergent bottles, milk jugs etc.) and industrial plastic waste under the brand name 'EcoTrax'. Similar types of sleepers are also manafuctured by IntegriCo Composites Inc., USA; I-Plas Ltd, United Kingdom; Tufflex Sleeper, South Africa; Lankhorst Mouldings, Netherland. Since its introduction, a variety of sleepers have been produced for different applications (Figure 3) including: switches; road crossing; bridges; passenger; and heavy duty track. This sleeper is more impervious to rot, insects, fungus and moisture, provides better resistance to plate wear with life span of about 50 years. Currently, these sleepers are introduced in Europe, Germany, Mexico, Australia, Netherland, New Zealand, Canada and Southeast Asia [1-7]. Figure 6 shows different types of recycled waste plastic sleeper.

Axion sleeper

I-Plas sleeper

KLP Lankhorst Mouldings sleeper

Tufflex Sleeper

Figure 6. Different types of recycled waste plastic sleeper [1-7]

Figure 7. Polyester fiber reinforced composite sleeper [1-7]

c) Fibre-reinforced Polymer (FRP) Sleeper

An FRP composite ones using fibre reinforcement and a resin matrix. The key features are: light weight (54 kg); longer service life (40 to 50 years); good resistance to corrosion; high electrical insulation; lower life cycle cost; and innovative design. From 2017 onwards, several have been installed in different locations in India for on regular purposes [1-7]. Figure 7 shows Polyester fiber reinforced composite sleeper.

d) Rubber composite Sleeper

Researchers in Thailand developed natural rubber based composite sleeper for vibration absorbers and blocks for the seismic protection of tall. They improved the mechanical properties of natural rubber using an ebonite system whereby the cross-link density of natural rubber was increased. They obtained better compressive modulus and hardness of the modified natural rubber than scrap rubber/tyres composites [1-7]. Figure 8 shows Rubber composite sleeper.

e) Hybrid Concrete Sleeper/ geopolymer concrete sleeper

The geopolymer concrete sleeper is now considered an alternative environmentally friendly railway sleeper as geopolymer concrete reduces landfill weights because it uses industrial by-product. Several researchers across the world have developed and investigated geo-polymer sleepers made using fly-ash, iron slag, waste rubber tyre, bamboo, waste concrete aggregate and found that durability of these eco-friendly sleeper is good which can be an alternative to conventional concrete sleeper with the additional advantage of low environmental impact [1-7]. Figure 9 shows Fly ash based concrete sleeper and Table 1 compares mechanical performance of different types of sleeper [1-23].

Figure 8. Rubber composite sleeper [1-7]

Figure 9. Fly ash based concrete sleeper [1-7]

Table 1. Mechanical performance of different types of sleeper [1-23]

	Timber	Concrete	FFU	Waste Plastic	FRP	Geo Polymer
Density, (kg/m^3)	850-1045	<2400	<820	850-1150	<1150	<2200
Screw Pull-out Force, (kN)	<40	<65	<65	31.6-35.6	<35.6	<65
Modulus of Elasticity, (GPa)	<16	<30	<8.1	1.5-1.8	<1.6	<30
Rail-Seat Compression, (MPa)	<60	<60	<55	15.0-20.6	<20.8	<60
Modulus of Rupture, (MPa)	<65	-	<142	17.0-20.6	<21.2	-
Shear Strength, (MPa)	<5	<25	<10	<5	<5	<23

6. Conclusions

Materials like cement and steel have not demonstrated so much dependable options in contrast to old wood sleepers thus specialists have zeroed in on elective materials for assembling railroad sleepers. For the most part various kinds of composite materials have been picked and tried to utilize them as sleeper materials. A large portion of the sleepers being used are made of prestressed concrete in Indian Railway. This paper additionally examines the cutting edge materials that can be utilized in the production of railroad sleeper that can represent the effect loads. It is important to examine the conduct of a sleeper under unique effect and weakness loadings, and its exhibitions in various ecological conditions, for example, UV radiation, dampness, high pH and raised temperatures before introducing it in a rail track. This examination proposed some possible methodologies of utilizing composite sleepers. The advancement of sleeper isn't just worthwhile regarding cost decrease yet additionally improves the parallel strength of rail track. The plan rules for composite sleeper is basic to set up for their broad acknowledgment to the basic fashioners and end clients. A decade ago observed headway in Indian Railways where utilization of various sorts of composite sleepers have discovered their utilization which will be useful sooner rather than later moreover.

References

[1] Shashikala A, Anilkumar P, Joseph G, John J and Lijith K 2015 Experimental Investigations on Use of Rubber Concrete in Railway Sleepers. In: *2nd RN Raikar Memorial International Conference & Bathia-Basheer International Symposium on 'Advances in Science & Technology of Concrete-2015',*

[2] Manalo A, Aravinthan T, Karunasena W and Ticoalu A 2010 A review of alternative materials for replacing existing timber sleepers *Composite Structures* **92** 603-11

[3] Ferdous M, Khennane A and Kayali O 2013 Hybrid FRP-concrete railway sleeper *University of New South Wales, Canberra, Australia*

[4] Ferdous W and Manalo A 2014 Failures of mainline railway sleepers and suggested remedies–review of current practice *Engineering Failure Analysis* **44** 17-35

[5] Xiao S, Lin H, Shi S Q and Cai L 2014 Optimum processing parameters for wood-bamboo hybrid composite sleepers *Journal of Reinforced Plastics Composites* **33** 2010-8

[6] Ferdous W, Manalo A, Aravinthan T, Van Erp G, Khennane A and Kayali O 2015 Composite railway sleepers: New developments and opportunities. In: *Proceedings of the 11th International Heavy Haul Association Conference: Operational Excellence (IHHA 2015), Perth, Australia,* pp 21-4

[7] Ferdous W, Manalo A, Van Erp G, Aravinthan T, Kaewunruen S and Remennikov A 2015 Composite railway sleepers–Recent developments, challenges and future prospects *Composite Structures* **134** 158-68

[8] Ferdous W, Manalo A, Aravinthan T and Van Erp G 2016 Properties of epoxy polymer concrete matrix: Effect of resin-to-filler ratio and determination of optimal mix for composite railway sleepers *Construction Building Materials* **124** 287-300

[9] Jain C, Khandelwal S, Mehrotra S and Gupta R 2016 A review paper on use of composite material for railway sleepers in railway track *SSRG Int J Civ Eng* **3** 104-8

[10] Zakeri J A, Esmaeili M, Mosayebi S A and Sayadi O 2016 Experimental investigation of the production of sleepers from concrete that contains blast furnace slag *Proceedings of the Institution of Mechanical Engineers, Part F: Journal of Rail Rapid Transit* **230** 77-84

[11] Gonzalez-Corominas A, Etxeberria M and Fernandez I 2017 Structural behaviour of prestressed concrete sleepers produced with high performance recycled aggregate concrete *Materials structures* **50** 94

[12] Silva É A, Pokropski D, You R and Kaewunruen S 2017 Comparison of structural design methods for railway composites and plastic sleepers and bearers *Australian journal of structural engineering* **18** 160-77

[13] Soehardjo K and Basuki A 2017 Utilization of bagasse and coconut fibers waste as fillers of sandwich composite for bridge railway sleepers. In: *IOP Conference Series: Materials Science and Engineering. Medan,* p 012036

[14] Stepanov V, Saldaev V and Tsvetkov V 2017 Composite material for railroad tie. In: *Solid State Phenomena*: Trans Tech Publ) pp 587-91

[15] Kaewunruen S, Li D, Chen Y and Xiang Z 2018 Enhancement of dynamic damping in eco-friendly railway concrete sleepers using waste-tyre crumb rubber *Materials Design* **11** 1169

[16] Kaewunruen S, Lopes L C and Papaelias M P 2018 Georisks in railway systems under climate uncertainties by different types of sleeper/crosstie materials *Lowland Technology International* **20** 77-86

[17] Khalil A A 2018 Mechanical testing of innovated composite polymer material for using in manufacture of railway sleepers *Journal of Polymers the Environment* **26** 263-74

[18] Stepanov V and Timerbaev N F 2018 Composite Railroad Ties Obtained by the Energy Efficient Recycle of Wooden Railroad Ties. In: *Solid State Phenomena*: Trans Tech Publ) pp 981-5

[19] Ferdous W, Manalo A, Muttashar M, Yu P and Kakarla R 2019 Composites for alternative railway sleepers

[20] Khalil A A, Bakry H M, Riad H S and Shnour A S 2019 " Analysis on railway sleepers manufactured from polymers and iron slag *Journal Archive* **14**

[21] Kondrashchenko V I, Jing G Q and Wang C 2019 Wood-Polymer Composite for the Manufacture of Sleepers. In: *Materials Science Forum*: Trans Tech Publ) pp 509-14

[22] Raj A, Nagarajan P and Shashikala A 2020 Investigations on Fiber-Reinforced Rubcrete for Railway Sleepers *ACI Structural Journal* **117**

[23] Raj A, Nagarajan P and Shashikala A 2018 A Review on the Development of New Materials for Construction of Prestressed Concrete Railway Sleepers. In: *IOP Conf. Ser. Mater. Sci. Eng,* p 012129

IOP Conf. Series: Materials Science and Engineering **970** (2020) 012003 doi:10.1088/1757-899X/970/1/012003

Fly-ash Enriched Earthen Un-Burnt Bricks for Low-cost Housing

Joygopal Jena, Chinmayanada Sahoo, Kalyani S. Singh

Department of Civil Engineering, Gandhi Institute for Technological Advancement, Bhubaneswar, Odisha, India

E-mail: jenajoygopal@gmail.com

Abstract. In spite of various major steps taken by the Governments, housing problem in rural India is becoming bad to worse due to population explosion and their low income. Various low-cost housing schemes has been floated through use of cheaper construction materials. Therefore, development of new cheaper materials has become need of the day. This can be possible by production of new materials produced from various by-products or wastes justifying the slogan converting 'Wastes into Wealth'. Fly ash, a coal combustion by-product available in power plants is one such material available in plenty in our surroundings those are posing problem to the environment. Presently, these are utilised by mixing various cementitious materials in production of bricks, those are substituting conventional bricks, for their cost effectiveness. But, the extreme poor people of tribal areas are deprived of using these bricks in their houses due to non-availability and financial constraints. Most of tribal houses in KBK districts of Odisha, Eastern and Central rural of Chhattisgarh, some parts of Jharkhand, Bihar and West Bengal are made of walls with twigs plastered with mud and Nauria tiled roofing. To bridge the gap a new product 'Fly ash Enriched Earthen Unburnt brick' has been proposed. Necessary laboratory tests for the proposed product has been carried and their suitability is presented in this paper. Poor people who can't afford for standard bricks for their houses, this can be an alternate cheaper solution for them.

1. Introduction

Most of tribal houses in KBK districts of Odisha, Eastern and Central rural chunk of Chhattisgarh, some parts of Jharkhand, Bihar and West Bengal are made of walls with twigs plastered with mud and Nauria tiled roofing. (Figure 1). General poor people who can't afford for burnt clay bricks for their houses, 'Fly ash Enriched Earthen Unburnt Brick' can be an easy solution for them. In arid zones of central India, KBK districts of Odisha, Jharkhand, Chhattisgarh having low rainfall these can use this material for construction [1-6]. Fly ash can be mixed with the local earth irrespective of its quality in production of bricks which can be used unburnt in construction of low-cost housing. These types of materials would be environment friendly. It is a simple technology i.e., only substitution of materials for brick and to be used un-burnt. These bricks shall be made of local earth enriched with fly ash. It can be produced anywhere, irrespective of availability of any type of earth and would be much cheaper material affordable by poor. The bricks can be manufactured fairly simply using local labour and will also have a positive impact on local employment. It can save huge numbers of trees every year though it is the substitute of trees [7-10].

Figure 1. Typical tribal house in KBK Districts in Odisha

If we look to the historical facts around the globe the case of the Taos Pueblo in New Mexico may be the most appropriate example, where the Pueblo group of people are residing since about last thousand years. They construct of their residential houses with the sun-dried bricks made of mud those are plastered with special mud prepared from local clayey soil, known as caliche mixed with finely-cut straw or some local suitable fibrous materials for smooth finish and that adds to the surface strength. Facility of additional one more floor is provided by making the walls thicker and the roofs with planks made from cedar trees, with added provision of wooden ladders from outer side of the house for the approach to higher floor. The statistical information reveals an estimated 180 billion tons clay burnt bricks are utilized annually for the above purpose. This figure is likely to increase with the demand for further housing units which may exceed the figure 80 million for lower and lower-middle class habitants. At present the housing sectors in our country is also facing a deficit of 20 million housing units for its lower and lower-middle class which may face a surge of about 22.5 million units in near future. In this situation to cope up the exponential growth of demand fly ash enriched mud brick can be a feasible solution to support the poorest of the poor ones. Dr Andrew Heath [4], a faculty of the University of Bath has presented an overview for field application of the said technology. Mauro Sassu, Ignasio Ngoma [10] have presented collective initiatives of the organizations, 'Earthquake Engineering Research Institute (EERI)' and 'International Association for Earthquake Engineering (IAEE)' in their comprehensive report of "World Housing Encyclopaedia". In their report they have emphasized that about 45% of housing of the country, MALAWI and in the adjacent countries, i.e., Zambia and Tanzania are constructed having thatched roof supported by unburnt mud brick wall with mud mortar. Mike Lawrence et.al. [6] has stated that unburnt bricks made of local clay can deliver workable and sustainable substitute as an alternate to conventional burnt brick masonry or concrete blocks for both load-bearing and non-load bearing walls. Sadek Deboucha and Rostan Hashim [9] have reviewed the state-of-the-art utilization of mud bricks and compressed earthen blocks. Daniel Maskell et.al. [11] have discussed in their paper regarding use of other cementing materials such as cement and lime along with the conventional masonry for stabilizing the earthen masonry. Andrew Heath et.al. [4] summarizes results of extensive testing on viability of mass-production of unburnt clay bricks commercially. The focus of the paper is to investigate the properties effecting the compressive strength of these products.

2. Materials and Methods

For production of the proposed fly-ash enriched bud brick two major materials required are earth and fly ash apart from water. Those are described briefly here.

2.1. Fly ash

It is needless to mention the status of accumulation of fly-ash, a bye-product of coal based thermal power plants and their possible utilization in various activities. It is a fact that about 72% of power plants of our country are coal fed with annual generation of approximately 40 million tons of fly ash. These are separated by use of the cyclone converter in almost all power plants and industries using coal as a fuel. This fly ash is a threat as a major atmospheric pollutant. Mostly these are disposed as land fill. Presently a small part of it is used as a raw material for manufacture of bricks, construction of roads etc.

2.1.1. Current Status of fly ash

Thermal Power Generation units are the major source energy supply addition to the present hydropower installations and other alternate energy sources in our country. These are serving as backbone of power capacity and are mostly Coal/Lignite based. It is an established fact that coals available in India are of lower quality having ash content of about 30% to 45 % and on the hand imported coals of having much lower ash content of nearly 10% to15%. Thus, large amount of ash is generated at these thermal Powerhouses which is real threat to environment as pollution of both air and water. Disposal these requires huge area of land. A very small part of it is being utilised in infrastructural activities. Therefore, use of these fly ashes for production of any new building materials shall help the infrastructural development and environmental protection.

2.1.2 General Characteristics of Indian Fly ash

Physical Characteristics:
Specific Gravity varies from: 2.54 to 2.65 gm/cc
Bulk Density is about : 1.12 gm/cc
Fineness varies from : 350 to 450 m^2/kg
Chemical Characteristics:
Silica content varies from : 35% to 59%
Alumina varies from : 23% to 33%
Calcium Oxide varies from : 10% to 16%
Sulphur varies from : 0.5% - 1.5%
Iron varies from : 0.5% - 2.0%

2.2. Earth for Bricks

For manufacture of the conventional earthen bricks quality of earth plays a major role. But the proposed bricks shall be independent of quality of earth. Normal local earth mixed with required proportion of fly ash and water for manufacture of the brick. Laboratory tests should be conducted for deciding the proportions of materials for mixing to obtain desired compressive strength.

2.3. Manufacturing of Bricks

Earth, after collection is gone under various processes such as weathering, blending by adding desired quantity fly ash, sand etc as per the designed mix. After proper formation of very sticky stuff, this is placed in moulds to form required shape and size. After removal from mould it is allowed to dry in sun for at least 3days. Traditionally the clay bricks are generally made by hand resulting variable dimensions and other properties. Mechanised moulding of bricks is not practicable when the bricks are produced in small numbers and remote localities.

3. Results

The bricks are moulded in the laboratory utilizing local earth and the fly ash available locally. Mechanical properties of earth and fly ash used have not been tested. Compressive strength of the bricks produced are tested to ensure the suitability of those in the building construction. The variation of compressive strength with different proportion of fly ash has been presented in next section.

3.1. Compressive strength of Bricks

Sample bricks with varying proportion of earth and fly ash manufactured for test purpose. Laboratory tests of the bricks are conducted for ascertaining the compressive strength and results of which are presented in Table 1.

Table 1. Compressive strength of the bricks.

Sl. No.	Replacement of fly ash%	Sample No	Dimension (mm)	Max Load (KN)	Comp. Strength N/mm^2	Avg. Comp. Strength (N/mm^2)
1	0%	1	247 x 118	32	1.09	1.196
		2	250 x 119	40	1.34	
		3	250 x 117	34	1.16	
2	10%	1	255 x 127	54	1.66	1.671
		2	255 x 125	55	1.72	
		3	255 x 125	53	1.63	
3	20%	1	255 x 127	84	2.59	2.543
		2	255 x 125	82	2.57	
		3	255 x 125	79	2.47	
4	30%	1	247 x 120	125	4.21	4.250
		2	250 x 119	123	4.13	
		3	250 x 117	129	4.41	
5	40%	1	247 x 118	146	5.01	4.856
		2	250 x 119	141	4.73	
		3	255 x 117	144	4.82	
6	50%	1	255 x 127	136	4.19	4.427
		2	255 x 125	148	4.64	
		3	255 x 125	142	4.45	
7	60%	1	255 x 125	81	2.54	2.432
		2	254 x 125	79	2.48	
		3	255 x 127	73	2.25	

The properties of these unburnt bricks thus produced are mostly dependent of quality of the earth. Unlike the burnt bricks which are produced with selected earth can't be compared with these as far as strength is concerned. It is therefore difficult assign single value of strength. The strength of unburnt brick masonry mostly depends on properties of these materials, wall thickness and the content of water in it.

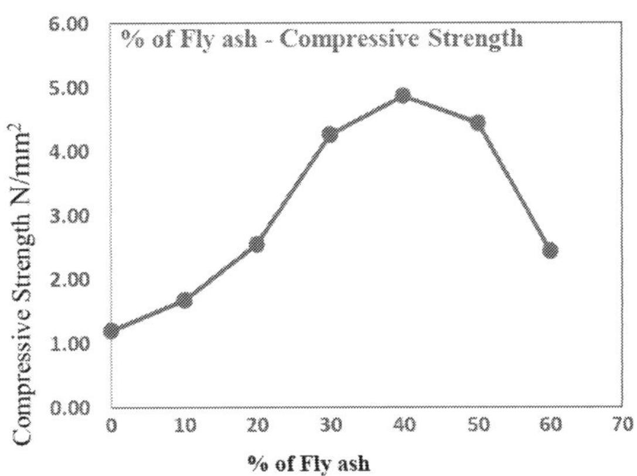

Figure 2. % of Fly Ash V/s Compressive strength of Brick

It is an established fact that with the increased water content in the masonry the compressive strength gets reduced. Therefore, it is desirable that the masonry be kept dry after its construction. The water content normally remains higher at the time of construction due to use of wet mortar and get stabilized afterwards during iuse. Figure 2 shows the difference of compressive strength of brick with the variation of fly ash content in the bricks. It is found that the compressive strength shows an optimum value at substitution of nearly 40% of fly ash for earth. This may vary from place to place and with quality of materials.

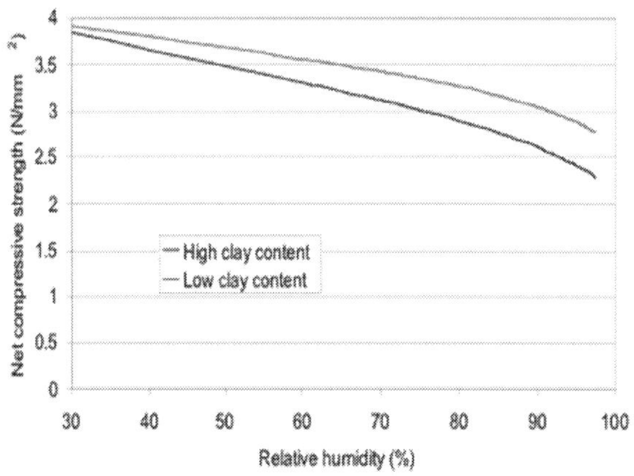

Figure 3. Result of Relative humidity V/s Compressive Strength [9]

3.2. Effect of varying Relative Humidity on compressive strength

The unburnt mud brick absorbs more amount of moisture from the during higher atmospheric humidity, that may affect its compressive strength. Much study has not been reported regarding the effect of relative humidity on the mud brick masonry. Figure 3 shows a typical test results on this issue. It shows that there is not great change in the strength even in the case of relative humidity varying widely from 30% to 97.5%. A case study of long-term monitoring of a structure built using unburnt clay brickwork in Dalguise, U.K., revealed the relative humidity of the structure was varying within 40% and 65% during the entire year, even in the case of continuous use of bathroom and with shower. It was found that in this case of change of relative humidity the strength only varied nearly 12% for the earth masonry having higher clay content, and about 8% in case of the lower content of clay used for the work.

4. Summary

The adaptability of this brick has been justified considering the various advantages, in spite of some major disadvantages. It can be substitution of burnt clay brick, it can be suitable for poor people who can't afford for standard bricks for their houses.

4.1. Advantages

- Safety: The prime concern for dwelling houses, especially, in remote rural areas is safety from wilds and rough weather, and for house hold valuables, those lack in the twig walled houses present in tribal areas. But use of these bricks adds to the safety at very lower cost.
- Strength: The compressive strength of these bricks are lower than the burnt clay bricks and fly ash bricks, it is higher than the ordinary unburnt mud bricks, which is suitable for walls of the sloped roofed houses and can used low-cost housing due to its cost effectiveness.
- Aesthetics: The aesthetics of fly-ash enriched mud bricks are very attractive due to their pleasing colour matching cement, uniformity in sizes and smoother finishing. They are free from cracks and bulges making the masonry better.
- Thermal conductivity: The thermal conductivity of these bricks varies from 0.9 to 1.05W/m^2. Less heat absorption keeps buildings cooler even in summer, hence suitable for varying Indian climates.

- Durability: Though the laboratory tests for durability of these bricks are not available, examples stated in earlier sections showed high durability and less permeable. The lower permeability of these bricks reduces their efflorescence effect. They are less porous which adsorbs less moisture and thus reduces dampness on the wall. These are also resistant to mild acid attack.
- Insulation for Sound: Masonry constructed with fly ash enriched mud bricks provide adequate insulation to sound for the building.
- Resistance to Fire: Obviously masonry constructed with these bricks are fire resistant as other bricks.
- Friendliness to Environment: As simple technology is adopted for manufacturing these bricks, these are eco-friendly. Avoiding the burning process emissions are automatically avoided and there is no discharge of effluents. Consequently, huge felling of trees is avoided protecting environment which otherwise needed for walling purpose.
- Conservation of Energy: Due to very low consumption of energy due to absence of burning process as in case of conventional bricks huge energy is conserved.
- Cost: The same number of fly ash enriched earthen bricks will cover more areas than clay bricks due to use of thicker mortar. Normally mud mortar is used for the masonry. Quality of water also does not play much importance. It requires less labour. The cost of this brick is approximately 70% lower than burnt clay brick.

4.2. Disadvantages

Major disadvantage of this brick is that it can be safely used for single storied buildings and construction of next storey is not advisable. Use of higher quality fly-ash is advisable for preventing negative effects. The defective construction shall result giving lesser strength and may not be suitable for construction. Bricks having poor quality will have negative impact on buildings. These are ideally suitable for arid zone. Mechanical strength being very low, this can be used by mixing other suitable ingredients. Availability and transportation of the fly ash at the brick manufacturing site may be the major constraint.

5 Conclusion

It can be concluded from the study that the proposed fly ash enriched brick can be used safely in the walls of the tribal houses providing safety and security to the habitant. It is quite safe from strength point of view. Local earth can be used and small quantity of bricks can be manufactured by the local labour at much lower cost. As it does not have the sufficient moisture resistance as burnt clay brick, it should be ensured for using in dry condition during and post construction. This brickwork has a lower strength than conventional brick masonry, it is not recommended to use thin-walled earth masonry. In spite of various advantages, the availability and transportation of fly ash for individual user may the major huddles. Governments/ NGOs should come forward to help in this regard for providing safe houses for tribal. Considering the various criteria described in details these bricks can be a utilized in the tribal houses substituting the present mud plastered twig walled houses.

References

[1] IS: 1077-1976, Bricks classification, bricks characteristics, bricks uses, bricks as per BIS, Bureau of Indian Standards, New Delhi, India
[2] Head K H 1998 *Manual of soil laboratory testing. Volume 3: effective stress tests*: John Wiley & Sons)
[3] Kamwanja G A 1988 Low Cost Building Materials in Malawi. University of Malawi
[4] Heath A, Lawrence M, Walker P and Fourie C 2009 The compressive strength of modern earth masonry. In: *11th international conference on non-conventional materials and technologies, NOCMAT*: Citeseer) pp 6-9
[5] https://www.greenspec.co.uk/building-design/unfired-clay-bricks
[6] Lawrence M, Heath A and Walker P 2008 Modern innovations with unfired clay masonry. In: *Proceedings of 8th International Seminar on Structural Masonry (ISSM 08),*

[7] Malaya C and Sreedeep S 2012 Critical evaluation of the drying water retention characteristics of a class F Indian fly ash *Journal of materials in civil engineering* **24** 451-9
[8] Minke G 2004 Building with earth: design and technology of a sustainable architecture. Birkhäuser-Publisher for Architecture)
[9] Deboucha S, Hashim R J S R and Essays 2011 A review on bricks and stabilized compressed earth blocks **6** 499-506
[10] Sassu M and Ngoma I 2002 HOUSING REPORT Unburnt brick wall building with pitched roof (nyumba ya zidina)
[11] Maskell D, Heath A and Walker P 2013 Suitable stabilisation methods for extruded earth masonry. In: *Central Europe towards Sustainable Building 2013 (CESB 13), Prague*: University of Bath

A Review on the Bacterial Concrete Properties

Shradha Jena[1], Bidyadhar Basa[1], Kishor Chandra Panda[2]

[1]Department of Civil Engineering, ITER, S'O'A (Deemed to be University), Bhubaneswar, Odisha, India
[2]Department of Civil Engineering, Government College of Engineering, Kalahandi, Bhawanipatna, Odisha, India

E-mail: sjena4203@gmail.com

Abstract. A very commonly used and widespread building material, concrete has various limits of using it, one of which is crack formation. Formation of larger cracks let water and CO_2 enter into the building materials which in turn react along with other chemicals causing an eye catching reduction in strength and durability. If immediate remedies are not taken to heal the cracks, it might lead to more serious problems like bigger cracks, more leakage of water, reduced strength, costlier to repair. Hence the best way to deal with such situation and to heal concrete even before it starts spreading is self-healing concrete. Self-healing concrete is the most recent and most current cutting-edge technological development in the field of building materials. Self-healing bacteria has the capacity to heal itself. The underlying principle of this mechanism is forming calcium carbonate precipitate with the help of bacteria as a blocking material for cracks. This new and innovative introduction of bacteria into the building materials show new light on production of sustainable, durable, strong concrete having prolonged life period. Due to the introduction of bacteria in concrete, the name becomes bacterial concrete. Bacterial/ microbial concrete is pollution free, economic and cost effective. This paper aims at defining bacterial concrete, mechanism and its effects on concrete properties.

1. Introduction

For ages now, concrete has been ruling the infrastructure world. It has gained a lot of recognition as the commonly used material for construction. Even though it has a lot of utilities, there are still two void spaces that concrete fails to fill. Firstly it has strong compressive strength and weak tensile strength, secondly crack formation is quite easy in concrete [1, 2]. Even smaller cracks in concrete result in further bigger problems, which include reducing life span of concrete [3]. Further weakening of reinforcement is resulted due to percolation of water, chloride attack, sulphate attack, carbon attack. In conventional method, crack formation is detected by doing examination manually, and usually repaired by filling it with cement [4]. But these repairs are costly and quite hard to do for deeper cracks [5, 6]. People prefer innovative ways over conventional ones and in these new methods, self-healing concrete is being used widely [7]. Self-healing concrete as its name describes, can heal cracks on its own. In this concrete, bacteria species are mixed with calcium and nutrient medium and injected while preparing concrete mixture. The cracks allow water into concrete which then form calcite precipitate around bacterial cells, which again in turn completely deal the cracks forming a permanent block. This review paper throws light on kinds of bacteria species, their morphology, mechanism of precipitate formation, and the effect of microbial organisms on strength of concrete.

2. About Bacteria

2.1. Bacteria

Bacteria species (singular: bacterium) are unicellular, simple and the most abundant microorganisms. Although bacteria are very simple in structure, they show extensive metabolic diversity. Bacteria are called prokaryotes due to absence of membrane bound structure inside their body. Bacteria have ability to survive in very adverse conditions. Bacteria are categorized into four types on the basis of their shape: the rod-shaped bacillus, the spherical coccus, the spiral Spirillum, the comma shaped Vibrium. Bacteria occur almost everywhere. They live in extreme conditions where survival of any other species is next to impossible. Some of the bacteria species also live inside bodies of other organisms as parasites. Bacteria also help in nutrient cycling, as some them are capable of nitrogen fixation, purification etc. bacteriology is a branch of science that includes study of bacteria in laboratory.

2.2. Morphology of Bacteria

Bacteria have varieties of shapes and sizes. Bacterial cells have a different kind of anatomy from that of eukaryotes. Their cells are comparatively smaller in size than of eukaryotic cells. They are normally 0.5 to 5 mm in size. A few bacterial species like Thiomargarita amibiensis and Epulopiscium fishelsoni are up to 0.5 mm long and are visible to the naked eye. Eubacteria are true bacteria due to presence of rigid cell wall. Archaebacteria are special because they live in harsh environments like extremely salty areas, hot springs, marsh areas. The mycoplasma are micro-organisms that lack a cell wall and are the smallest organisms (0.3 micrometres) that can live without oxygen. One of the situations that help an organism in surviving in adverse conditions is less expenditure of energy. In bacteria, large surface area to volume ratio helps in achieving this goal, it helps in environments having less nutrient supply. The cell wall and cycloskeleton of bacteria helps in deciding the shape of bacteria. The cycloskeleton which includes flagella, cilia help bacteria in locomotion, avoiding preys and pili help in conjugation.

3. Self-Healing Process and Basic Mechanism

3.1. Self-Healing Process

The main idea behind using self-healing concrete in construction is to automate the detection of crack formation in its initial stage only, so that it becomes easy to heal them primarily. When a mixture of concrete and bacteria is prepared, it results in forming calcite precipitation layer, which helps in healing crack [8, 9]. A general criteria for selection of the bacterial species to be added is that it should be highly alkaline in nature considering the fact that concrete itself is alkaline and has adverse conditions[10, 11]. The calcite precipitation layer blocks the path of and acts binds sand and gravel [12]. *Bacillus sphaericus* is a species of bacteria that acts in very alkaline medium and can convert urea into ammonia and carbonate [13]. Cracks of size greater than 0.2 mm cause main problem to concrete as they allow chemicals and water to enter. Cracks of lesser size can be healed themselves. In self-healing concrete, a small crack formation leads way to the entry of water, air, chemicals into concrete causing dormant bacteria to be active. These bacteria chemically produce a hard calcite layer which blocks the path of the cracks. After that, bacteria again become dormant. As the bacterial species are involved, this process is better called Microbiologically Induced Calcium Carbonate Precipitation abbreviated as MICP.

3.2. Reactions Involving in Self-Healing

The way bacterial concrete works is by forming calcium carbonate precipitation and forming a tight bind around otherwise loose bound particle. It heals concrete. Most common bacteria are urease producing bacteria [14, 15].

1 mole of urea when hydrolysed, gives rise to 1 mole ammonia and 1 mole of carbamic acid (NH_2COOH), which is shown Eq. 1 [16].

$$CO(NH_2)_2 + H_2ONH_2 \longrightarrow COOH + NH_3 \qquad (1)$$

These products form 1 mole of bicarbonate and 2 moles of ammonia, which is shown in Eq. 2.

$$NH_2COOH + H_2O \longrightarrow 2\,NH_3 + 1\,H_2CO_3 \qquad (2)$$

Bicarbonate decomposes into bicarbonate ion and H^- ion, which is shown Eq. 3.

$$H_2CO_3 \longleftrightarrow HCO_3^- + H^+ \qquad (3)$$

This reaction causes an increased pH, for which the reaction shifts towards right (law of mass action), which is seen in Eq. 4.

$$HCO_3^- + H^+ + 2NH_4^+ + 2OH^- \longleftrightarrow CO_3^{2-} + 2NH_4^+ + 2H_2O \qquad (4)$$

Bacterial cell wall attracts Ca^{2+} cations from the surrounding environment, which is shown in Eq. 5. The previously deposited CO_3^{2-} ions react with these Ca^{2+} ions to form $CaCO_3$ precipitation, shown in Eq. 6.

$$Ca^{2+} + Cell \longrightarrow Cell\text{-}\ Ca^{2+} \qquad (5)$$

$$Cell\text{-}Ca^{2+} + CO_3^{2-} \longrightarrow Cell\text{-}CaCO_3\downarrow \qquad (6)$$

4. Impact of Bacteria on the Properties of Concrete

4.1. Compressive Strength

Compressive strength is one of the determining factors for concrete durability. Therefore, the application of bio concrete to find out its true potential is quite a complicated area for research purpose. Inclusion of bacteria in concrete and mortar caused a significant amount of progress in compressive strength of concrete. The underlying principle is that crystals of calcium carbonate would precipitate on the bacterial cell surface and gradually form a blockage to pores and cracks which would lead to shortening the supply of oxygen, water and other nutrients, making it an unfavourable condition for bacteria to grow. The cells of the bacteria either become dead or turn into endospores to act as organic fibres which also help in improvement of concrete strength [17]. Various tests have been conducted to study the effect of bio concrete on compressive strength of concrete and mortar [18-21]. In all these experiments, various kinds of microorganisms have been added into the concrete mixture and all significant changes have been marked. Experiments were carried out by including varying concentrations of the species *Bacillus pasteurii* in the sample cement mortar specimens. A remarkable 18% increment in compressive strength was observed at an interval 28 days and it was due to the presence of sufficient quantity of organic substances in the mixture [17]. Positive potential of a microorganism was experimented that belong to *Shewanella* species (thermophilic anaerobic bacteria) on compressive strength of concrete. They reported a 25-30% increment in the compressive strength after a time period of 28 days [22]. *Bacillus pseudofirmus* was chosen along with *Bacillus cohnii* to add into the concrete and a 10% increment in the compressive strength was noticed [20]. When *Bacillus cohnii* bacteria species is added, it increases concrete strength up to certain cell concentration which agains shows decreasing levels with passing time [23]. However, 36% increment in compressive strength was observed by mixing of *Bacillus sp. CT-5* to mortar specimen [14]. *Bacillus Pasturii* now known as *Sporosarcina Pasturii* was used (both live and dead bacteria cell) which shows an increment in compressive strength upto 10% [11]. 22% increment in the compressive strength of concrete containing fly ash was observed when the concrete specimens were treated with the bacterial species *Sporosarcina pasteurii*. In this experiment, three percentages (10, 20 and 30) of fly ash by weight acted as a replacement for cement and three disparate bacterial cell concentration 10^3, 10^5 and 10^7 cells/ml were used. This increment in compressive strength was mainly due to deposition of bacterial cells which blocked the pores [24]. By using *Bacillus Sphaericus* microorganisms which is in dormant condition, there is 30.76%, 46.15% and 32.21% increase in compressive strength and 13.75%,14.28% and 18.35% increase in split tensile strength in a period of 3,7 and 28 days respectively [25]. By utilizing and comparing two types of bacteria like *Bacillus Cereus* and *Bacillus Pasturii*, it is observed that with *Bacillus Cereus* compressive strength increases up to 38% and with *Bacillus Pasturii* increment in strength was 29% [26]. *Bacillus subtilis* was used with cell concentration from $10\text{-}10^6$ cell/ml and highest strength was achieved with 10^5

cells/ml. It may be due to its ability to form calcite which block the pores of concrete structure [27]. *Bacillus subtilis* bacteria was used with two carrier compound which are Graphite Nano platelets and Light weight aggregate. As compared to control specimen (Concrete without bacteria), there is increase of 12% of compressive strength in case of Light weight aggregate and 9.8% of compressive strength in case of Graphite Nano platelets [2].

4.2. Water Permeability

Permeability of concrete is its physical property that controls the rate of absorption of liquid by it. Durability of concrete mainly depends on this aspect as aggressive and hazardous materials mixed in water can penetrate it if the concrete mass is more permeable and vice versa. Porosity in concrete is formed due to improper compaction leaving air pockets, space left due to evaporation of water, inadequate water/cement (w/c) ratio, improper curing and formation of micro cracks [28]. Deposit of Calcium Carbonate ($CaCO_3$) is found to be helpful in reducing permeability of concrete specimens [29]. Studies revealed a marked decrease in porosity and there by permeability in test samples by adding *S. Pastteurii* bacteria in fly ash concrete. Concentration of bacteria in water at 10^5cells/ml resulted in fourfold reduction in water absorption in concrete. Presence of bacteria in bacterial concrete causes Calcium Carbonate precipitation which fills the pores in it [25]. *Bacillus Halodurans strain KG1* was used in concrete with cement kiln dust. It was observed that there is 20% less water absorption in comparison to concrete without bacteria [30]. Water absorption is found to be less than one third in the specimens that are cast by adding *Bacillus Megaterium* along with its nutrients and they do so by depositing microbial calcite [18]. Injection of *Bacillus Aerius* bacteria also result in the same effect and helps in increased durability of concrete structures by reducing the permeability [31]. Concrete specimens with *AKKR5 bacterial* (10^5 cells/ml) showcase moderate to very low permeability as the calcium carbonate precipitate fills the concrete pores [32]. Microbial precipitation improves the quality of recycled aggregates with reduction in water absorption [33, 34].

4.3. Chloride Permeability

Reinforced concrete structures get deteriorated by chloride attack resulting in corrosion of reinforcing steel. Internal pore network of concrete plays a major role of accessing ion of chloride into it. The pore structure depends on factors like mix design, use of admixtures, construction practices, degree of hydration and curing process etc. Chloride permeability test is carried out by calculating how much electrical current is passing across a concrete specimen. On the basis of the charge passing across it, concrete's permeability is assessed. Adding bacteria to the concrete, its resistance to chloride permeation can be enhanced. The effect of bacterial carbonate precipitation on the durability of concrete with different porosity was observed. *Bacillus Sphaericus* bacteria was used for this research work. The result shows that the surface deposition of carbonate crystals decreased the chloride migration about 10 to 40% [35]. It was noticed that the average number of coulombs in concrete that contains bacteria was 11.7% less in comparison to the concrete that has no bacterial addition. Use of *Sparcious Pasteurii* and *Bacillus Subtilis* are found to decrease the chloride penetration in concrete [36]. In rapid chloride tests, injection of *Bacillus Aerius* bacteria in concrete can cause effective reduction in the net amount of charge passing through RHA and control concrete specimens at all curing ages. At the age of 7, 28 and 56 days it was found that charge passed in concrete specimen having bacteria to be decreased by 55.8%, 49.9% and 48.4% respectively, in comparison to nonbacterial concrete [32]. The addition of *Sparcious Pasteurii* (10^5 cells/ml) in concrete with 10% silica fume resulted in optimum resistance for chloride penetration (380 coulombs) [24]. In case of fly ash concrete, it was observed that, the chloride ion concentration was reduced by maximum amount by adding *Sporoscarcina Pasteurii* (10^5 cells/ml concentration) i.e. concrete having 30% fly ash content gave 762 coulombs penetration which is a quite less amount. The activity period of concrete specimens that are allowed to come in contact with de-icing salts or marine environments is measured by the capability of concrete to prevent chloride ions penetration [24]. *Bacillus Cereus* and *Bacillus Pasturii* of various cell concentrations were utilized in cement concrete. By utilizing bacteria, the chloride entrance of bacterial sample decreases as compared to control concrete [26].

4.4. Corrosion

Large infrastructures fail mainly cause of corrosion, which affect steel when reinforced concrete is mainly exposed to chloride ions, which essentially react with steel, eating it inside and creating void. For this reason, necessary attention needs to pay to reduce concrete's permeability. In order to avoid corrosion, sufficient measures are needed to be taken to reduce the permeability of concrete. Because, higher permeability of concrete allows unnecessary water, chloride ions along with chemicals which cause corrosion of steel within. When calcium carbonate precipitation occurs, it blocks entry of substances and reduces [20]. In a trial method, when bacteria species like *Sporosarcina pasteurii* and *Bacillus* sp. CT-5 [18] were mixed with concrete, it showed large decrement in steel's corrosion. Bacteria, blocks the water absorption by cracks by forming calcite precipitation, hence provides longer average self-life to reinforcement bars [22, 37-49].

5. Conclusions

This research paper provides brief introduction on the bacteria types, their morphology and properties of self-healing concrete. One of the main goals was to know the underlying mechanism of how urease producing bacteria can help in healing of cracks. It highlights how permeability and compressive strength of concrete are highly affected by the presence of bacterial specimen. The percentage of water and chloride permeability of concrete is highly reduced when bacterial specimens are present in the concrete, the reason being their ability to block the cracks. Biological reactions which involve bacteria protect concrete from freeze-thaw-attack and it in turn enhances concrete's properties positively. The self-healing concrete reduces crack formation and heals already existing ones, thus it can be safely assumed that this concrete can effectively act as a replacement of conventional concrete.

References

[1] Wiktor V and Jonkers H M 2011 Quantification of crack-healing in novel bacteria-based self-healing concrete *Cement Concrete Composites* **33** 763-70

[2] Khaliq W and Ehsan M B 2016 Crack healing in concrete using various bio influenced self-healing techniques *Construction Building Materials* **102** 349-57

[3] Jonkers H M and Schlangen E 2007 Crack repair by concrete-immobilized bacteria. In: *Proceedings of the first international conference on self healing materials,* p 20

[4] Neville A 1996 Properties of concrete. Pearson higher education

[5] Wang J, Van Tittelboom K, De Belie N and Verstraete W 2012 Use of silica gel or polyurethane immobilized bacteria for self-healing concrete *Construction building materials* **26** 532-40

[6] Muhammad N Z, Shafaghat A, Keyvanfar A, Majid M Z A, Ghoshal S, Yasouj S E M, Ganiyu A A, Kouchaksaraei M S, Kamyab H and Taheri M M 2016 Tests and methods of evaluating the self-healing efficiency of concrete: A review *Construction Building Materials* **112** 1123-32

[7] Huang H and Ye G 2014 A review on self-healing in reinforced concrete structures in view of serving conditions. In: *3rd International Conference on Service Life Design for Infrastructure, Zhuhai, China, 15-17 Oktober 2014*: Citeseer)

[8] Pei R, Liu J, Wang S and Yang M 2013 Use of bacterial cell walls to improve the mechanical performance of concrete *Cement Concrete Composites* **39** 122-30

[9] Jonkers H M, Thijssen A, Muyzer G, Copuroglu O and Schlangen E 2010 Application of bacteria as self-healing agent for the development of sustainable concrete *Ecological engineering* **36** 230-5

[10] Siddique R and Chahal N K 2011 Effect of ureolytic bacteria on concrete properties *Construction building materials* **25** 3791-801

[11] Erşan Y Ç, Da Silva F B, Boon N, Verstraete W and De Belie N 2015 Screening of bacteria and concrete compatible protection materials *Construction Building Materials* **88** 196-203

[12] Dhami N K, Reddy M S and Mukherjee A 2012 Improvement in strength properties of ash bricks by bacterial calcite *Ecological Engineering* **39** 31-5

[13] Van Tittelboom K, De Belie N, De Muynck W and Verstraete W 2010 Use of bacteria to repair cracks in concrete *Cement Concrete Research* **40** 157-66

[14] Achal V, Pan X and Özyurt N 2011 Improved strength and durability of fly ash-amended concrete by microbial calcite precipitation *Ecological Engineering* **37** 554-9

[15] Dhami N K, Reddy S M and Mukherjee A 2012 Biofilm and microbial applications in biomineralized concrete *Advanced topics in Biomineralization* 137-64

[16] Burne R A and Chen Y-Y M 2000 Bacterial ureases in infectious diseases *Microbes Infection* **2** 533-42

[17] Ramachandran S K, Ramakrishnan V and Bang S S 2001 Remediation of concrete using micro-organisms *ACI Materials Journal-American Concrete Institute* **98** 3-9

[18] Achal V, Mukherjee A and Reddy M S 2011 Microbial concrete: way to enhance the durability of building structures *Journal of materials in civil engineering* **23** 730-4

[19] Bang S S, Galinat J K and Ramakrishnan V 2001 Calcite precipitation induced by polyurethane-immobilized Bacillus pasteurii *Enzyme microbial technology* **28** 404-9

[20] Jena S, Basa B and Panda K C 2020 Effect of Bacillus Cohnii Bacteria on the Properties of Concrete *Recent Developments in Sustainable Infrastructure: Select Prceedings of ICRDSI 2019* Springer pp 597-606

[21] Jena S, Basa B and Panda K C 2020 Bacterial Concrete for the Development of Sustainable Construction - A Review *Recent Trends in Civil Engineering: Select Prceedings of ICRTICE 2019* Springer 587-600

[22] Ghosh P, Mandal S, Chattopadhyay B and Pal S 2005 Use of microorganism to improve the strength of cement mortar *Cement Concrete Research* **35** 1980-3

[23] Jonkers H M 2007 Self-healing concrete: a biological approach *Self healing materials*: An Alternative Approach to 20 Centuries of Materials Science Springer pp 195-204

[24] Chahal N, Siddique R and Rajor A 2012 Influence of bacteria on the compressive strength, water absorption and rapid chloride permeability of fly ash concrete *Construction Building Materials* **28** 351-6

[25] Gavimath C, Mali B, Hooli V, Mallpur J, Patil A, Gaddi D, Ternikar C and Ravishankera B 2012 Potential application of bacteria to improve the strength of cement concrete *International journal of advanced biotechnology research* **3** 541-4

[26] Maheswaran S, Dasuru S, Murthy A R C, Bhuvaneshwari B, Kumar V R, Palani G, Iyer N R, Krishnamoorthy S and Sandhya S 2014 Strength improvement studies using new type wild strain Bacillus cereus on cement mortar *Current Science* 50-7

[27] Jena S, Basa B, Panda K C and Sahoo N K 2020 Impact of Bacillus subtilis bacterium on the properties of concrete *Materials Today: Proceedings*

[28] Phung Q T, Maes N, De Schutter G, Jacques D and Ye G 2013 Determination of water permeability of cementitious materials using a controlled constant flow method *Construction Building Materials* **47** 1488-96

[29] Chahal N, Siddique R and Rajor A 2012 Influence of bacteria on the compressive strength, water absorption and rapid chloride permeability of concrete incorporating silica fume *Construction Building Materials* **37** 645-51

[30] Siddique R and Rajor A 2014 Influence of bacterial treated cement kiln dust on the properties of concrete *Construction building materials* **52** 42-51

[31] Siddique R, Nanda V, Kadri E-H, Khan M I, Singh M and Rajor A 2016 Influence of bacteria on compressive strength and permeation properties of concrete made with cement baghouse filter dust *Construction Building Materials* **106** 461-9

[32] Siddique R, Singh K, Singh M, Corinaldesi V and Rajor A 2016 Properties of bacterial rice husk ash concrete *Construction Building materials* **121** 112-9

[33] Qiu J, Tng D Q S and Yang E-H 2014 Surface treatment of recycled concrete aggregates through microbial carbonate precipitation *Construction Building Materials* **57** 144-50

[34] Grabiec A M, Klama J, Zawal D and Krupa D 2012 Modification of recycled concrete aggregate by calcium carbonate biodeposition *Construction Building Materials* **34** 145-50

[35] De Muynck W, Cox K, De Belie N and Verstraete W 2008 Bacterial carbonate precipitation as an alternative surface treatment for concrete *Construction Building Materials* **22** 875-85

[36] Nosouhian F, Mostofinejad D and Hasheminejad H 2016 Concrete durability improvement in a sulfate environment using bacteria *Journal of Materials in Civil Engineering* **28** 04015064

[37] Ghosh S, Biswas M, Chattopadhyay B and Mandal S 2009 Microbial activity on the microstructure of bacteria modified mortar *Cement Concrete Composites* **31** 93-8

[38] Ranganathan N, Macherone A, Mehta B P R, Marczely J and Dickstein J 2000 Urea Hydrolysis And Ammonia Uptake by Bacillus Pasteurii, Kibow Biotech, Inc. Philadelphia, USA. In: *10th International Congress on Nutrition and Metabolism in Renal Disease-poster# P-19, France from,* pp 6-8

[39] Rowan N J, Caldow G, Gemmell C G and Hunter I S 2003 Production of diarrheal enterotoxins and other potential virulence factors by veterinary isolates of Bacillus species associated with nongastrointestinal infections *Applied environmental microbiology* **69** 2372-6

[40] Baumann P, Clark M A, Baumann L and Broadwell A H 1991 Bacillus sphaericus as a mosquito pathogen: properties of the organism and its toxins *Microbiology Molecular Biology Reviews* **55** 425-36

[41] Stevens D A, Hamilton J R, Johnson N, Kim K K and Lee J-S 2009 Halomonas, a newly recognized human pathogen causing infections and contamination in a dialysis center: three new species *Medicine* **88** 244-9

[42] Andalib R, Abd Majid M Z, Hussin M W, Ponraj M, Keyvanfar A, Mirza J and Lee H-S 2016 Optimum concentration of Bacillus megaterium for strengthening structural concrete *Construction Building Materials* **118** 180-93

[43] Cappitelli F, Toniolo L, Sansonetti A, Gulotta D, Ranalli G, Zanardini E and Sorlini C 2007 Advantages of using microbial technology over traditional chemical technology in removal of black crusts from stone surfaces of historical monuments *Applied environmental microbiology* **73** 5671-5

[44] De Muynck W, Debrouwer D, De Belie N and Verstraete W 2008 Bacterial carbonate precipitation improves the durability of cementitious materials *Cement concrete Research* **38** 1005-14

[45] Ghosh P, Mandal S, Pal S, Bandyopadhyaya G and Chattopadhyay B 2006 Development of bioconcrete material using an enrichment culture of novel thermophilic anaerobic bacteria

[46] Park S-J, Park Y-M, Chun W-Y, Kim W-J and Ghim S-Y 2010 Calcite-forming bacteria for compressive strength improvement in mortar *Journal of microbiology biotechnology* **20** 782-8

[47] Kumar V R, Bhuvaneshwari B, Maheswaran S, Palani G, Ravisankar K and Iyer N R 2011 An overview of techniques based on biomimetics for sustainable development of concrete *Current Science* 741-7

[48] Kota K P, Kota R K, Dulla J B and Karlapudi A P 2014 Bioconcrete Enhancement from Biofilm Producing Marine Bacterium *Int. J. Pharm. Sci. Rev. Res* **25** 276-9

[49] Kumar S S, Krishna A M and Dey A 2014 Dynamic soil properties of Brahmaputra sand using Cyclic Triaxial tests. In: *North East Students Geo-Congress On Advances In Geotechnical Engineering, 1*: Citeseer

Compressive Strength of Mortar with Alkali Activated Fly ash and Ground Granulated Blast Furnace Slag

Sagarika Panda[1], Ramakanta Panigrahi[2], M.L. Narasimham[1]

[1]Department of Civil Engineering, Centurion University of Technology and Management, Bhubaneswar, Odisha, India
[2]Department of Civil Engineering, Veer Surendra Sai University of Technology, Burla, Odisha, India

E-mail: sagarikapanda1988@gmail.com

Abstract. The geopolymer concrete is a new technology that replace ordinary cement concrete with cement less concrete using the different industrial by-products considered as waste to the society. These materials create a hazardous environmental condition. To overcome the same, many researchers have reported replacement of these materials with cement in concrete work. These materials do not possess much binding properties. Hence, an alkaline solution is utilized to activate the same. In the present study, the compressive strength of the geopolymer mortar prepared with Class F fly ash and ground granulated blast furnace slag in different ratios are examined. It is observed that the compressive strength achieved from elevated temperature cured samples is higher than the ambient temperature cured sample. Increase in ground granulated blast furnace slag content also increases the compressive strength parameters of the geopolymer mortar sample.

1. Introduction

Geopolymer is the recent type of technique used in construction industry to develop concrete without cement content to reduce the CO_2 emission and hence protect our planet. It further conserves energy consumption in the world for future [1]. Davidovit's study states that the annual CO_2 emission worldwide is around 0.184 tonnes and OPC includes 80% of the total. The reuse of waste recycled industrial by-products decline manufacturing process of raw materials of OPC [2-4]. The industrial by-products used for the development of geopolymer concrete should be rich in alumina-silicates like Fly ash (FA) and ground granulated blast furnace slag (GGBS) which are further activated with an alkaline solution [5].

The properties and applications of geopolymer concrete such as fire resistance, thermal insulation, and thermal shock refractories is highly appreciated in many fields including engineering process technologies [3]. To use this geopolymer technology in construction industry, the fresh and hardened properties of the mixture shall be investigated. Consistency, workability, setting times of the concrete mix, type of curing and compressive strength are the important parameters to be considered [5]. The strength of concrete mainly depends on the mortar strength which is a factor of great importance. From recent studies, it is revealed that the fly ash emits 80%- 90% less greenhouse gases than that of OPC manufacturing. Similarly, the production of binder with use of GGBS reduces the greenhouse gas emission rate by 80% [6]. According to A.Z. Warid Wazien *et al.* [7], when the solid particles of aluminosilicate raw materials dissolute with an alkali activator solution, polycondensation process forms a 3D network structure known as geopolymerization. Curing conditions have a major role in the strength development of geopolymer mortar/concrete. The strength parameters for geopolymer concrete are much higher as compared to ordinary conventional concrete [8]. Geopolymer concrete also have good flexural [9] as well as shear [10] strength.

In the present work, geopolymer mortars are prepared with Fly ash and GGBS as binders and Sodium Hydroxide and Sodium Silicate as alkali activator solution along with some extra water for better workability. The samples are well compacted and kept for curing in both ambient and elevated temperatures for 7 days and tested for compressive strength under Universal Testing Machine (UTM). A comparative explanation between the strength parameters of the samples at both temperature curing conditions is presented.

2. Materials

Class F Fly ash and GGBS, used as the raw materials, are collected from Jindal Steel and Power Limited (JSPL), Jajpur, Odisha and Mideast Integrated Steel Limited (mesco), Jajpur, Odisha respectively. Table 1 shows the characterization of the Fly ash and GGBS carried out with help of X-ray Fluorescence test (XRF). The sand of Daya river is used as fine aggregates for the manufacturing of geopolymer mortar. The grading confirms to Zone II as per IS 383. The specific gravity and fineness modulus of the fine aggregates are recorded as 2.62 and 2.6% respectively.

Alkali Activator solution is prepared with NaOH and Na_2SiO_3. Sodium hydroxide of 14 M molarity is prepared by mixing 560 grams of Caustic Soda pellets in one litre of total solution and stirring continuously for two minutes and slowly mixing Sodium Silicate solution (Water Glass) into it. The alkaline solution is kept for 24 hrs. After which it is suitable for use. Hence the solution is needed to be prepared a day before experiment. It gives best results when used hot.

Table 1. Chemical composition of Fly ash and GGBS in XRF test ((in percentage)

	Al_2O_3	SiO_2	P_2O_5	Cl	K_2O	CaO	TiO_2	V_2O_5	CrO_3	MnO	Fe_2O_3	SO_3
Fly ash	28.80	59.48	0.95	0.11	1.78	1.66	2.04	429.6	259.0	409.5	4.78	-
GGBS	18.92	29.58	-	0.14	0.91	45.71	1.36	92.1	29.9	0.28	0.97	1.71

3. Experimentation and Results

Geopolymer mortar samples are prepared in different ratios of binders i.e. Class F Fly ash and GGBS. The alkaline solution-to-binder ratio is considered as 0.25 and addition water-to-binder ratio is taken as 0.2. The mix proportions for the geopolymer mortar are shown in Table 2.

Table 2. Mix proportions (kg/m³)

Samples	Fly ash	GGBS	Fine Aggregates	NaOH	Na_2SiO_3	AAR	Extra Water
M1	580	-	1740	41	104	145	116
M2	551	29	1740	41	104	145	116
M3	522	58	1740	41	104	145	116
M4	493	87	1740	41	104	145	116
M5	464	116	1740	41	104	145	116
M6	435	145	1740	41	104	145	116
M7	406	174	1740	41	104	145	116
M8	377	203	1740	41	104	145	116
M9	348	232	1740	41	104	145	116
M10	319	261	1740	41	104	145	116
M11	290	290	1740	41	104	145	116

The Fly ash and GGBS based geopolymer mortar samples are prepared with 5%, 10%, 15%, 20%,

25%, 30%, 35%, 40%, 45%, 50% GGBS content. Six cubes, each of dimension 70.6 mm x 70.6 mm x 70.6 mm, are casted for 7 days' compressive strength of mortar. Out of these cubes, three set of samples are taken for ambient temperature curing and other three set of samples are used for oven curing. The ratio of binder-to-sand is kept 1: 3. For each sample, 200 gm binder and 600 gm of sand is taken as per Indian Standards. All the samples are well compacted and finished in fresh state.
Before preparing the geopolymer mortar mix, the moulds are properly cleaned and tightened. After cleaning, the moulds are well greased to achieve smooth demoulding. In each and every mould the freshly prepared mix is filled in three layers of equal depths and in each layer 25 blows of tamping rod are provided for compaction of the sample. The top surface of the sample is smoothened with help of a trowel removing excess mixture. After setting of the samples, they are demoulded and cured for 7 days.

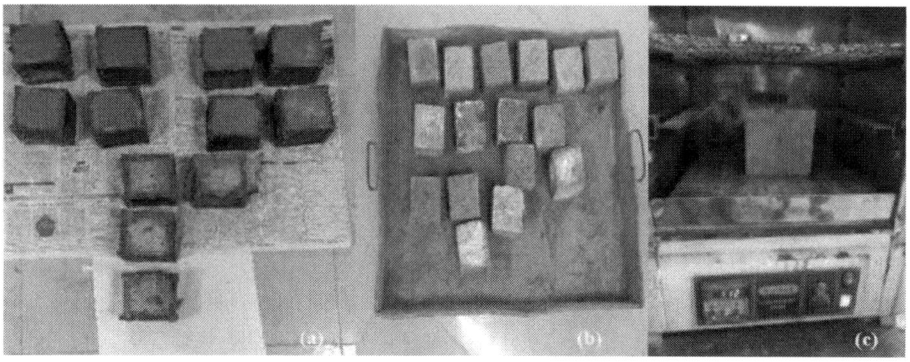

Figure 1. (a) Cubes after casting, (b) Cubes kept for ambient curing, (c) Cubes kept in Oven for elevated temperature curing

The samples once set are demoulded and divided for different curing process. Three of the samples are kept at room temperature curing (ambient curing) and other three samples are kept in oven at 112 °C for 24 hrs and then at room temperature for the rest of the period. Figure 1 shows the before and after curing samples of geopolymer mortar.

3.1. Consistency Test

The procedure considered to find the normal consistency, compressive strength of geopolymer is similar like cement. In these tests, cement is replaced by fly ash and GGBFS. The geopolymer mortar in fresh state is tested in Vicat apparatus and the standard consistency is found in the range of 30-32 percentage.

3.2. Compressive Strength Test

The compressive strength test is carried out for geopolymer mortar with the standard cube of size 70.6mm x 70.6mm x 70.6mm. For both ambient and heat curing samples,7 days compressive strength is found in CTM and recorded. Table 3 shows compressive strength of 11 trial mix ratios at 7 days of curing.
The compressive strength of the Fly ash-GGBS ratio 50:50, cured in elevated temperature condition, is recorded as the highest i.e. 64.325 MPa and the lowest compressive strength i.e., 27.349 MPa is observed with 100% Fly ash geopolymer mortar sample cured in ambient temperature. Figure 2 represents the compressive strength of different ratios after 7 days curing.
The compressive strength of ambient temperature curing is observed to be less than that of oven curing. The change in ratio of Fly ash and GGBS has a great effect over the compressive strength. GGBS plays an important role in strength gain in geopolymer mortar. Increase in GGBS content improves the strength of the mortar. This shows that the compressive strength of the geopolymer mortar varies with the type and quantity of binder used.

Table 3. Compressive strength (MPa)

Test samples	7 days compressive strength in MPa	
	Elevated Temperature Curing	Ambient Temperature Curing
M1	11.682	7.247
M2	13.450	8.756
M3	14.235	9.237
M4	16.862	11.854
M5	19.275	13.143
M6	26.423	15.364
M7	31.918	17.928
M8	39.254	19.864
M9	48.487	22.680
M10	57.873	25.483
M11	64.325	27.349

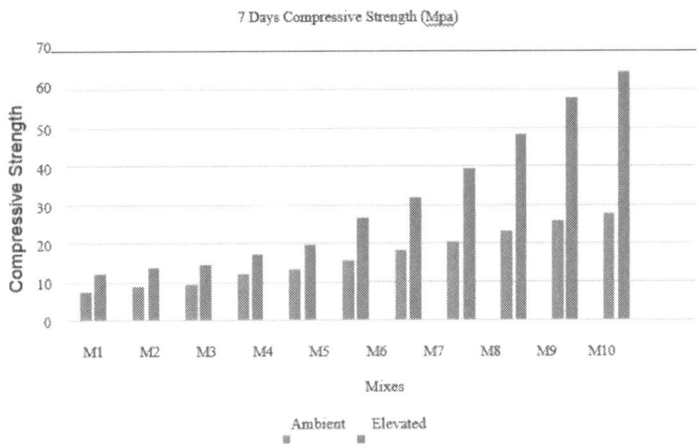

Figure 2. Compressive strength of geopolymer mortar after 7 days

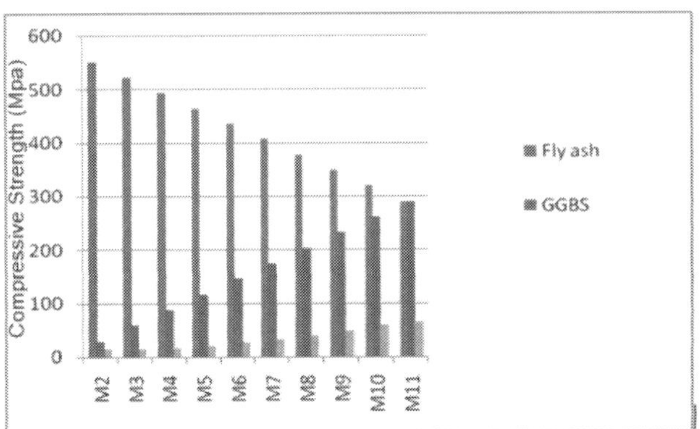

Figure 3. Compressive strength of geopolymer mortar at elevated temperature with variation of Fly ash and GGBFS content

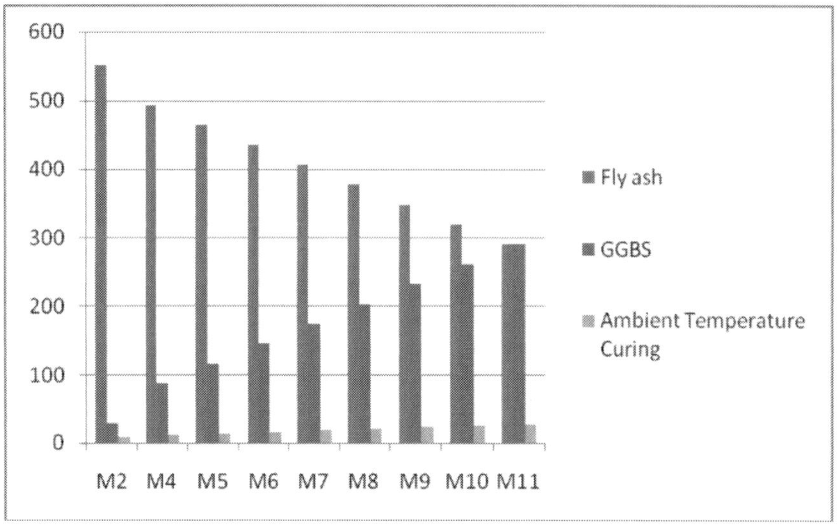

Figure 4. Compressive strength of geopolymer mortar at elevated temperature with variation of Fly ash and GGBFS content

4. Conclusions

This paper has presented the compressive strength of geopolymer mortar with different ratios of Fly ash and GGBS as binder material along with river sand as fine aggregate. An alkaline solution of NaOH and Na_2SiO_3 is considered with constant ratio and concentration. The following conclusions are made from this study:

- The strength achieved for oven cured mortar sample is higher as compared to ambient cured geopolymer mortar sample.
- With increase in the percentage of GGBS in the geopolymer mix, the compressive strength increases.
- Fly ash based geopolymer mortar shows less compressive strength compared to the samples with GGBS replacement.

References

[1] Narayanan A and Prabavathy S 2018 Study on geopolymer mortar using hydrogen peroxide as foaming agent *Journal of Structural Engineering (India)* **45** 139-47

[2] Lee W-H, Wang J-H, Ding Y-C and Cheng T-W 2019 A study on the characteristics and microstructures of GGBS/FA based geopolymer paste and concrete *Construction Building Materials* **211** 807-13

[3] Saint-Quentin 2011 *Geopolymer: chemistry and applications* (Institut Géopolymère World Congress Geopolymer: Institut Géopolymère)

[4] Kabir S A, Alengaram U J, Jumaat M Z, Yusoff S, Sharmin A and Bashar I I 2017 Performance evaluation and some durability characteristics of environmental friendly palm oil clinker based geopolymer concrete *Journal of cleaner production* **161** 477-92

[5] Samantasinghar S and Singh S P 2019 Fresh and Hardened Properties of Fly Ash–Slag Blended Geopolymer Paste and Mortar *International Journal of Concrete Structures Materials Today: Proceedings* **13** 47

[6] Pradeep U, Purushotham R S and Markandeya R P 2016 The Effect of Age on Alkali-Activated Geo Polymer Mortar at Ambient Temperature *SSRG International Journal of Civil Engineering* **3** 16-8

[7] Wazien A W, Abdullah M, Abd Razak R, Rozainy M and Tahir M F M 2016 Strength and density of geopolymer mortar cured at ambient temperature for use as repair material. In: *IOP Conference Series: Materials Science and Engineering,*

[8] Manjunath G and Giridhar C 2011 Compressive strength development in ambient cured geo-polymer mortar *International journal of earth sciences engineering* **4** 830-4

[9] Bouaissi A, Li L-y, Abdullah M M A B and Bui Q-B 2019 Mechanical properties and microstructure analysis of FA-GGBS-HMNS based geopolymer concrete *Construction Building Materials* **210** 198-209

[10] Yacob N S, ElGawady M A, Sneed L H and Said A 2019 Shear strength of fly ash-based geopolymer reinforced concrete beams *Engineering Structures* **196** 109298

IOP Conf. Series: Materials Science and Engineering **970** (2020) 012006 doi:10.1088/1757-899X/970/1/012006

Advances in Seismic Strengthening Materials and Techniques

Sanjivan Mahadik[1], Lomesh Mahajan[2], S. R. Bhagat[1]

[1]Department of Civil Engineering, Dr. Babasaheb Ambedkar Technological University, Lonere, Maharashtra, India –402104
[2] Department of Civil Engineering, R. C. Patel Institute of Technology, Shirpur, Maharashtra, India

E-mail: sanjivanmahadik@gmail.com

Abstract. Seismic strengthening of reinforced concrete (RC) members is a method of improving their existing capacity to withstand seismic forces. The process typically includes strengthening of RC structural members such beams, columns, slabs, footings, beam-column junctions, etc. to improve the axial, flexure, and shear capacity. Experiences during the recent earthquakes and changes in Indian Standards made it essential to carry out structural audit of buildings. The buildings, which do not comply with the recent Code provisions have to be reconstructed or to be strengthened to achieve the desired capacity level. Reconstruction of such buildings will consume time and money. In addition, evacuation of such buildings for reconstruction is a difficult task. In this paper an attempt has been made to focus on the seismic strengthening of structural members of RC buildings which are designed to only gravity loads or the structures designed prior to release of new building Codes. This paper covers a brief introduction of conventional methods of strengthening followed by an introduction of advanced FRP composite materials and techniques of retrofitting for existing RC buildings. Finally, it is concluded that the advanced FRP composite materials and recent innovative methods can more advantageously used for seismic strengthening of RC buildings.

1. Introduction

Many buildings in India are vulnerable due to inadequacy in their strength due to some reason though they are coming under high intensity zone of earthquake as per study of Giuseppe Oliveto and Massimo Marletta [1]. In seismic strengthening seismic load resisting capacities of buildings are improved such that these become safe as per existing Codal provisions. Different factors affecting strength of structures are its dimensions, materials used for their construction and material properties, arrangement and percentage of reinforcement etc. Structures requires strengthening due to several reasons like weakening of structure due to accident, damage in earthquake or changes in standards due to change of Code of practice. Sometimes changes in structure are made due changes of uses of building or changes in environment. The main aim in strengthening of structure is to improve its strength, stiffness and sometimes ductility or durability. Rehabilitation is requires sometimes for old buildings which had lost their strength due to old edge. Retrofitting is essential to meet current Code requirement. Different conventional technics and some innovative technics available and are in practice for seismic strengthening. Use of composite has become popular in recent days due to their added properties and simplicity in applications. Work done by previous researchers on seismic strengthening of RC structure with conventional and innovative methods is reviewed and special attention is made on use of FRP composite materials and other combinations used in seismic resistance of existing structure, strengthened structure, structure repaired after damage and review of research made by many researchers in regards to strength, stiffness and effectiveness. Use of FRP was studied by Asfa M et al. [2] for seismic strengthening of shear wall with openings using different configuration of FRP around opening and found that, FRP composite can be used effectively for improving behaviour of shear walls with openings. And it was also investigated that, there is remarkable enhancement in behaviour of shear wall having opening by making FRP strengthening. Different techniques of retrofit or strengthening are available but their choice depend on expected failure mode, consequences of wall retrofit, physical limitations, and budget available as investigated by Galall K. and El-Sokkary H [3]. Application of CFRP strips and its usefulness in applying as a of the externally bonded CFRP around structural member in improving the capacity of RC member compared to externally bonded CFRP sheets is made by Mofidi Amir and Chaallal Omer [4]. Use of FRP in RC structure was studied considering parameters like point of application of these strips, width of strip and location of strips with respected to shear reinforcement. It was observed that, use of FRP in the form of strips is more advantages than to use it as continuous sheets. Placing CRP strip between reinforcement found beneficial than that of placing them along same location. Use of jacket having high performance was tried for strengthening of existing shear wall by Marini A. and Meda A. [5]. Performance of shear wall found improved significantly in term of strength, as well as ductility and deformability. Use of cement composite with fiber reinforcement was tried by Parra- Montesinos and Gustavo J. [6] for seismic strengthening of structures like beam to column joint, shear wall having low height, coupling beam etc. It was observed that, use of this type of composite significantly improves behaviours like strength, deformation capacity and overall behaviour of structural element. Marini A. and Meda A. [5] had presented use of jacket made with steel mesh for retrofitting of shear wall and observed that, use of this jacket having very good performance in safeguarding structures from earthquake and in improving behaviours like strength, deformation capacity, ductilityetc. Layssi Hamed and Mitchell Denis [7] made tests on two walls repaired with jacket of steel fibers concrete having self-consolidation property and retrofitted with warp of CFRC and found that retrofitting of wall make it to behave with improved ductility. Repair with above jacket improved displacement behaviour of walls and flexural behaviour. Galall K. and El-Sokkary H [3] made use of carbon fiber polymer in rehabilitation of shear wall improving its behaviour and observed that, performance of shear wall improved remarkably in shear strength, flexure and ductility. Lee Kang Seok et al [8] in their investigation on column found that, use sprayed FRP in strengthening of column improved shear strength and deformation capacity markedly.

2. Conventional Strengthening Methods

IST Group (2004) presented conventional strengthening used in structure to improve performance of structural members. These methods consists of addition of structural member where required, placing a new shear wall, enlarging size of existing structural element or bracing of existing structural elements etc. Addition of shear wall and bracing system found most efficient and it has became more popular due to their ease in construction, lower in cost and having good effect compared to other methods. Shear wall helps to resist most of the probable lateral load acting on structure keeping structure safe [9, 10].

Following conventional methods are in present practices.

- Supporting with steel sections
- Addition of precast walls
- Strengthening with additional foundation
- Jacketing with suitable high performance material
- Addition of extra column
- Grouting weaker section

Use of steel section in the form of cross bracing and as a steel plate shear wall is common method since past many years. These are easy to apply and causing minimum disturbance to occupants while installations and have good strength. Some other methods are also in use like shotcreting, grouting and epoxy injection, and prefabricated reinforced concrete thin walls panels or steel plate bracing fixed to one side of brick wall etc. are also work effectively in strengthening of structures. Results of conventional methods of repair and retrofitting are not up to mark of satisfaction. Existing practices of repair of buildings go little beyond cosmetic treatment of the structure. Such methods neither strengthen the structure nor extend its life as observed by Mukherjee Abhijit and Joshi Mangesh V. [11]. Recently new and innovative methods are come in practice, which includes use of composite materials, use of energy dissipaters and isolators in seismic resisting system.

3. Retrofit of Structures using Innovative Materials and Approaches

The innovative methods of seismic retrofitting may be grouped into the following classes as per Gustavo J. Parra-Montesinos et al. [9] had differentiated these methods in to following classes.

- Base isolation
- Energy Dissipation System
- Use of Composite
- Combination of above methods

3.1. Base isolation (Stiffness reduction)

The base isolation is effective method of stiffness reduction approach. Base isolation is separation of super structure of buildings from sub structure introducing base isolators in between for the purpose of reducing seismic effect on buildings. Though technology is still developing, it is being used in many structures in worldwide. Each structure has its own natural frequency of vibration depending on various factors like mass, stiffness and height. Whole building oscillates in earthquake violently with frequency of earthquake. Base isolators are available in different variety like roller bearing isolators, springs isolators, sliding Bearing, elastomeric bearings. Each type has its own advantages and dis advantages or limits to use. Figure1 shows all these type of isolators.

(A) Roller bearing (B Spring isolators) (C) Sliding Bearing (D) Elastomeric bearings

Figure 1. Type of Base Isolators

3.2. Confinement (Ductility increase)

Confinement of weak or failed element can be done with FRP composites. It improves ductility of member. In this method, structural members are strengthened with application or warping of structural members where essential for improving seismic resisting capacity of them. These materials in new form are easy to apply and are more efficient.

3.3. Energy Dissipation System (damage controlled structures)

In this method of seismic retrofitting new elements called dissipaters are introduced with diagonal braces to damp seismic forces by absorbing seismic energy striking to structures at the time of occurrence of earthquake.

Dampers works like hydraulic shock absorbers in vehicles, which absorbs most of the jerks due to acceleration or rough roads. Commonly used seismic dampers are shown in Figure 2 and are explained below.

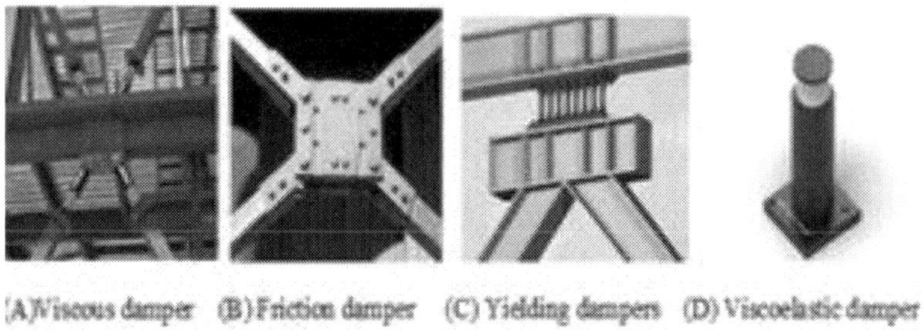

Figure 2. Type of Seismic Dampers

a) Viscous Dampers: in this system energy of seismic wave is absorbed with fluid passing between piston and cylinder arrangements. Piston head utilizes specially shaped passages and damp the energy by altering the flow.
b) Friction Dampers: in this system, seismic energy is absorbed with friction between rubbing surfaces. Use of friction between plates is made for controlling seismic waves.
c) Yielding dampers: In this system, seismic energy is absorbed by yielding of secondary components inserted between two main components during shock or vibration.
d) Viscoelastic dampers: In this system, seismic energy is absorbed by utilizing the controlled shearing of solids converting seismic energy to heat energy.

3.4. Use of Composite Materials

Composite materials are made with a addition of two or more materials for getting available a new material having required additional properties. Use of composite materials is familiar and aware to human being from ancient days for improving properties of materials to be used in homes or making articles as per study of Pandey P.C. [10]. Use of straw in mud bricks in old days is one of the best examples for this thing. CFRP, GFRP, AFRP are the mostly used polymers in construction industry for strengthening or retrofitting. These materials are superior in quality, easy to apple and reasonable in cost so had preferred for strengthening or retrofitting by most of the peoples from last few decades. Each ingredient of the composite material do not dissolve each other though they are working in same mixture and for same work. One material is material mix with other in the form of reinforcement. Reinforcing material is generally available in fibers, particles or sheets and they are mixed to matrix which is a bulk material in which reinforcing materials are dispersed as per nptel.ac.in [12]. As per study of Susana Cabral-Fonseca [13] four main fibers used in the construction industry are: (i)bcarbon fiber, (ii) glass fiber, (iii) basalt and (iv) aramid. Composite materials are used generally for local strengthening of structural elements either in compression, flexure of in shear.

These materials are applied generally for
(a) Strengthening of RC members for flexural
(b) Strengthening of RC members shear
(c) Confinement of RC members

4. Flexural Strengthening of RC Members

Structural members can be strengthened using FRP in the region of their maximum tension, compression or shear stresses. FRP is applied in the form of strips or sheets as per requirement. Epoxy is used as adhesive generally [4]. For proper application of FRP standers are available. Flexural strengthening of beam is shown in Fig. 3.

Figure 3. Girder strengthened for flexure [14].

5. Shear Strengthening of RC Members

Shear strengthening of structural member is done as shown in Figure 4. For shear strengthening FRP is applied on beam surface in inclined manner with beam axis to resist shear stresses in beam cross section[13] Strengthening of column is usually done as shown in Figure 5

Figure 4. Shear strengthening of bridge [14]

Figure 5. Confinement of RC Column Applying CFIR [14]

Peoples are familiar to use of composite materials from ancient days for improvement of quality of material and for innovative works as per study of Pandey P.C. [10]. Advanced composite materials are more durable, easy to apply and having good strength. Strengthening of structural members has become easier due to introduction of FRP in strengthening application. Commonly CFRP, GFRP, and AFRP are used in most of situations like flexural, shear or compression strengthening. Columns are generally strengthened for compression with confinement. For beams, bonding of FRP is made on externally for flexure or shear at specific locations.

6. Summary

Retrofitting of structures can be made with different way according to suitability of use for particular type of failures. Table 1 gives information of different types of retrofitting.

Table 1. Summary of retrofitting methods

Method of retrofit	Arrangement used	Applications, use and nature
Conventional methods	Steel bracing or steel plate, jacketing, grout etc.	Applied for existing as well as failed structures
Base isolation	Base isolators	Generally applied to heavy structures. Separates sub and super structures. Reduce seismic energy by stiffness reduction using isolating pads or bearings.
Energy Dissipation System	Dampers	Suitable for short as well as tall building. Applied as cross bracing. Damps seismic energy to safeguard structure.

7. Conclusion

Different innovative methods are now available for strengthening and retrofitting technique for improving behaviour of structures and make them safe for use as per standards. These different techniques consists of strengthening of existing structures by addition of member, application of shotcreting, ferrocement filling, use of advanced composite materials and different techniques for seismic retrofitting. RC structures can successfully be strengthen or retrofitted with advanced composite materials and by using different techniques. FRP composites are most efficient easier in application material for retrofitting of structural element. Composites can be used effectively for retrofitting. Addition of new shear walls and steel bracing elements can be suitably done as per requirement. Retrofitting can be done successfully by strengthening applications mainly due to the relative ease of installation. It is essential to retrofit the buildings, which are seismically unsafe using suitable materials and techniques. The retrofitted structural members are able to develop their nominal axial, flexural, shear capacities and ductility. Finally, it can be concluded that the advanced composites materials can be used for seismic strengthening of structures with better performance than conventional methods. Innovative methods like base isolation, energy dissipation or use of composite materials are new, easier, economic and effectives techniques over conventional strengthening methods.

References

[1] Oliveto G and Marletta M 2005 Seismic retrofitting of reinforced concrete buildings using traditional and innovative techniques *ISET Journal of earthquake technology* **42** 21-46

[2] Asfa M, Mostofinejad D and Abdoli N 2011 *Advances in FRP Composites in Civil Engineering*: Springer) pp 837-40

[3] Galal K and El-Sokkary H 2008 Recent advancements in retrofit of RC shear walls. In: *Proceedings of the Fourteenth World Conference on Earthquake Engineering, Beijing, China,*

[4] Mofidi A and Chaallal O 2011 Shear strengthening of RC beams with externally bonded FRP composites: Effect of strip-width-to-strip-spacing ratio *Journal of Composites for Construction* **15** 732-42

[5] Marini A and Meda A 2008 Seismic retrofitting of existing shear walls by means of high performance RC jacket. In: *Proc. of the 14th World Conference on Earthquake Engineering, Beijing,*

[6] Parra-Montesinos G J 2005 High-performance fiber-reinforced cement composites: an alternative for seismic design of structures *ACI Structural Journal* **102** 668

[7] Layssi H and Mitchell D 2012 Experiments on seismic retrofit and repair of reinforced concrete shear walls. In: *Proceedings of the 6th International Conference on FRP Composites in Civil Engineering (CICE),* pp 13-5
[8] Lee K S, Lee B Y and Seo S Y 2016 A seismic strengthening technique for reinforced concrete columns using sprayed FRP *Polymers* **8** 107
[9] Parra-Montesinos G J, Canbolat B A and Jeyaraman G 2006 Relaxation of confinement reinforcement requirements in structural walls through the use of fiber reinforced cement composites. In: *8th National Conference on Earthquake Engineering,*
[10] 2004 Methods for Seismic Retrofitting of Structures-technical *IST Group*
[11] Mukherjee A and Joshi M V 2002 Seismic retrofitting technique using fibre composites *Indian concrete journal* **76** 496-502
[12] https://nptel.ac.in/courses/105/108/105108124/#
[13] Custódio J and Cabral-Fonseca S 2013 *Advanced Fibre-Reinforced Polymer (FRP) Composites for Structural Applications*: Elsevier) pp 814-82
[14] Motavalli M and Czaderski C 2007 FRP composites for retrofitting of existing civil structures in Europe: State-of-the-art review. In: *International Conference of Composites & Polycon*: American Composites Manufacturers Association Tampa, FL, USA) pp 17-9

IOP Conf. Series: Materials Science and Engineering **970** (2020) 012007 doi:10.1088/1757-899X/970/1/012007

Studies on influence of mineral admixture on mechanical properties of CC and SCC

Shashwati Soumya Pradhan[1], Kishor Chandra Panda[2]

[1] Department of Civil Engineering, National Institute of Technology, Agartala, Tripura, India
[2] Department of Civil Engineering, Government College of Engineering, Kalahandi, Bhawanipatna, Odisha, India

E-mail: saswatipradhan09@gmail.com

Abstract. Now a day's research is more concern on the successful utilization of different mineral admixture so as to cut down the environmental pollution caused by cement industries throughout cement production. Self-Compacting Concrete (SCC) is a new generation concrete has been developed without compromising the strength characteristics with minimum time and labour cost. So to make more economic and environmental friendly concrete, two mineral admixtures (waste products of the industries) i.e. Silpozz and Lime Stone Powder (LSP) were used during this experimental work. The aim of this study is to evaluate the fresh and hardened properties of the conventional concrete (CC) and Self Compacting Concrete (SCC) containing mineral admixtures. Cement is partially replaced with Silpozz and LSP as 0%, 10% and 20% for both CC and SCC. Total ten numbers of concrete mixes have prepared for M30 grade of concrete. Fresh concrete properties of CC have evaluated by slump test and for SCC, slump flow, T500, V-funnel and J-ring test were conducted. Hardened concrete properties were evaluated after 7 and 28 days for compressive, split tensile and flexural strength test. The desired amount of superplasticizer (SP) is used in SCC to make the concrete workable. The test results indicate that the 20% replacement of Silpozz influences the mechanical properties of concrete as compared to LSP replacement.

1. Introduction

Japan introduced Self-Compacting Concrete (SCC) in 1988 due to deficiency in skilled workers [1]. SCC is gaining popularity due to its nature of compact itself in congested stirrups and rebar without any vibration. It decrease the labour cost, reduces noise level and pouring the concrete time get reduced during construction [2]. The key features of SCC are increase in segregation resistance, high flowability and high passing ability which make it advantages over other concrete in presence of superplasticizer and other additives [3]. The higher cost of superplasticizer reduces its application [4]. This problem can be solved by utilizing mineral admixtures as partial replacement of cement as reported by Kannan [5].

Due to industrialization the waste product generated from the industries created the disposal problem which also pollute soil, water and air [6]. Researches were using these mineral admixtures as partial substitution for cement to mitigate the problem [7]. Mineral admixtures help to enhance the strength and overall performance of concrete. An agro-industrial waste is Rice Husk Ash (RHA) which is generated by the combustion of rice husk in a control temperature to form husk [8]. This husk has

beneficial effects such as improve strength and durability, reduce heat of hydration and reduces the cement consumption when utilizing it as partial cement substitute [6]. The production of rice husk is about 148 million tons in 2015 in worldwide. The final product of RHA contain about more than 90% of amorphous silica[4]. In SCC, RHA with urea improve the flowability in mass concrete and the concrete temperature is get reduced [9]. RHA provides denser microstructures based on chemical interaction among calcium hydroxide and water and form C-S-H gel in SCC [10]. Inclusion of RHA in conventional concrete (CC) shows better performance as compared to control specimen [11]. Silpozz addition enhances the Compressive, split tensile and flexural strength at all ages of curing [12]. He et al. [13] reported that the waste Lime Stone Powder (LSP) was generated during the quarrying process of crushing and grinding Lime Stone which is a non pozzolanic material. The mineral composition of LSP is present in various forms such as calcite, aragonite, amorphous calcium carbonate and vaterite etc. Crystalline, clastic, massive or granular are some of the form of LSP based on the formation[14]. It generally used in cement production due to its filler effect. LSP is very fine and caused the environmental pollution as well as health hazards. It increases the workability, total shrinkage but reduces the setting time as compared to control specimen [1]. Cement paste containing high LSP has come across different stages of hydration such as dissolution, acceleration, dynamic balance and hardened stage [13]. It induces high early strength by increasing the hydration [15]. LSP can be used as low cost ultra-high performance concrete and reduces the carbon dioxide emission [16] without compromising the strength characteristics [17] and improves the sulphate resistance in CC [18]. Using wood fiber waste with limestone powder waste help to produce light weight concrete [19]. In cement paste, the $CaCO_3$ consumption increases due to the formation of carboaluminate when the lime stone powder increases. LSP has low cost and wider availability make it more popular to use in concrete [20]. RHA and LSP substitution decreases the unit weight of concrete [1]. The combine effect of slag and pulverized LSP helps to reduce the slump loss by providing better workability and reduces depth of carbonation, drying shrinkage for better durability properties [16]. Matsagar [21] concluded that the porosity of transition zone among both cement paste and aggregate can be reduced through substitution of Silpozz with 10% FA. FA and Silpozz replacement in concrete up to certain limit improve the resistance against deterioration [22]. The combine effect 10-30% FA and 10-20% of Silpozz affect the fresh and hardened characteristics of concrete [23].
It is evident that the past studies are showing good performance by the utilization of mineral admixtures. Here the effort has been done to investigate of mineral admixtures such as Silpozz and LSP utilization in CC and SCC as partial cement replacement to minimise the contamination generated by the manufacture of cement and the efficient use of waste industrial byproducts, which creates the disposal problem. In order to achieve this aim, various experiments on fresh and hardened concrete have been carriedout to determine the effect of raw materials on the finished products.

2. Experimental Program

Table 1. Chemical composition of raw materials

Oxide	SiO_2	Al_2O_3	Fe_2O_3	CaO	MgO	SO_3	LOI	Carbon	K_2O	Others	Moisture
OPC	20.50	5.05	2.99	62.00	2.07	2.40	3.10	-	-	-	-
Silpozz	88.18	1.61	0.56	1.59	1.63	-	-	2.67	1.67	2.09	0.79
LSP	10.50	0.82	0.29	49.00	0.58	0.10	38.71	-	-	-	-

2.1. Materials

Ordinary Portland Cement (OPC) 43 grade is used and it's specific gravity is 3.15. Silpozz is the trade name for RHA which is an agricultural by products which contain more than 85% of silica. Its particle size of Silpozz is 25 microns and specific gravity is 2.71 [24]. LSP is a non pozzolanic material which can act as filler materials and it has specific gravity 2.8. The chemical compositions of OPC, Silpozz and LSP are presented in Table 1.

River sand is used as fine aggregate passing through 4.75 mm sieve conform to zone III is used having water absorption 0.40, specific gravity 2.67. Coarse aggregate has specific gravity 2.86 and water absorption 0.20. CERA HYPERPLAST XR W-40 superplasticizer is used in SCC to obtain good workability [24]. This conforming to IS 9103, ASTM C 494-03 and BS 5075.

2.2. Mix proportions and test procedure

For M30 grade concrete with w/c ratio 0.43 with mixing proportions of 1:1.44:2.91 as per IS: 10262-2009, the mix design was prepared. By varying the Silpozz and LSP content in OPC, five concrete mixes for CC is prepared. Variations of fine and coarse aggregate, variation of Silpozz and LSP replacement in cement along with the addition of superplasticizer (SP) is used to prepare five SCC mixes as represented in Table 2. Fresh concrete workability was assessed immediately after completion of the mixing. For CC, slump test is conducted to determine the fresh concrete properties. For SCC the fresh concrete is accessed by Slump test for filling ability, T_{500} test and V-Funnel test for flow ability by following EFNARC 2005 guidelines and J-Ring test measure the passing ability by following EFNARC 2002 guidelines. The mechanical characterization of hardened concrete was evaluated by compressive strength test in a cube of size 150 × 150 × 150 mm, split tensile strength test in a cylinder of size 100 × 200 mm and flexural strength in the prism of size 100 × 100 × 500 mm for both CC and SCC after the curing period of 7 and 28 days. Three specimens were cast and examined for each day of testing. The average value is taken as final value for the specific test conducted.

Table 2. Concrete Mix proportions of CC and SCC

Mix Identity	Concrete Mix Proportion
CL0S0	OPC 100%
CL10S0	OPC 90%+ LSP 10%
CL20S0	OPC 80%+ LSP 20%
CL0S10	OPC 90%+ Silpozz 10%
CL0S20	OPC 80%+ Silpozz 20%
SCL0S0	OPC 100%+ SP 0.35%
SCL10S0	OPC 90%+ LSP 10% + SP 0.35%
SCL20S0	OPC 80%+ LSP 20% + SP 0.35%
SCL0S10	OPC 90%+ Silpozz 10% + SP 0.45%
SCL0S20	OPC 80% + Silpozz 20% + SP 0.65%

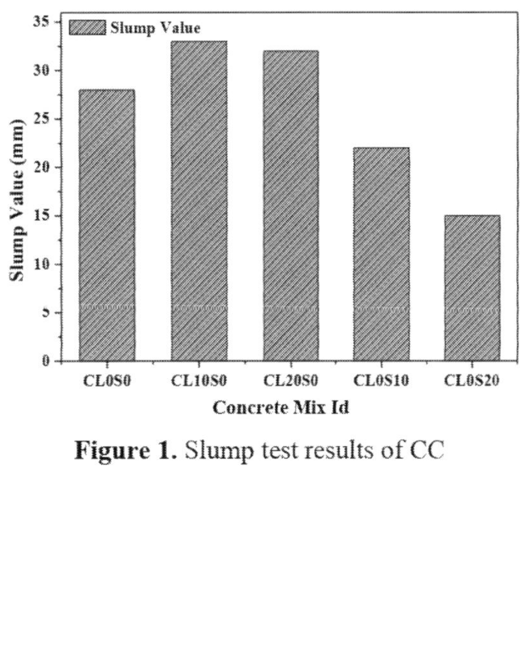

Figure 1. Slump test results of CC

3. Results and Discussion

3.1. Fresh Concrete Test Results

Slump test is performed to evaluate the fresh concrete properties of CC. Figure 1 represents the slump test results of CC. The use of LSP in concrete has been found to improve workability in concrete. Concrete containing 10% LSP shows good workability as comparision to other CC mixes. This is related to the changes occurs in morphology in presence of finer particles of LSP which reduce water bleeding and segregation [16]. But Silpozz addition decrease the workability of CC due to higher surface area of Silpozz [11]. In absence of superplasticizer addition, the more will be the Silpozz content in concrete the lesser will be the workability [25].

Table 3. Slump flow and T_{500} Test results of SCC

Mix Identity	SP (%)	Slump Flow		T_{500} Test	
		Result (mm)	EFNARC(2005)	Result (sec)	EFNARC(2005)
SCL0S0	0.35	560	550-650	15	>2
SCL10S0	0.35	573	550-650	12	>2
SCL20S0	0.35	645	550-650	4	>2
SCL0S10	0.45	580	550-650	9	>2
SCL0S20	0.65	672	660-750	5	>2

In the case of SCC, the results of workability by slump flow, T_{500} test is represented in Table 3 and J-Ring Test, V-Funnel test results is represented in Table 4. Constant amount of superplasticizer i.e. 0.35% required in control specimen and 10%, 20% of LSP addition and it shows higher slump flow[26]. Increasing the Silpozz content in SCC tends to decreases the workability and required more superplasticizer for better workability [8] due to higher viscosity is caused by increase in particle friction [2]. T_{500} test results satisfy EFNARC (2005) guidelines and come under VS2 category which provides good flowability in SCC. Finer the Silpozz particle the more will be the water requirement [7].

Table 4. J-Ring and V-Funnel Test results of SCC

Mix Identity	J-ring Test			V- Funnel Test	
	Height(mm)	Flow(sec)	EFNARC(2002)	Result	EFNARC(2005)
SCL0S0	12	21	0 – 10	20	7 – 27
SCL10S0	07	15	0 – 10	14	7 – 27
SCL20S0	05	7	0 – 10	9	7 – 27
SCL0S10	07	8	0 – 10	12	7 – 27
SCL0S20	05	9	0 – 10	14	7 – 27

Passing ability of SCC is measured through J-Ring and all concrete mix fulfills EFNARC (2000) criteria except the control specimen. Increase in porosity of Silpozz due to 20% replacement in cement increases the water requirement and required more superplasticizer [9]. Flowability of SCC measured by V-Funnel test fall under EFNARC (2005) criteria VF2 [4]. All SCC mixes shows acceptable flow ability as specified by EFNARC criteria. Addition of 20% LSP shows better performance as compared to other mixes in SCC by incorporation of higher amount of superplasticizer.

3.2. Compressive Strength Test Results

The compressive strength results of CC and SCC is represented in Figure 2. The results shows that the increase in compressive strength of CC and SCC in case of 10% LSP addition and more the addition shows decrease in strength as comparision with Silpozz addition [1]. Adding LSP has both positive and negative effect on compressive strength [26]. LSP had detrimental impact on ultra-high performance concrete at 20% replacement [17]. It also reduces the hydration and increase concrete porosity [27]. Due to the nucleation and filling effect, LSP affects compressive strength at an early stage [16]. The strength decrease in LSP based concrete may be due to the dilution effect of LSP [20]. Author reported that the use of LSP in CC shows good strength as compared to control specimen [19]. Inclusion of Silpozz has beneficial effect on both CC and SCC [2]. After 28 days of curing time, pozzolanic reaction start to proceed in addition of RHA in CC improve the densification by decreasing the amount of CH as reported by the researcher [11]. In SCC, 10% of Silpozz has higher strength than control specimen and 20% of Silpozz shows small reduction in strength. More the addition of Silpozz in concrete more will be the compressive strength [7]. The strength increment caused by the micro filling effect of RHA that enhances the microstructure and provide denser matrix phase [1, 4]. The strength is also get influenced by increasing the curing time from 7 days to 28 days [12]. CC prepared with 25% RHA shows highest strength than other as reported by the researchers [19].

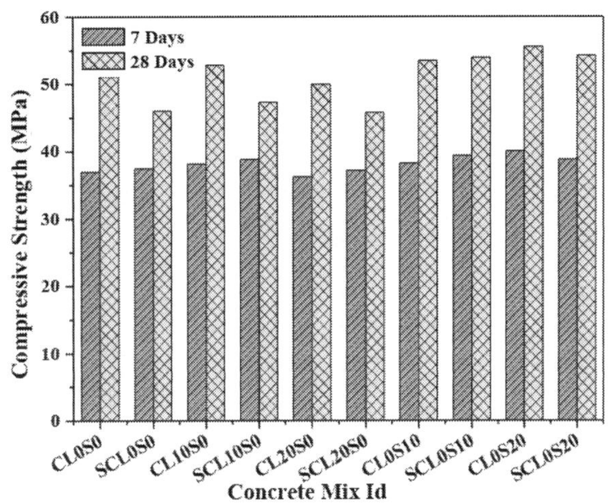

Figure 2. Compressive Strength Results of CC and SCC

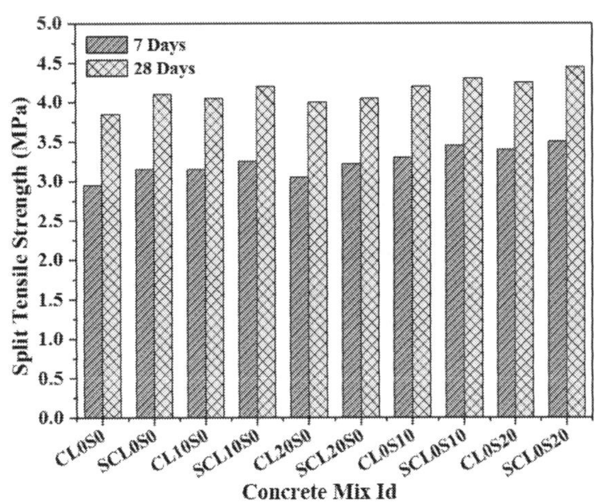

Figure 3. Split Tensile Strength Results of CC and SCC

3.3. Split Tensile Strength Test Results

The results of split tensile strength results for all mixes of CC and SCC are shown in Figure 3. Better results are observed by the substitution of Silpozz in both CC and SCC as compared to LSP addition at 7 and 28 days of curing. The improvement of strength may be related to the higher pozzolanic impact of Silpozz addition. Lesser amount of Silpozz addition in SCC increases the split tensile strength as reported by Gill and Siddique [8]. Increasing in concrete age increases the process of hydration which help to improve the split tensile strength. At all curing period inclusion of 10% Silpozz significantly increases the split tensile strength as comparision with reference sample [12]. More than 10% of LSP addition shows decrement in split tensile strength due to lesser hydration product with lower bonding and weak ITZ determined by Demirhan et al. [26]. Pradhan and Panda [28] reported that the utilization of 20% LSP shows lower split tensile strength as compared to 20% of Silpozz addition in SCC.

3.4. Flexural Strength Test Results

Flexural strength test results of CC and SCC concrete mixes are represented in Figure 4. SCC shows comparatively better performance as compared to CC. LSP addition of 20% has least strength value as compared to all other concrete mixes. Similar results were obtained by Pradhan and Panda [28]. The calcium carbonate in LSP has a very little reaction with cement hydrates create the filler effect in concrete [1]. Silpozz replacement shows better performance as compared to control mix in SCC and this leads to resistant against microcracks propagation with lower porosity [4]. Highest flexural strength is observed by 20% Silpozz addition in both CC and SCC.

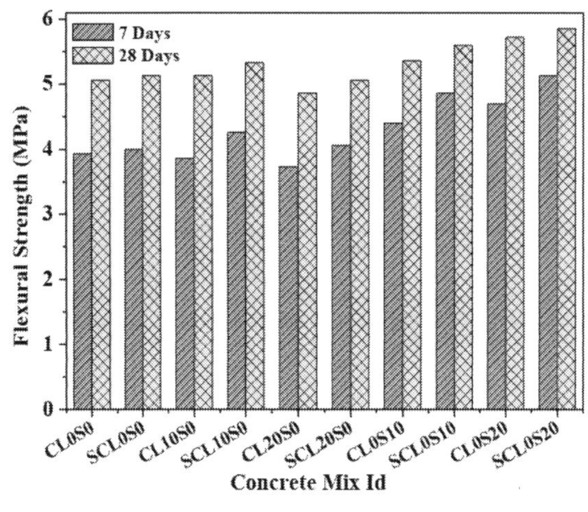

Figure 4. Flexural Strength Results of CC and SCC

4. Conclusions

Performance of mineral admixtures significantly influences the fresh and hardened concrete properties of CC and SCC. Silpozz and LSP help to produce green concrete by reducing cement consumption and solve disposal problem. The following conclusion can draw from this experimental assessment:

- Incorporation of superplasticizer in concrete plays a critical role in workability of SCC.
- When Silpozz is used in CC it absorbs water and decreases the workability. In SCC, the addition of a superplasticizer facilitates to get desired workability.
- Adding LSP in CC enhances the slump flow and a fixed amount of superplasticizer shows remarkable results.
- Compressive strength test result the higher strength is achieved by CC as compared to SCC.
- Compressive strength test, the replacement of 10% LSP shows better performance than 20% LSP and control specimen for both CC and SCC.

- Increasing the substitution of Silpozz in cements as partial replacement shows an increment in compressive strength in both CC and SCC.
- SCC improves the split tensile and flexural strength test results as compared to CC. It also shows that both 7 and 28 days of curing have a similar pattern of strength gain.
- Among all concrete mixes, split tensile and flexural strength test results show LSP as 10% replacement and Silpozz as 20% replacement shows good performance in both CC and SCC.

References

[1] Sua-Iam G and Makul N 2013 Utilization of limestone powder to improve the properties of self-compacting concrete incorporating high volumes of untreated rice husk ash as fine aggregate *Construction Building Materials***38** 455-64

[2] Gholhaki M, Hajforoush M and Kazemi M 2018 An investigation on the fresh and hardened properties of self-compacting concrete incorporating magnetic water with various pozzolanic materials *Construction Building Materials***158** 173-80

[3] Le H T and Ludwig H-M 2016 Effect of rice husk ash and other mineral admixtures on properties of self-compacting high performance concrete *Materials Design***89** 156-66

[4] Safari J, Mirzaei M, Rooholamini H and Hassani A 2018 Effect of rice husk ash and macro-synthetic fibre on the properties of self-compacting concrete *Construction Building Materials***175** 371-80

[5] Kannan V 2018 Strength and durability performance of self compacting concrete containing self-combusted rice husk ash and metakaolin *Construction Building Materials***160** 169-79

[6] Sandhu R K and Siddique R 2017 Influence of rice husk ash (RHA) on the properties of self-compacting concrete: A review *Construction Building Materials***153** 751-64

[7] Gill A S and Siddique R 2018 Durability properties of self-compacting concrete incorporating metakaolin and rice husk ash *onstruction Building Materials***176** 323-32

[8] Gill A S and Siddique R 2017 Strength and micro-structural properties of self-compacting concrete containing metakaolin and rice husk ash *Construction Building Materials***157** 51-64

[9] Makul N and Sua-iam G 2018 Effect of granular urea on the properties of self-consolidating concrete incorporating untreated rice husk ash: Flowability, compressive strength and temperature rise *Construction Building Materials***162** 489-502

[10] Panda K and Prusty S 2015 Influence of silpozz and rice husk ash on enhancement of concrete strength *Advances in concrete construction***3** 203

[11] Chao-Lung H, Le Anh-Tuan B and Chun-Tsun C 2011 Effect of rice husk ash on the strength and durability characteristics of concrete *Construction building materials***25** 3768-72

[12] Raisi E M, Amiri J V and Davoodi M R 2018 Mechanical performance of self-compacting concrete incorporating rice husk ash *Construction Building Materials***177** 148-57

[13] He Z, Cai R, Chen E and Tang S 2019 The investigation of early hydration and pore structure for limestone powder wastes blended cement pastes *Construction Building Materials***229** 116923

[14] Wang D, Shi C, Farzadnia N, Shi Z, Jia H and Ou Z 2018 A review on use of limestone powder in cement-based materials: Mechanism, hydration and microstructures *Construction Building Materials***181** 659-72

[15] Jung S-H, Saraswathy V, Karthick S, Kathirvel P and Kwon S-J J I J o C S 2018 Microstructure characteristics of fly ash concrete with rice husk ash and lime stone powder *International Journal of Concrete Structures Materials***12** 1-9

[16] Li C and Jiang L 2020 Utilization of limestone powder as an activator for early-age strength improvement of slag concrete *Construction Building Materials***253** 119257

[17] Li P, Brouwers H, Chen W and Yu Q 2020 Optimization and characterization of high-volume limestone powder in sustainable Ultra-high Performance Concrete *Construction Building Materials***242** 118112

[18] Hu L and He Z 2020 A fresh perspective on effect of metakaolin and limestone powder on sulfate resistance of cement-based materials *Construction Building Materials* **262** 119847
[19] Wang D, Shi C, Farzadnia N, Jia H, Zeng R, Wu Y and Lao L 2019 A quantitative study on physical and chemical effects of limestone powder on properties of cement pastes *Construction Building Materials* **204** 58-69
[20] Torkaman J, Ashori A and Momtazi A S 2014 Using wood fiber waste, rice husk ash, and limestone powder waste as cement replacement materials for lightweight concrete blocks *Construction building materials* **50** 432-6
[21] Matsagar V 2015 *Advances in structural engineering*: Springer)
[22] Jena T and Panda K 2015 *Advances in structural engineering*: Springer) pp 1863-73
[23] Jena T and Panda K 2017 Compressive strength and carbonation of sea water cured blended concrete *International Journal of Civil Engineering Technology* **8** 153-62
[24] Pradhan S and Panda K 2017 Assessment of fresh and hardened properties of concrete using supplementary cementitious materials *Materials Today: Proceedings* **4** 9837-41
[25] Sabat L and Pradhan S S 2020 Experimental evaluation of a composite material using GFRC *Materials Today: Proceedings* **21** 1330-4
[26] Demirhan S, Turk K and Ulugerger K 2019 Fresh and hardened properties of self consolidating Portland limestone cement mortars: Effect of high volume limestone powder replaced by cement *Construction Building Materials* **196** 115-25
[27] Souza A T, Barbosa T F, Riccio L A and dos Santos W J 2020 Effect of limestone powder substitution on mechanical properties and durability of slender precast components of structural mortar *Journal of Materials Research Technology* **9** 847-56
[28] Pradhan S and Panda K 2014 Effect of silpozz and LSP on fresh and hardened properties of SCC *Carbon* **2** 1.67

Geosynthetic as Sustainable Materials for Earthquake resistant of Masonry Structures

Hasim Ali Khan, Biplab Behera, Radhikesh Prasad Nanda

Department of Civil Engineering, National Institute Technology, Durgapur, West Bengal, India

E-mail: hiihasim@gmail.com

Abstract. When exposed to simultaneous in-plane and out-of-plane behaviour generated by earthquakes or high wind pressure, unreinforced masonry (URM) walls are vulnerable to failure. The paper presents the performance of brick masonry panels using non-woven geotextile in-plane shear and out-of-plane bending experimentally tested. The panels were reinforced with a diagonal geometric pattern on both sides. From the laboratory assessment, it was determined that geosynthetic enhancement significantly improved the capacity of load carrying, displacement capacity, bending moment, stiffness, diagonal shear, in-plane shear, and flexural strength with higher performance for both cases in the case of the cross pattern. In contrast to the un-strengthened panel, experimental findings have demonstrated a less fragile behaviour. Geosynthetics can therefore be preferably suited for strengthening the brick masonry panel in order to defend brick constructions in seismic populated places.

1. Introduction

Masonry is the major form of building structures for habitation in developing countries. It has innumerable advantages, including thermal insulation, sound resistivity, the option of addition and alternation after construction, less formwork, easy and low-cost repair, use of local and eco-friendly materials, need of less-skilled labor, and less interference in engineering over other present-day reinforced concrete and steel construction. Masonry buildings are constructed via a combination of non-engineered bricks and mortar [1-4]. Bricks have good compressive strength, perform well under gravity loading, and act vertically on the structures. However, such unreinforced brittle structures are not strong enough for resisting lateral thrust generated due to earthquake ground motion; hence they cause major hindrance in seismically active regions. The history of post-earthquakes has resulted in maximum damages to structures and has also accounted for the maximum loss of lives and properties. Moreover, field surveys of past earthquake affected areas have illustrated that low-strength failure dwellings are mainly responsible for the loss of lives and properties. Figure 1 presents the critical in-plane collapse strategies of URM walls exposed to seismic behaviour by Eigawady et al [1]; Kalali and Kabir [2, 3]. In various civil engineering building ventures, the application of geosynthetic materials has been commonly used. Walls of retention, embankments, soil backfill [7].

Most of the experimental and numerical work concentrate on obtaining capacity of strengthened masonry wallettes under either in-plane or out of plane loading. Since during seismic action, a wall is subjected to loading in all the direction and undergoes simultaneous in plane and out of plane loading, the accuracy depends upon the simultaneous effect [4-9]. Present investigation focuses the performance of masonry wallettes strengthened by geosynthetic material viz. geotextile with different patterns subject to in-plane shear and out-of-plane bending loading activities, separately.

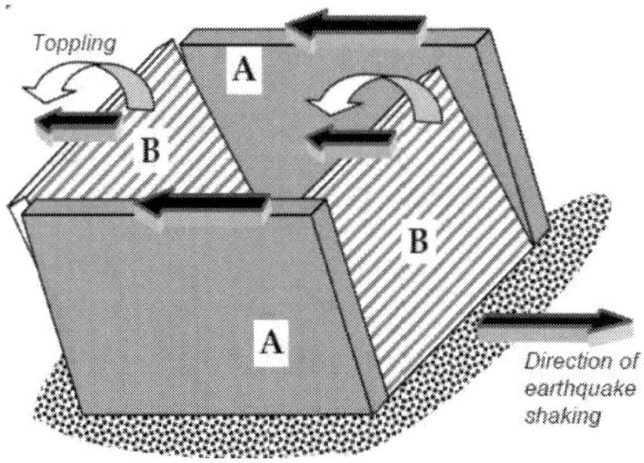

Figure 1. In-plane shear (wall A) and out-of-plane bending (wall B)

2. Experimental Programme

2.1. Introduction

To assess the in-plane shear and out-of-plane bending for masonry wallettes, a series of tests was aimed at wallettes. Twenty-four masonry wallettes, 600 mm in length, 600 mm in width, and 125 mm in depth, were cast for the experiments in this report. Masonry bricks, manufactures by a local supplier in Durgapur, were used in this experimental project. A cement mortar with a mixed proportion of 1:5 (cement: sand) was adopted, and a proper quantity of water was used to guarantee the mortar's good working ability. The water-cement ratio was maintained at 0.45. Six mortar cubes were prepared and cured for 28 days. The compressive strength of the mortar was evaluated by dividing the cross-sectional area with the maximum load. The descriptions of masonry wallets before and after geosynthetic reinforcement are shown in Figure 2.

Table 1. Component of the brick-masonry

Properties	Brick	Mortar
Density	1750	2150
Elasticity modulus	2020	4050
Poisson's ratio	0.15	0.22
Ultimate tensile strength	1.66	0.86
Ultimate compressive strength	8.93	3.49

Table 2. Composition of nonwoven polypropylene geotextile

Property	Unit	Value
Tensile strength	MPa	0.18
Young's modulus	MPa	15850
Poison's Ratio	-	0.35
Thickness (at 2 kPa)	mm	1.5
Mass per unit area	g/m^2	262
Elongation	%	85

2.2. Parameters taken in brick-masonry

In Table 1 and Table 2, the physicochemical characteristics of the brick, mortar and non-woven geotextiles used throughout the presenting research are summarized.

Nine single-leaf masonry Wallette, having dimensions of 600 mm × 600 mm × 125 mm, were constructed in a casting yard at normal temperature. Each brick masonry wallette was cast with seven layers of bricks via English bond. The schematic of a masonry bond pattern is illustrated in Figure 2. Out of thesis nine wallettes, three wallettes were kept un-strengthened, named as UR (un-reinforced), three wallettes were kept as diagonal and cross pattern.

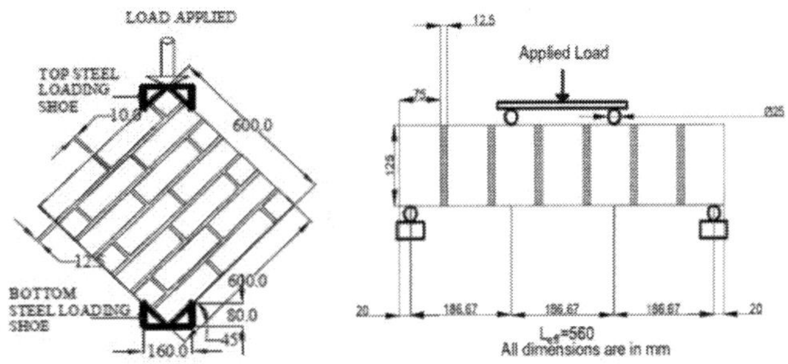

(a) (b)

Figure 2. Masonry specimen (a) Diagonal Compression test (b) Four-point bending test

2.3. Test setup

All wallettes were tested at the Hydraulic servo Dynamic Actuator Laboratory of NIT Durgapur. A servo-controlled actuator with a load-carrying capability of 100 kN was utilized for static and dynamic loading operations. In compliance with ASTM E-519 and ASTM E518, two steel loading shoes and steel rods were produced for diagonal compression and four-point bending tests of brick-masonry, respectively.

3. Results and Discussion

The deformed image of the panels is presented in Figure 4. For both in-plane and out-of-plane, deformation is expanded from UR to GRXSS.

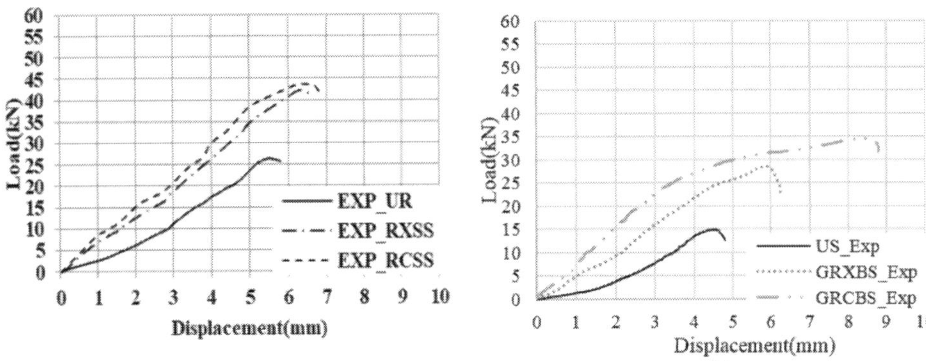

(a) (b)

Figure 3. (a) Diagonal compression of masonry specimen (b) Four-point bending test

(a) UR_IN-PLANE

(b) GRXSS_IN-PLANE

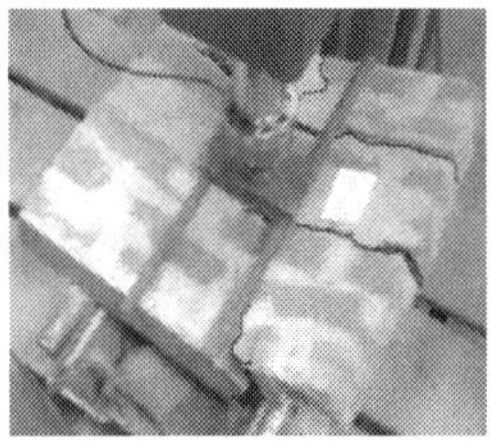
(c) UR_OUT-OF-PLANE

(d) GRXSS_OUT-OF-PLANE

Figure 4. Collapse shapes of tested brick wallettes

The graphs of load-deformation assessed along the compressed diagonal are described. After strengthening with the highest in the case of diagonal pattern, the capabilities of load deformation increase. Shear stress and flexural strength for masonry specimens are measured in accordance with ASTM E 518 and ASTM E 519 and are shown in Table 3.

Table 3. Evaluated of diagonal strength and flexural strength

Strengthening arrangement	Diagonal shear strength	Increase in Diagonal shear strength over UR (%)	Flexural Strength	Increase in Flexural strength over UR (%)
UR	0.23	-	0.99	-
RWSS	0.3128	36	1.76	73.73
RWSS	0.3197	39	1.87	88.89

Diagonal shear strength and bending strength increased from 36% to 39% and from 73.73% to 88.89%. In addition, the panel with diagonal reinforcement was also seen to have more power.

4. Conclusions

The diagonal and flexural strength of brick masonry wallettes, with diagonal reinforcement patterns, utilising geosynthetic, were experimentally tested under diagonal compression and four-point bending tests. The remarks below are illustrated based on the current report:

- For both in-plane and out-of-plane, the strengthened specimen enhanced the failure load and deformation from un-strengthened to diagonal reinforced.
- Diagonal shear strength and flexural strength are improved from 36% to 39% and from 73.73% to 88.89%. Diagonal reinforcement, in addition, has more rigidity.
- Brittle failure demonstrated an increased deformation potential for the un-strengthened panel when strengthening it.

It is also noted that the capacity of load bearing, the capacity of deformation and diagonal shear strength are substantially improved from un-reinforced to diagonal.

References

[1] ElGawady M A, Lestuzzi P and Badoux M 2007 Static cyclic response of masonry walls retrofitted with fiber-reinforced polymers *Journal of composites for Construction* **11** 50-61
[2] Kalali A and Kabir M 2012 Cyclic behavior of perforated masonry walls strengthened with glass fiber reinforced polymers *Scientia Iranica* **19** 151-65
[3] Kalali A and Kabir M Z 2012 Experimental response of double-wythe masonry panels strengthened with glass fiber reinforced polymers subjected to diagonal compression tests *Engineering Structures* **39** 24-37
[4] Nanda R P, Agarwal P and Shrikhande M 2012 Base isolation by geosynthetic for brick masonry buildings *Journal of Vibration Control* **18** 903-10
[5] Palmeira E M, Tatsuoka F, Bathurst R J, Stevenson P E and Zornberg J G 2008 Advances in geosynthetics materials and applications for soil reinforcement and environmental protection works *Electronic Journal of Geotechnical Engineering* **13** 1-38
[6] Khan H, Roy P and Nanda R 2016 Retrofitting of Brick Masonry Panels with Glass Fibre Reinforced Polymers *IOSR Journal of Mechanical Civil Engineering* , 11-8
[7] Zucchini A and Lourenço P B 2004 A coupled homogenisation–damage model for masonry cracking *Computers Structures* **82** 917-29
[8] ASTM 2010 Standard test method for diagonal tension (shear) in masonry assemblages
[9] Khan H A and Nanda R P 2020 Out-of-plane bending of masonry wallette strengthened with geosynthetic *Construction Building Materials* **231** 117198

An Experimental Study on Self-Compacting Concrete Using Tyre Rubber and Recycled Aggregate

M. Mishra[1], K. C. Panda[2]

[1]Department of Civil Engineering, Gandhi Institute for Technical Advancement, Bhubaneswar, Odisha, India
[2]Department of Civil Engineering, Government College of Engineering, Kalahandi, Bhawanipatna, Odisha, India

E-mail: mishramadhusmita1986@gmail.com

Abstract. The present study carried out for an experimental investigation on both fresh and hardened properties of self-compacting concrete (SCC) containing waste tyre rubber and recycled aggregate (RA). The tyre rubber replaced 5 % and 10 % with coarse aggregate by its weight. The recycled aggregate replaced 5 % and 10 % with coarse aggregate by its weight. Totally 7 mix were prepared including the combination of 5 % tyre rubber with 5 % recycled aggregate and 10 % tyre rubber with 10 % recycled aggregate. The 5 mm and 10 mm size of rubber chips are used. The M30 grade concrete mix design was adopted for the present work. For SCC preparation, the coarse aggregate amount decreased by 35 % and fine aggregate amount increased 35 % with water cement ratio (w/c) 0.43 and super plasticizer (SP) 0.4%. To obtain fresh properties slump flow test, T500 test, J - ring test, V - funnel test, L - box test and U - box test is conducted. For hardened concrete properties the compressive strength, flexural strength and split tensile strength tests are conducted. An attempt has been taken to identify the combine potential effect of tyre rubber and recycled aggregate in concrete and necessity of fresh and hardened properties for the SCC mix.

1. Introduction

Due to modernization and maintaining of day-to-day life, people are modified their life style. According this the old motor vehicles and buildings were replaced with new vehicles and building respectively. This modernization causes the growth of waste material in worldwide. In this research two waste materials were adopted i.e. waste tyre rubber from motor vehicles and recycled aggregate (RA) from demolished building. There was no other way to discard the tyre rubber and building waste completely. To avoid the increment of the waste amount the researcher started reused it in construction field. Both the waste was used as the replacement of the coarse aggregate. This helps to save the natural resources. The conclusion drawn from previous research studies that when tyre rubber replaced with coarse aggregate it having less mechanical strength, increase ductile nature in concrete and absorb vibration. Similarly, the concrete made from recycled aggregate (RA) gives less in strength, more in drying shrinkage, creep and less resistance for chloride ion penetration as compare to normal concrete using natural coarse aggregate.

According to Kheder and Al-Windawi [1] due to the w/c ratio of the SCC mix the compressive strength of recycled aggregate concrete varies. Levy and Helene [2] stated that the workability and compressive strength have same value for both the concrete made with recycle coarse aggregate (RCA) with 20 %, 50 % and 100 % replacement and natural coarse aggregate (NCA) concrete in the range of 20 - 40 MPa at 28 days. When RCA replaced 20 % with NCA then the strength result was equal and sometimes better than the control mix made with 100 % NCA. According to Mishra and Panda [3] the strength reduces in CRC as compare to the SCRC. As the rubber chips amount increases in SCRC the performance of concrete for workability test was better in less time period. The strength of SCC enhances after addition of super plasticiser in early age. The 5 % replacement of rubber chips in SCC has more strength than other replacement. When strength comparison occurs in between 5 % replacement and control mix the variation of strength was very less in both CRC and SCRC. According to Mishra and Panda [4] they used 0 %, 5 %, 10 %, 15 % and 20 % of rubber chips as the replacement of coarse aggregate in SCRC (self-compacting rubberized concrete). 40 % and 60 % of rubber volume used from 5 mm and 10 mm size respectively for the concrete mix. The M30 grade concrete mix design was adopted for the work. After analyzing the test results of fresh and hardened concrete the quality of strength was depending on the quantity of rubber chips. As the quantity of rubber chips increases the strength decreases. Up to 15 % of rubber chips replacement was adoptable for construction work beyond that the strength decreases. Mishra and Panda [5] stated that, recycled coarse aggregate (RCA) has been partially replaced by 0 %, 10 % and 20 % with NCA (natural coarse aggregate). Similarly the silpozz also partially replaced by 10 % with cement of its weight. The M30 grade concrete was designed with w/c ratio 0.43 and super plasticizer (SP) of 0.35 and 0.5 %. The results obtained from this experimental work that due to use of RCA the strength decreased but after adding silpozz the strength increased. RCA can be used as a partially replacement of NCA hence proved. Najim and Hall [6] studied the mechanical and dynamic properties of self-compacting crumb rubber modified concrete (SCCRMC). The advantages of using SCCRMC in vibrator plant are productivity rate increased, manpower decreased and destruction of noise and fuel consumption. As compare with plain concrete, the SCCRMC has fully self-compacted ability under its own weight. It also has high flowing and filling ability. It reduces the blocking in congested reinforced areas, resist in high segregation, maximum durability, less permeability and more compressive strength. Current research work focuses on the increment of compatibility when crumb rubber particles used as the replacement of fine aggregate in self-compacting concrete due to its high compacted and dense microstructure, which results in high mechanical strength and brittle failure modes. According to Padhi and Panda [7] in this experimental work fine rubber (FR) particles from scrap rubber tires has partially replaced with the fine aggregate (FA) and Silpozz has been used 5 %, 10 % and 15 % as the partial replacement of cement in addition of super plasticiser (SP) which help to increase the strength of concrete at early age. The fresh concrete test results indicate that the workability for both CRC (conventional rubberized concrete) and SCRC decreases as FR quantity increases. It has been observed that upto 5 % FR replacement with natural fine aggregate (NFA) leads the decrease of compressive strength (CS) with the increase of flexural strength (FS). In SCRC the strength enhances due to the replacement of silpozz in cement. Panda et al. [8] investigated on the influence of the rubber content on fresh and hardened properties of rubberized concrete with 0 % (i.e. without rubber) 3 %, 6 %, 9 % and 12 % rubber content with the partial replacement of coarse aggregate in the M20 grade concrete. As the replacement percentage of rubber increased the compressive strength decreased but in 3 % rubber replacement the concrete gives slightly high compressive strength than without rubber replacement concrete. According to Panda and Bal [9] the RCA was replaced partially i.e. 10 %, 20 %, 30 % and 40 % with NCA in the M25 grade concrete. As the RCA replacement ratio increases the strength of SCC decreases. In SCC mix up to 30 % RCA replacement achieves the acceptable compressive strength. Rehman et al. [10] performed experiment on the fundamental properties of rubber modified self-compacting concrete (RMSCC) rubber particles obtained from the waste tyre was partially replaced with fine aggregate. The test was conducted to obtain the fresh and hardened properties of the RMSCC. Normally concrete incorporate with rubber particles because of its poor bonding. If the quantity of rubber kept within 20 - 30 % replacement of fine aggregate then it may be the preferred material over the applications. The mixture may provide an advantage of absorbing vibration without compromising its strength. According to Sami and Akmal [11] the RCA has higher toughness and soundness percentage. As compare with NCA the value was less but it remained within the permissible limits. The objective of this experiment is to study the strength of self-compacting concrete after replacement of rubber chips and recycled aggregate.

2. Material Used

For this experimental work OPC 43 grade cement, zone III fine aggregate, natural coarse aggregate (NCA) (20 mm downgraded), waste Tyre rubber chips (5 mm and 10 mm size), RA (20 mm downgraded), water reducing high end super plasticizer (SP) and normal water. The tyre rubber chips replaced 5 % and 10 % with NCA by its weight. The recycled aggregate also replaced 5 % and 10 % with NCA by its weight in SCC mix with water cement ratio (w/c) 0.43. The details properties of fine, natural and recycled coarse aggregate tests were conducted in laboratory and presented in Table 1. The used sample details for this study were represented in Figure 1.

Table 1. The details properties of fine aggregate, natural and recycled coarse aggregate

Characteristics of test	As per the value obtained from IS 383-1970		
	FA	NCA	RCA
Fineness modulus	3.03(Zone-III)	7.00	6.87
Specific Gravity	2.67	2.86	2.55
Water Absorption	0.40	0.20	0.10
Bulk Density (kg/m^3)	1568	---	---
Abrasion Value (%)	---	34.78	55.42
Impact Value (%)	---	24.00	29.46
Crushing Value (%)	---	23.30	32.90

Figure 1. Tyre rubber chips (5mm and 10mm size), Recycled coarse aggregate (RCA 20 mm downgraded) and water reducing high-end super plasticizer (SP)

Table 2. The details of mix proportion along with identifications for SCC mix

Mix Identity	NCA(%)	Tyre Rubber (%)	RA(%)
SCT0R0	100	0	0
SCT5R0	95	5	0
SCT10R0	90	10	0
SCTOR5	95	0	5
SCTOR10	90	0	10
SCT5R5	90	5	5
SCT10R10	80	10	10

3. Mix Proportions

For this experimental work, the SCC mix has prepared by increasing 35 % of fine aggregate quantity and by decreasing 35 % of coarse aggregate quantity. The M30 grade SCC mix was design as per

EFNARC guidelines 2002 and 2005. The SCC mix was prepared in three batches, in first batch the replacement of RA was zero and the replacement of tyre rubber was 0 %, 5 % and 10 % with NCA. In second batch mix, the tyre rubber was zero and the replacement of RA was 0 %, 5 % and 10 % with NCA. In third batch mix, the combination of tyre rubber and RA equally replaced 5 % and 10 % with NCA. The details of mix proportion along with identifications for SCC mix are given in Table 2. The details of mixing quantity for SCC mix in Kg/m3 were presented in Table 3.

Table 3. The details of mixing quantity for SCC mix in Kg/m^3

Mix Identity	SCC mix quantity for different ingredients in Kg/m^3							
	Cement (kg)	FA (kg)	NCA (kg)	Tyre Rubber (kg)		RCA (kg)	SP (kg)	Water (kg)
				5 mm 60 %	10 mm 40 %			
SCT0R0	395	1043.55	787.15	-	-	-	1.58	158
SCT5R0	395	1043.55	747.75	23.64	15.76	-	1.58	158
SCT10R0	395	1043.55	708.45	47.22	31.48	-	1.58	158
SCT0R5	395	1043.55	747.75	-	-	39.4	1.58	158
SCT0R10	395	1043.55	708.45	-	-	78.7	1.58	158
SCT5R5	395	1043.55	708.35	23.64	15.76	39.4	1.58	158
SCT10R10	395	1043.55	629.75	47.22	31.48	78.7	1.58	158

4. Mixing and Casting

The mentioned quantity of cement, fine aggregate, natural and recycled coarse aggregate was placed in concrete mixture for one minute dry mix in batch wise and then add the calculated amount of water and super plasticizer to form the SCC mix paste. Then the workability test was conducted to find the properties of fresh concrete. For SCC mix six test were conducted in laboratory i.e. slump flow test, T_{500} test, J - ring test, V - funnel test, L - box test and U - box test. To find the mechanical properties of SCC mix the compressive strength, flexural strength and split tensile strength test were conducted after 7 and 28 days of curing on the cube, prism and cylinder specimens respectively. The details of lab set up for workability test were shown in Figure 2 and Figure 3. The details of lab set up for the compressive strength, split tensile strength and flexural strength test was shown in Figure 4.

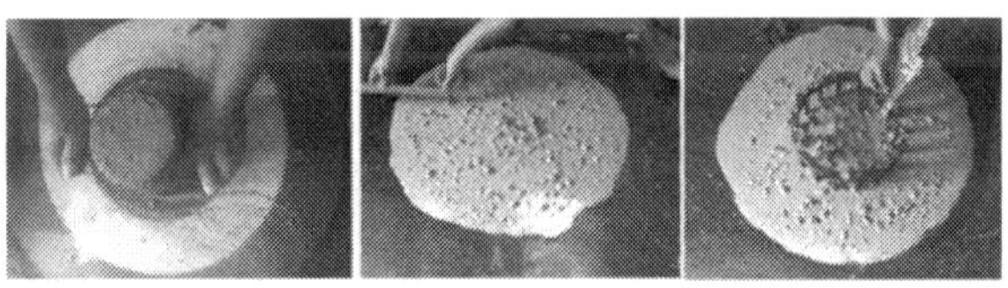

Figure 2. Lab set up for slump cone, T_{500} and J – ring test

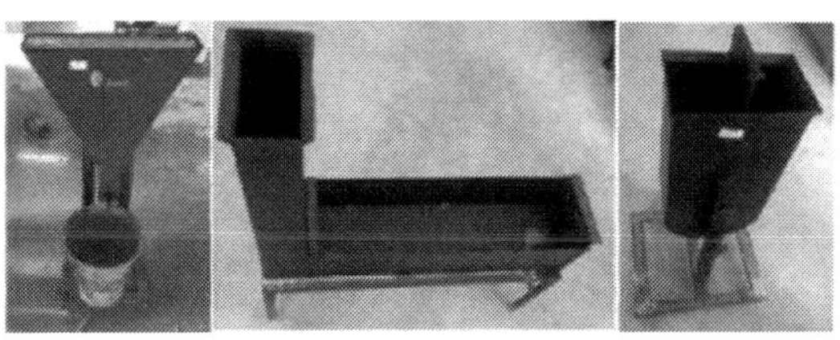

Figure 3. Lab set up for V - funnel, L - box and U - box test

Figure 4. Lab set up for compressive strength, split tensile strength and flexural strength test

5. Result and Discussion

To know about the flowing, passing and filling ability of SCC mix the workability test was performed in laboratory. The test conducted in fresh concrete after preparation of each batch mix. To check the flowing ability performance of SCC mix the slump cone test and T_{500} test were performed. Similarly, to check the passing ability performance of SCC mix the J- ring test and V- funnel test were performed. Again to check the filling ability performance of SCC mix the L- box test and U- box test were performed. After performing the entire tests for workability in fresh SCC mix then the test continues for hardened concrete. To analyses the hardened properties of SCC mix the cube, prism and cylinder specimen were cast and curing for 7 and 28 days in order to determine the compressive strength, flexural strength and split tensile strength respectively [12-15].

Table 4. The test results for the flowing ability of SCC mix with EFNARC guidelines (2005) criteria

Mix Identity	Slump Flow (mm)		T_{500} Test (Seconds)	
	TestResult	EFNARC(2005)Criteria	TestResult	EFNARC(2005)Criteria
SCT0R0	565	550-650	15	>2
SCT5R0	625	550-650	12	>2
SCT10R0	690	660-750	9	>2
SCT0R5	560	550-650	18	>2
SCT0R10	555	550-650	21	>2
SCT5R5	595	550-650	14	>2
SCT10R10	660	660-750	16	>2

Table 5. The test results for the passing ability of SCC mix with EFNARC guidelines (2002) and EFNARC guidelines (2005) Criteria

Mix Identity	V- funnel (Seconds)		J - ring Test		
	TestResult	EFNARC(2005) Criteria	Step height result(mm)	Total flow result(sec)	EFNARC(2002) Criteria
SCT0R0	19	7 - 27	13	15	0 – 10
SCT5R0	14	7 - 27	9	11	0 – 10
SCT10R0	12	7 - 27	8	9	0 – 10
SCT0R5	18	7 - 27	11	16	0 – 10
SCT0R10	21	7 - 27	12	18	0 – 10
SCT5R5	16	7 - 27	10	13	0 – 10
SCT10R10	17	7 - 27	9	14	0 – 10

The test results for the flowing ability of SCC mix with EFNARC guidelines (2005) criteria was presented in Table 4. Similarly, the test results for the passing ability of SCC mix with EFNARC guidelines (2002) and EFNARC guidelines (2005) criteria was presented in Table 5. Also the test results for the filling ability of SCC mix with EFNARC guidelines (2005) criteria was presented in Table 6.

To determine the mechanical properties of SCC mix the compressive strength test, flexural strength test and split tensile strength test were conducted for concrete specimen after 7 and 28 days of curing. Cube concrete specimen was adopted for compressive strength test. Prism specimen was adopted for flexural strength test. Similarly, cylinder specimen was adopted for split tensile strength test. The comparison of compressive strength with days for SCC mix was represented in Figure 5. The comparison of flexural strength with days for SCC mix was represented in Figure 6. The comparison of split tensile strength with days for SCC mix was represented in Figure 7.

Table 6. The test results for the filling ability of SCC mix with EFNARC guidelines (2005) Criteria

Mix Identity	U - Box (mm)		L - Box			
	TestResult	EFNARC (2005)Criteria	TestResult H_2/H_1	T_{20} Sec	T_{40} Sec	EFNARC(2005) Criteria
SCT0R0	77	≤ 80	0.76	10	15	≥ 0.75
SCT5R0	65	≤ 80	0.84	6	10	≥ 0.75
SCT10R0	60	≤ 80	0.86	4	8	≥ 0.75
SCT0R5	73	≤ 80	0.77	8	12	≥ 0.75
SCT0R10	75	≤ 80	0.78	9	13	≥ 0.75
SCT5R5	70	≤ 80	0.80	7	11	≥ 0.75
SCT10R10	67	≤ 80	0.82	5	9	≥ 0.75

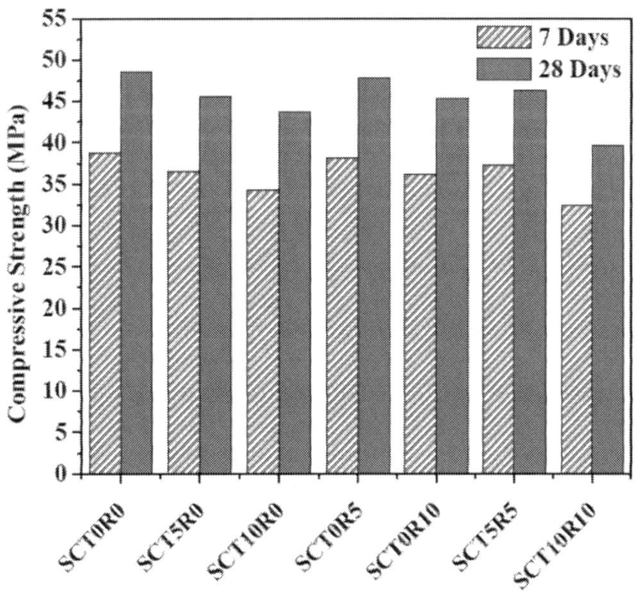

Figure 5. The comparison of compressive strength with days for SCC mix

The observation was made from Figure 5 that the control mix (SCT0R0) gives highest compressive strength at all age. When the tyre rubber replaced the NCA the strength started decrease as compare to the control mix. As the replacement quantity increase the strength started decrease due to its poor bonding. At 5 % replacement of tyre rubber (SCT5R0), the compressive strength was high from 10 % tyre rubber replacement (SCT10R0). When the RA was replaced the NCA that to 5 % and 10 %, the 5 % replacement (SCT0R5) gives more compressive strength than 10 % replacement of RA (SCT0R10). As compare with control mix (SCT0R0) the compressive strength of 5 % replacement of RA (SCT0R5) was minimum. When the combine effect of tyre rubber and RA was considered, whatever strength was decreased due to addition of rubber chips the concrete gradually regain the strength after addition of RA. The 5 % replacement combination of tyre rubber and RA (SCT5R5) gives more strength than the 5 % and 10 % replacement of tyre rubber with NCA. When the replacement amount of tyre rubber chips and recycled aggregate amount increases in SCC mix the bonding become weak and the failure in specimen occurs at the inter face of the rubber chips and recycled aggregate. Up to 15 % of rubber chips replacement in coarse aggregate was acceptable after that the strength started decreases to a non-acceptable limit.

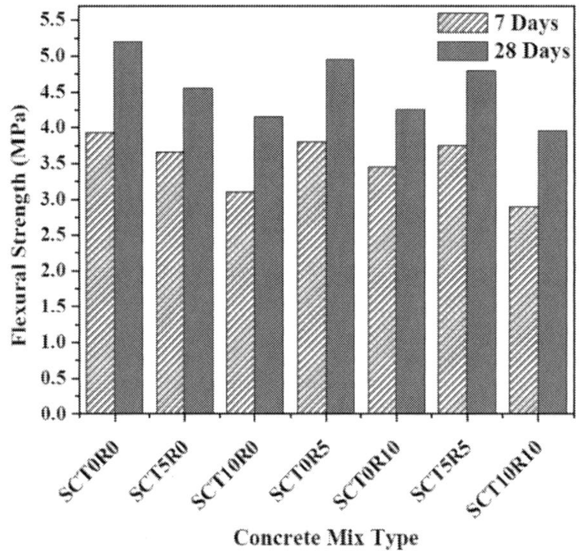

Figure 6. The comparison of flexuralstrength with days for SCC mix.

The observation was made from Figure 6 that the control mix (SCT0R0) gives highest flexural strength at all age. When the tyre rubber replaced the NCA the strength started decrease as compare to the control mix. As the replacement quantity increase the strength started decrease. At 5 % replacement of tyre rubber (SCT5R0) the flexural strength was high from 10 % tyre rubber replacement (SCT10R0). When the RA was replaced the NCA that to 5 % and 10 %, the 5 % replacement (SCT0R5) gives more flexural strength than 10 % replacement of RA (SCT0R10). As compare with control mix (SCT0R0) the flexural strength of 5 % replacement of RA (SCT0R5) was minimum. When the combine effect of tyre rubber and RA was considered, whatever strength was decreased due to addition of rubber chips the concrete gradually regain the strength after addition of RA. The 5 % replacement combination of tyre rubber and RA (SCT5R5) gives more strength than the 5 % and 10 % replacement of tyre rubber with NCA. The 10 % combination of tyre rubber and RA (SCT10R10) having less strength as compare to all the mix. When the replacement amount of tyre rubber chips and recycled aggregate amount increases in SCC mix the bonding become weak and the failure in specimen occurs at the inter face of the rubber chips and recycled aggregate. Up to 15 % of rubber chips replacement in coarse aggregate was acceptable after that the strength started decreases to a non-acceptable limit.

The observation was made from Figure 7 that the control mix (SCT0R0) gives highest split tensile strength at all age. When the tyre rubber replaced the NCA the strength started decrease as compare to the control mix. As the replacement quantity increase the strength started decrease. At 5 % replacement of tyre rubber (SCT5R0) the split tensile strength was high from 10 % tyre rubber replacement (SCT10R0). When the RA was replaced the NCA that to 5 % and 10 %, the 5 % replacement (SCT0R5) gives more split tensile strength than 10 % replacement of RA (SCT0R10). As compare with control mix (SCT0R0) the split tensile strength of 5 % replacement of RA (SCT0R5) was minimum. When the combine effect of tyre rubber and RA was considered, whatever strength was decreased due to addition of rubber chips the concrete gradually regain the strength after addition of RA. The 5 % replacement combination of tyre rubber and RA (SCT5R5) gives more strength than the 5 % and 10 % replacement of tyre rubber with NCA. The 10 % combination of tyre rubber and RA (SCT10R10) having less strength as compare to all the mix. When the replacement amount of tyre rubber chips and recycled aggregate amount increases in SCC mix the bonding become weak and the failure in specimen occurs at the inter face of the rubber chips and recycled aggregate. Up to 15 % of rubber chips replacement in coarse aggregate was acceptable after that the strength started decreases to a non-acceptable limit.

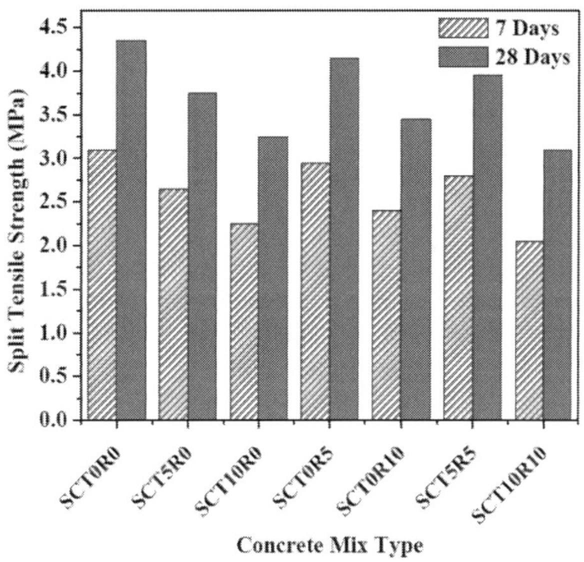

Figure 7. The comparison of split strength with days for SCC mix.

6. Conclusion

The flow ability of SCC mix was more and it satisfied the second class criteria as mentioned EFNARC guide lines 2005 when the tyre rubber quantity was increased. From the seven mix the SCT10R0 and SCT10R10 satisfied the second class criteria as per EFNARC guidelines and the rest mix satisfied the first class criteria as per EFNARC guide lines. In T_{500} test, the replacement of rubber takes less time to complete the test. Because the rubber chips were not properly bond with concrete so it separated and flow with water. Although the RA had the same problem of bonding but it tooks little more time to complete the test. For this T_{500} test, all the mix satisfied the second class criteria of EFNARC guidelines (2005). The passing ability of SCC mix was more in rubber chips replacement mix due to poor bonding. As rubber chips replacement quantity increases the segregation occurs and the mix passes quickly. In the case of RA replacement, the mix takes time to pass through small opening and gap between rods. To know the filling ability, the rubber replaced mix filled the compartment with small height variation. The RA replaced concrete take more time for filling and the height variation between two compartments was more than rubber replaced mix. All the SCC mix with different mix proportion satisfies the EFNARC guidelines (2002 & 2005). The compressive strength was more in control mix as the tyre rubber replaced quantity increases the compressive strength decreases. When RA was replaced with NCA, the difference of compressive strength with control mix was very minimum in percentage. The 5 % replacement of RA (SCT0R5) gives more compressive strength at all age with comparison of all other replacements. The combine effect of 5 % replacement of tyre rubber and RA (SCT5R5) was acceptable for construction field. This gives strength as well as ductility and vibration absorbing quality. The combine effect of 10 % replacement of tyre rubber and RA (SCT10R10) gives minimum strength at all age as compare with all other mixes. In comparison between SCT5R5 and SCT10R10 the compressive strength, flexural strength and split tensile strength was more in SCT5R5. As compare with SCT5R0 and SCT0R5 the strength was more in SCT0R5 at all age. The concrete with rubber chips gives very good vibration absorbing quality. Although this type of concrete is not suitable for high strength carrying structure but it can be used for small work like footpath, pedestal etc.

References

[1] Kheder G and Al-Windawi S 2005 Variation in mechanical properties of natural and recycled aggregate concrete as related to the strength of their binding mortar *Materials Structures* **38** 701-9

[2] Levy S M and Helene P 2004 Durability of recycled aggregates concrete: a safe way to sustainable development *Cement concrete research* **34** 1975-80

[3] Mishra M and Panda K 2015 *Advances in Structural Engineering*: Springer) pp 1785-94

[4] Mishra M and Panda K 2015 An experimental study on fresh and hardened properties of self compacting rubberized concrete *Indian Journal of Science Technology* **8** 1-8

[5] Mishra M and Panda K 2020 *Advances in Structural Engineering and Rehabilitation*: Springer) pp 189-99

[6] Najim K B and Hall M R 2012 Mechanical and dynamic properties of self-compacting crumb rubber modified concrete *Construction building materials* **27** 521-30

[7] Padhi S and Panda K 2016 Fresh and hardened properties of rubberized concrete using fine rubber and silpozz *Advances in Concrete Construction* **4** 49-69

[8] Panda K, Parhi P and Jena T 2012 Scrap-Tyre-Rubber replacement for aggregate in cement concrete: Experimental study *Int. J. Earth Sci. Eng* **5** 1692-701

[9] Panda K and Bal P 2013 Properties of self compacting concrete using recycled coarse aggregate *Procedia Engineering* **51** 159-64

[10] Rahman M, Usman M and Al-Ghalib A A 2012 Fundamental properties of rubber modified self-compacting concrete (RMSCC) *Construction Building Materials* **36** 630-7

[11] Tabsh S W and Abdelfatah A S 2009 Influence of recycled concrete aggregates on strength properties of concrete *Construction building materials* **23** 1163-7

[12] EFNARC 2002 Specification and guidelines for self-compacting concrete *European federation of specialist construction chemicals concrete system, February*

[13] 2005 The European Guidelines for Self-Compacting Concrete *BIBM* **22**

[14] 1970 IS: 383-1970, Indian Standard Specification for course and fine aggregates from natural sources for concrete, (second revision), Bureau of Indian Standards, New Delhi, India

[15] IS: 8112-1989, Indian Standard, 43 Grade Ordinary Portland Cement Specifications, (First Revision), Bureau of Indian Standards, Manak Bavan, 9 Bahadur Shah Zafar Marg, New Delhi, India.

Thermal Power Plant Waste By-product and Its Utility in Concrete

S. R. Bhagat[1], Lomesh Mahajan[2]

[1] Department of Civil Engineering, Dr. Babasaheb Ambedkar Technological University Lonere, Maharashtra, India
[2] Department of Civil Engineering, R. C. Patel Institute of Technology, Shirpur, Maharashtra, India

E-mail: srbhagat@dbatu.ac.in

Abstract. Nearly 117 coal based thermal power station has working conditions in India. About 71% (169 x 103 MW) electricity is generated from four Indian regions i.e. Western, Northern, Eastern, and Southern. All these thermal power plant has produces by-product after burning coal as an energy inputs. The by-product obtained in the form of fine ash which has been known as fly ash and bottom ash. These by-products directly dump on the nearby area or thrown into low lying areas. The utility of these finer materials has proven a feasible substance in the concrete constituents. The paper presents possible advantages to incorporate it into concrete. Fly ash, bottom ash and pond ash has potential to act as a binding and filler agent in concrete.

1. Introduction

India is the 3rd biggest coal handling country in the globe and about 70% thermal power stations (TPP) consuming energy source as coal[1]. Coal is the prime necessary fuel source utilized in TPP for era of developing power. The basic few reasons for most of electricity generated from coal based TPP are as (i) coal based TPP can react to quickly changing loads easily. (ii) A parcel of the steam created can be utilized as a prepare steam in most of the industries. (iii) Steam motors and turbines can work beneath 25 % of over-burden continuously and (iv) Fuel utilized is cheaper in generation taken a toll in comparison with that of diesel consuming TPS. These four reasons unknowingly promotes the use of coal for electricity generation But at the other end, All TPP has been generated enormous amount of coal burning ash in the granular and slurry form. The coal pond has limited capacity to store these unwanted ash and therefore coal ash transported to nearby dumping area. The transfer of ash from a huge capacity TPP is of also consuming vehicle fuel and cost. Due to unavailability of land for dumping ash, transportation distance increases upto 50-100km from TPP. Unfortunately, coal ash dealing with a major issue in society due to their huge amount storage in local residential areas. The coal ash generated by TPP categorized into two sorts, i.e., fly ash and bottom ash. Amid burning process of pulverized coal the tiny and light weighted particles extricated by the electrostatic precipitators. This ash trap at bottom of electrostatic precipitators generally identified as fly ash (FA). The fly ash has major part of burning ash as it weightage nearly 80% of total ash generation. Few heavier particle having coarser nature trapped downword direction at the base of furnace hopper and generally identify by the name as coal bottom ash (CBA). The quality of FA and CBA are depends upon the quality of raw coal properties.

FA and CBA is unwanted by product of TPP but it can used as successful supplementary construction material. Since starting of 21st century, many researchers has been took effort for searching applicability of it into construction material. As of now, more than 50% of the fly ash concrete used in United States. Many times the rates change as per class of fly ash and their reactivity chances. Ordinarily, Class 'F' FA utilized at 15% to 25% by mass of cementitious material, and Class 'C' at 15% to 40%. FA broadly utilized in generation of Pozzolanic Portland cement (PPC) and in fabricating mortar mix. FA also potential in concrete making as partial cement substitution [2-5]. Supplementary cementing fabric in generation of FA concrete moves forward its quality and toughness properties. While, CBA identified due to their coarser and more intertwined particles, where the FA appears less pozzolanic property. For the most part FA is considered as an idle fabric for cement replacement and unacceptable for utilize as supplementary coarser aggregate formation fabric. Since molecule measure conveyance of CBA is comparative to natural sand, it pulls in to utilize as substitution of sand in a few gracious designing applications. The various available modes and consumption of FA utilization in the year 2016-17 are mention in the following Table no.1

Table 1. Sector wise Fly Ash Utilization in India during the Year 2016-17 (CEA, New Delhi, 2017)

Sr. No	Sectors	Quantity of Fly Ash utilized (%)
1	Cement	23.98
2	Mine filling	6.96
3	Bricks & Tiles	8.81
4	Reclamation of low lying area	6.52
5	Ash Dyke Raising	7.02
6	Roads & flyovers	3.66
7	Agriculture	1.14
8	Concrete	0.45
9	Hydro Power Sector	0.01
10	Others	4.72
11	Unutilized Fly Ash	36.72
	Total	100.00

2. Use of Fly ash (FA) material in Concrete

The many researchers successfully used TPP by product FA for alternative replacement option to cement for concrete. These forms and applications include expansion in binding materials which backbone of concrete products and cover fabric, roadway and asphalt to develop good source of low-density aggregate, infiltration boundary and eco-friendly material development. Cement is costlier and vitality seriously binder in concrete. The unit cost of concrete decreased by halfway substitution of FA. The utilization of FA remains pozzolana for halfway substitution of cement, and mostly since of its advantageous impacts, minimize water component for comparative workability, diminished dying, and advancement of warm. It has been utilized especially in huge mass concrete applications situation to control extension due to heat-hydration additionally makes a difference in diminishing splitting at 1-7days curing. The major disadvantage of fiber-strengthened concrete is its moor workability. To solve this inadequacy, a fabric is required, make strides the workability, and no comprising quality. The utilize of FA improves the workability of concrete and being broadly suggested as halfway substitution of cement. As per the details of Bureau of Indian Guidelines FA upto 35% can be utilized in preparation of PPC, whereas around the world there are cases in other nations that allow upto 55% usage of FA in

PPC generation. Setting aside 25% of cement generation for OPC for such applications, the adjust 75% can be PPC with a normal FA substance of 30%. FA, produced due to coal vitality used for power generation, is a mechanical by-product as a natural toxin. Since of the natural issues displayed by the FA, impressive inquire about has been embraced around the world [2-3]. Antiohos et al. [3] examined FA concrete and found satisfactory advantages in strength scenario. A significant sum of investigate has been conducted utilizing FA for investigation on organic compounds, mercury, colors and other natural compounds in waters. The investigation suggested the fly ash is a promising adsorbent. FA, as a mixed cement component offers a for both crude material and coordinate concrete admixture use. Since FA inevitably in concrete, its characteristics need to be reasonable for that purpose. Be that as it may, since small or no alteration can be given at the concrete blending arrange, FA debris for utilize in mixed cements, which possess reliable and uniform FA characteristics.

3. Use of Bottom ash (BA) material in Concrete

A few authors have detailed the advantage of BA concrete as halfway substitution of portland cement [5-10] or as a fractional substitution of fine aggregates [11-14]. At the stage of beginning, strength gaining slower in bottom ash concrete specimens but as the curing time increases the strength also improve at some extent. In the initial phase of hydration of BA concrete, BA did not familiar with the property of calcium hydroxide. The pozzolanic characteristics does not create in early curing. A smaller size BA particles provide high density and ultimately good in packing density [16-18]. The plausibility of substituting normal fine aggregate with industrial by-product such as waste foundry sand and BA offers specialized, financial benefits, which are of extraordinary significance in the show setting of maintainability in infrastructural sector [14]. The nearness of BA expanded the amount of water loss and water discharge rate. The higher BA amount in the concrete, the higher impact. Researcher [15] reported that the distinctive shapes of BA blend with physical properties within the initial state of concrete BA is utilized as a substitution for characteristic aggregates[16-18]. In brief the FA and BA can be used to reduce environmental pollution and burden of concrete ingredient. Now days, FA-BA concrete gripped on replacement possibilities experienced. Author [19] verified the compressive strength for BA mixed concrete, the result found nearly half and full proportions about 40% lower strength. The strength of concrete thoroughly related to proportioning and properties of its inert materials.

The researcher [20] investigated that presence of BA for fine aggregate in concrete and recommended the use of BA. The reduction seen in strength of concrete because of porous nature of BA and higher water consumption. The strength can been controlled by controlling the water demand with the help of super plasticizers. Researchers [21, 22] experienced the same observation for modulus of elasticity (E) too. Inversely proportional behavior found with BA content. For 100% sand replacement, the range of (E) found 15% to 16% lower. The split strength (ft) of BA concrete have potential to get satisfactory range. Also detected that (ft) of specimens 20% BA for fine aggregate more results (CC) specimens were. The (ft) observations seen decremented effect with increase bottom ash by 20%, whereas researcher [13] is witnessed that no variation in the (ft) up to 10% fine aggregate replacement.

4. Advantages in Fresh Concrete

The particles of fly ash (FA)being spherical and uniform in size similar to a cement particle, the water required for preparation and placing of concrete is lesser compared to concrete prepared without using FA. In the case of precast concrete, the use of FA gives better workability which results in sharp, distinctive corners and edges with a smooth surface appearance. This also makes it possible to prepare elements with intricate shapes and patterns. The use of FA in precast concrete elements reduce the overall permeability, thereby making it more durable. The finer particles of FA reduces bleeding and segregation, and improves pumpability and finishing, especially in lean mixes.

5. Advantages in Hardened Concrete

The compressive strength of concrete depends on many factors, the most important of which is the ratio of water to cement. Good quality of FA generally improves the workability of concrete or for the same workability, the required water-cement ratio is less. The reduction in the water-cement ratio leads to improved strength. Good quality of FA possesses reactive characteristics that take part in the hydration process along with portland cement leading to better strength. It has been observed that the ordinary FA when partially replaced with cement influence the early strength gain of concrete, however at a later age the pozzolanic reaction between FA and the calcium hydroxide from cement takes place leading to improve in compressive strength. Properly proportioned concrete containing FA will lead to the overall economy in producing concrete. The improved impermeability of FA concrete leads to a reduction in steel reinforcement corrosion. Sulfate and other chemical attacks reduced when FA is added in producing concrete [17-18].

6. Experimental Evaluation of Concrete Specimens

The compressive strength (CS) of FA mixed concrete evaluated with considering the water cementitious ratio 0.5, 0.55, and 0.6. The size of a 20 mm sieve used for coarse aggregates selection. The FA contents varied according to a 10 percent interval from 0 to 50%. For getting greater accuracy of work, sample specimens cured and tested in the control condition at 28 days. Lean mix to high performance concrete evaluation took into account by considering binder content 300kg/m3, 375kg/m3, and 450kg/m3. CTM machine used for testing the 150mm x 150 mm size cube specimens. Class F (IS: 3812 (2003) Part – I)[23] FA used with a specific gravity (SG) 2.15 for concrete making. OPC 53 grade cement with SG 3.15 was used. The effect of replacement of FA to cement for concrete making is graphically recorded in the following Figure 1, 2, and 3.

7. Result and Discussion

The experimental investigation was made for computing CS of 28 days for fly replacement at 0, 10, 20, 30, 40, and 50% by OPC. Considering the result of mechanical property, i.e. CS of FA-based concrete specimens for an average of three cubes at every interval of replacement shows that as the FA amount increases the strength decreases considerably after 30%. For w/cm ratio 0.4 and 0.5 the results are quite better than 0.55 as the C-S-H gel formation process more effectively developed between these phases. All the specimens show the permissible range for 20 and 30% FA incorporation but as FA concentration proportionally affects final CS results. Results are higher in lower binder content specimens compared to 450kg/m^3.

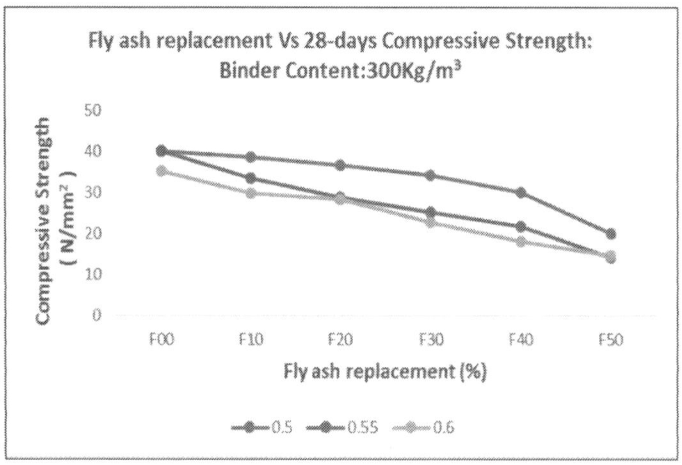

Figure 1. Fly ash replacement (%) Vs 28-day Compressive strength at binder content of 300 kg/m^3.

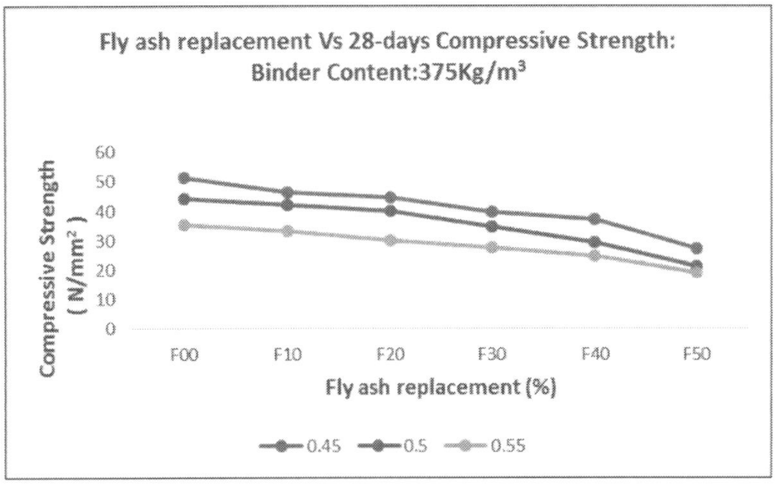

Figure 2. Fly ash replacement (%) Vs 28-day Compressive strength at binder content of 375 kg/m^3.

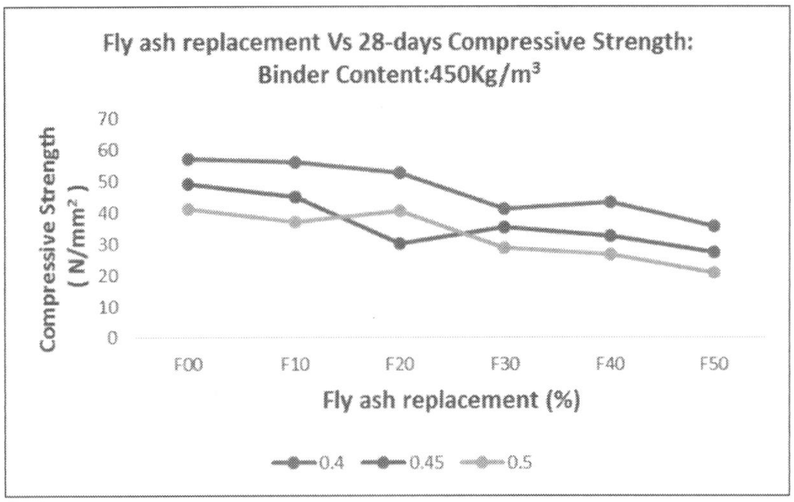

Figure 3. Fly ash replacement (%) Vs 28-day Compressive strength at binder content of 450 kg/m^3.

8. Conclusion

It has a known fact around the world that the utilization of an enormous sum of fossil powers has made different antagonistic effects on the environment, counting corrosive rain and worldwide warming. This paper has endeavoured wide range of information with small experimental investigation so that the perused fly ash utilization has potential to better mix of concrete with satisfactory strength approach. For Binder content 300 kg/m^3 and 375 kg/m^3 is more applicability to consume high fly ash with satisfactory results. The w/cm ratio 0.4 and 0.45 has found similar results in all concrete specimens. At the 30% replacement of FA found 23% reduction as compared with fully OPC used concrete specimens. As considering economy and mass concrete places, FA mixing is good solution at partial level. Fly ash, even though posturing natural contamination, it is a significant crude fabric for different applications. Among the regulation and legitimate obstructions is the need for knowledge of potential cinder employments, scattered information on environmental and wellbeing impacts, compositional irregularities within the products, belief that other raw thermal power plant waste materials are promptly accessible. The topics motivates us for thermal ash utilization in various structural application and increase the integrated FA and BA amount in the concrete.

References

[1] Sahu S, Bhangare R, Ajmal P, Sharma S, Pandit G and Puranik V 2009 Characterization and quantification of persistent organic pollutants in fly ash from coal fueled thermal power stations in India *Microchemical Journal* **92** 92-6

[2] Bouzoubaa N, Zhang M-H and Malhotra V 2001 Mechanical properties and durability of concrete made with high-volume fly ash blended cements using a coarse fly ash *Cement Concrete Research* **31** 1393-402

[3] Antiohos S, Papageorgiou D, Chaniotakis E and Tsimas S 2007 Mechanical and durability characteristics of gypsum-free blended cements incorporating sulphate-rich reject fly ash *Cement Concrete Composites* **29** 550-8

[4] Mahajan L and Bhagat S 2020 Hardened behaviour of fly ash incorporated concrete with bottom ash as partial replacement of fine aggregate *International journal of innovative technology and exploring engineering* **9** 2898-907

[5] Kurama H and Kaya M 2008 Usage of coal combustion bottom ash in concrete mixture *Construction building materials* **22** 1922-8

[6] Lomesh Mahajan and S.R.Bhagat 2020 The Contribution of Bottom Ash toward Filler Effect with Respect to Mortar *Lecture Notes in Civil Engineering Springer* **72** *145-154*

[7] Kula I, Olgun A, Sevinc V and Erdogan Y 2002 An investigation on the use of tincal ore waste, fly ash, and coal bottom ash as Portland cement replacement materials *Cement Concrete Research* **32** 227-32

[8] Yüksel İ and Bilir T 2007 Usage of industrial by-products to produce plain concrete elements *Construction Building Materials* **21** 686-94

[9] Yüksel İ, Bilir T and Özkan Ö 2007 Durability of concrete incorporating non-ground blast furnace slag and bottom ash as fine aggregate *Building Environment* **42** 2651-9

[10] Bai Y, Darcy F and Basheer P 2005 Strength and drying shrinkage properties of concrete containing furnace bottom ash as fine aggregate *Construction Building materials* **19** 691-7

[11] Bai Y and Basheer P 2003 Influence of furnace bottom ash on properties of concrete *Proceedings of the Institution of Civil Engineers-Structures Buildings* **156** 85-92

[12] Ghafoori N and Cai Y 1998 Laboratory-made roller compacted concretes containing dry bottom ash: Part ii—long-term durability *Materials Journal* **95** 244-51

[13] Ghafoori N and Buchole J 1997 Properties of High-Calcium Dry Bottom Ash for Structural Concrete *Materials Journal* **94** 90-101

[14] Aggarwal Y and Siddique R 2014 Microstructure and properties of concrete using bottom ash and waste foundry sand as partial replacement of fine aggregates *Construction Building Materials* **54** 210-23

[15] Andrade L, Rocha J and Cheriaf M 2009 Influence of coal bottom ash as fine aggregate on fresh properties of concrete *Construction Building Materials* **23** 609-14

[16] Siddique R 2004 Performance characteristics of high-volume Class F fly ash concrete *Cement Concrete Research* **34** 487-93

[17] Mahajan L and Bhagat S 2020 Fly Ash Effects on Fresh Concrete *International journal of innovative technology and exploring engineering* **9** 1247-1252

[18] Mahajan L and Bhagat S 2020 Hardened behaviour of fly ash incorporated concrete with bottom ash as partial replacement of fine aggregate *International journal of innovative technology and exploring engineering* **9** 2898-2907

[19] M. Singh, R. Siddique 2013 Effect of coal bottom ash as partial replacement of sand on properties of concrete, *Resour. Conserv. Recycl.* 72 , 20– 32.

[20] Mahajan Lomesh, Bhagat S.R. 2020, The Contribution of Bottom Ash Toward Filler Effect with Respect to Mortar, *Lecture Notes in Civil Engineering, Springer Nature*, Vol. 72, 145-154.

[21] K. Arumugam R Ilangovan, M.D. James 2011 A study on characterization and use of pondash as fine aggregate in concrete. *Int. J. Concr. Struct. Mater.* 2 , 466–74.

[22] I. Yuksel, A. Genc 2007 Properties of concrete containing non ground ash and slag as fine aggregate. *ACI Materials Journal* 104(4), 397–403.

[23] IS 3812 Part 2 (2003), Specification for pulverized fuel ash part 2. For use as admixture in cement mortar and concrete, *Bureau of Indian Standar*d, New Delhi

Investigation of Fly Ash Concrete by Slump Cone and Compaction Factor Test

Lomesh Mahajan[1], Sanjivan Mahadik[2], S. R. Bhagat[3]

[1] Department of Civil Engineering, R. C. Patel Institute of Technology, Shirpur, Maharashtra, India
[2,3] Department of Civil Engineering, Dr. Babasaheb Ambedkar Technological University Lonere, Maharashtra, India

E-mail: loms786@gmail.com

Abstract. The supplementary or alternative material to cement has been an emerging field in civil engineering. The concrete ingredients have become modern due to the need for reducing global warming and material scarcity problem. Fly ash is used to replace cement partially in concrete, therefore the concrete mix has changed their characteristics as obtained by normal concrete. The proper investigation always required for performance measurement and to measure the fresh concrete property, the slump cone test, and compaction factor test are the tool. The Slump Test has become the most frequently performed due to the practicality of the recommended equipment and the experiment protocol. The slump test involves the cone's behavior under the action of gravitational forces. The slump check is a realistic way to gauge the workability. The concrete slump test and compaction factor test are used to determine the consistency of fresh concrete before it sets. This paper has focused on testing the 54 mixes having various water-cm ratios like 0.5, 0.55, 0.6, 0.45, and 0.4. Fly ash material possesses satisfactory workability properties due to their similar oxide compositions. Due to the fineness of fly ash, less bleeding observed than control concretes.

1. Introduction

As the request for foundation advancement increases, it moreover increments the possible applicability of concrete in various emerging construction activities. The need of populations and their infrastructural demand has a responsible cause of more cement consumption. In all the stages of infrastructural growth, cement manufacturing has created gigantic impacts on environmental sustainability, economy, and utilization [1,2]. Cement is the energy of worldwide infrastructure. It evaluated that the worldwide generation of cement is approximately 4200 million metric tons in 2017. With the help of innovative progress, the concrete industry always trying to find supplementary cementitious fabric in creating esteem included items conjointly decreasing the strong waste disposal issue [2]. There is a got to economize cement utilization and one of the practical solutions is to in part supplant cement with pozzolanic materials. Pozzolona may be a characteristic or counterfeit fabric containing silica in a receptive frame. It could be a siliceous and aluminous fabric, which possesses small or no cementitious esteem but in the partitioned shape, and within the nearness of dampness chemically responds with CH at standard temperatures to create natural cementitious character to the concrete. Fly ash (FA) is among the foremost unremarkably used alternative cementing material as a result of it may be act a neglected item coming about from the coal combustion[3]. The pozzolana characteristics of FA materials could be the best suitable place in concrete for both within the new and solidified state [4-6]. The participation of FA in fresh concrete mass has potential to get several benefits on the environment and improves the fresh and long period properties of concrete like easy to place, strength, and durable. It conjointly assist to slow down heat extension and water utility. In addition, the formation of surrey in the concrete mix minimized with the help of the FA material as constituents. The analysts have found that the substitution of FA by 15% to 40% of the full volume of cement [7-10]. Fresh properties of concrete have always affected the results of hardened properties in terms of its quality and durability. Sometimes, the delayed in fresh concrete placement on site tends to solidify the mix, and this phenomenon reflected in the process of workability [1]. The direct field measurement method embraced concrete quality, and this process is to consider a complex one due to its uncommon necessity of test estimate. If the fly ash used as cement substitution fabric into the concrete, then the workability of blended mix has been improved few extents. This is occurs due to circular surface and shape of FA glossy particles. The circular shape of FA debris acts as a frame of a ball bearing affect, which helps the grease between sticky natures in concrete mix [1-3].

2. Experimental Evaluation

2.1 Materials used in Manufacture of Fly Ash Concrete

Cement of OPC-53 grade, and specific gravity 3.15 used for testing purpose. River sand was used as fine aggregate with the Specific gravity (SG) is 2.68. The F class fly ash used for partial replacement in the experimentation. Used coarse aggregate the maximum size was 20 mm, and SG 2.7.

2.2 Slump Cone Apparatus Test

The slump cone apparatus test is the foremost straightforward workability measurement for new liquid phase concrete, includes very cost effective and fast way. This reality broadly push to utilized for workability tests since 19th century. The slump test followed methods mentioned in ASTM C143 (United States), IS: 1199 – 1959 in India and EN 12350-2 in Europe. The slump test is the foremost common strategy for evaluating the stream properties of new concrete; the slump gives a degree of workability [11]. Utilizing this test, the slump can be inferred by measuring the drop from the top of the slumped new concrete. Shape for slump cone, non permeable base plate, measuring scale, temping bar. The shape for the test is in the shape of the frustum-cone having 30 cm high, foot distance across 20 cm and top distance across 10 cm. The tamping bar is of steel 16 mm breadth and 60cm long and adjusted at one end. The compacting factor test is planned essentially for utilize within the laboratory but it can also be utilized within the field.

It is more exact and delicate than the slump test, is especially valuable for concrete blends of exceptionally low workability, and regularly utilized when concrete is to be compacted by vibration. The strategy applies to plain and air-entrained concrete, made with lightweight, typical weight or overwhelming aggregates having a nominal greatest estimate of 38 mm or less but not to circulate air through concrete or no-fines concrete. The test of concrete to be tried should be put delicately within the upper container, utilizing the hand scoop. The container might be filled level with its brim and the trap door might be opened so that the concrete falls into the lower container. Instantly after the concrete has come to rest, the barrel should be revealed, the trap door of the lower container opened, and the concrete permitted to drop into the barrel. The abundance of concrete remaining over the level of the beat of the barrel might at that point be cut off by holding a trowel in each hand, with the plane of the edges level, and moving them at the same time one from each side over the best of the barrel, at the same time keeping them squeezed on the best edge of the barrel. The compacting factor is characterized as the proportion of the weight of mostly compacted concrete to the weight of completely compacted concrete. It might regularly be expressed to the closest second decimal place. As the phase of consistency assessment with considering experimental execution, the most popular methods i.e. Slump Cone and Compaction factor (C.F) carried out as per guidelines of IS: 1199:1959 [12]. For the performance of workability of concrete, the slump cone apparatus conforming to IS 7320:1974 [13] and Compacting Factor Apparatus were used throughout the experimentation. Binding material content 300kg/m^3,

375kg/m^3, 450kg/m^3 and water-cementitious combinations like (0.5), (0.55), (0.6), (0.45), (0.4) are considered for experimentation. The fly ash amount were taken as 0% (F00), 10% (F10), 20% (F20), 30% (F30), 40% (F40), and 50% (F50). The detail results of slump test and compaction factor test are record in the following Table 1 and Table 2 respectively.

Table 1. Slump Cone test results for fly ash replacement

Sr. No	Mixes	Binder Content	w/c ratio	Slump
1	S1-F00	300	0.5	5
2	S2-F10	300	0.5	5
3	S3-F20	300	0.5	5
4	S4-F30	300	0.5	5
5	S5-F40	300	0.5	5
6	S6-F50	300	0.5	10
7	S7-F00	300	0.55	7
8	S8-F10	300	0.55	8
9	S9-F20	300	0.55	9
10	S10-F30	300	0.55	9
11	S11-F40	300	0.55	10
12	S12-F50	300	0.55	10
13	S13-F00	300	0.6	19
14	S14-F10	300	0.6	18
15	S15-F20	300	0.6	17
16	S16-F30	300	0.6	16
17	S17-F40	300	0.6	15
18	S18-F50	300	0.6	15
19	S19-F00	375	0.45	5
20	S20-F10	375	0.45	8
21	S21-F20	375	0.45	8
22	S22-F30	375	0.45	9
23	S23-F40	375	0.45	10
24	S24-F50	375	0.45	10
25	S25-F00	375	0.5	28
26	S26-F10	375	0.5	27
27	S27-F20	375	0.5	25
28	S28-F30	375	0.5	22
29	S29-F40	375	0.5	20
30	S30-F50	375	0.5	18
31	S31-F00	375	0.55	122
32	S32-F10	375	0.55	95
33	S33-F20	375	0.55	95
34	S34-F30	375	0.55	77
35	S35-F40	375	0.55	90
36	S36-F50	375	0.55	118
37	S37-F00	450	0.4	16
38	S38-F10	450	0.4	15
39	S39-F20	450	0.4	12
40	S40-F30	450	0.4	10
41	S41-F40	450	0.4	10
42	S42-F50	450	0.4	11
43	S43-F00	450	0.45	70

44	S44-F10	450	0.45	55
45	S45-F20	450	0.45	42
46	S46-F30	450	0.45	40
47	S47-F40	450	0.45	38
48	S48-F50	450	0.45	36
49	S49-F00	450	0.5	230
50	S50-F10	450	0.5	210
51	S51-F20	450	0.5	195
52	S52-F30	450	0.5	190
53	S53-F40	450	0.5	185
54	S54-F50	450	0.5	185

Table 2. Compaction factor test results for fly ash replacement

Sr. No	Mixes	Binder Content	w/c ratio	C.F.
1	CF1-F00	300	0.5	0.78
2	CF2-F10	300	0.5	0.78
3	CF3-F20	300	0.5	0.79
4	CF4-F30	300	0.5	0.80
5	CF5-F40	300	0.5	0.81
6	CF6-F50	300	0.5	0.81
7	CF7-F00	300	0.55	0.81
8	CF8-F10	300	0.55	0.82
9	CF9-F20	300	0.55	0.83
10	CF10-F30	300	0.55	0.84
11	CF11-F40	300	0.55	0.85
12	CF12-F50	300	0.55	0.85
13	CF13-F00	300	0.6	0.91
14	CF14-F10	300	0.6	0.91
15	CF15-F20	300	0.6	0.89
16	CF16-F30	300	0.6	0.88
17	CF17-F40	300	0.6	0.88
18	CF18-F50	300	0.6	0.88
19	CF19-F00	375	0.45	0.85
20	CF20-F10	375	0.45	0.85
21	CF21-F20	375	0.45	0.84
22	CF22-F30	375	0.45	0.85
23	CF23-F40	375	0.45	0.85
24	CF24-F50	375	0.45	0.85
25	CF25-F00	375	0.5	0.95
26	CF26-F10	375	0.5	0.93
27	CF27-F20	375	0.5	0.89
28	CF28-F30	375	0.5	0.89
29	CF29-F40	375	0.5	0.87
30	CF30-F50	375	0.5	0.88
31	CF31-F00	375	0.55	0.99
32	CF32-F10	375	0.55	0.98
33	CF33-F20	375	0.55	0.97
34	CF34-F30	375	0.55	0.97
35	CF35-F40	375	0.55	0.95
36	CF36-F50	375	0.55	0.96

37	CF37-F00	450	0.4	0.90
38	CF38-F10	450	0.4	0.89
39	CF39-F20	450	0.4	0.87
40	CF40-F30	450	0.4	0.85
41	CF41-F40	450	0.4	0.83
42	CF42-F50	450	0.4	0.83
43	CF43-F00	450	0.45	0.98
44	CF44-F10	450	0.45	0.97
45	CF45-F20	450	0.45	0.96
46	CF46-F30	450	0.45	0.94
47	CF47-F40	450	0.45	0.93
48	CF48-F50	450	0.45	0.93
49	CF49-F00	450	0.5	0.98
50	CF50-F10	450	0.5	0.98
51	CF51-F20	450	0.5	0.98
52	CF52-F30	450	0.5	0.97
53	CF53-F40	450	0.5	0.97
54	CF54-F50	450	0.5	0.96

3. Conclusion

Few conclusions can be drawn after this research:

- Fly ash particles being fine & siliceous in nature improves fresh state of concrete.
- Fly ash particles being round and fine improves workability and helps getting desired slump.
- Helps in reducing water demand.
- Prevents bleeding and drying shrinkage.
- Reduced heat of hydration.
- If replacing cement by fly ash upto 50 percentage then it should not affect the workability much according to compaction factor test recorded results.

In a few cases of slump test, the results varied according to binder content and w/cm ratio. The slump reduction seen higher side in the binder content 300kg/m^3 comparative to others.

References

[1] Shaikuthali, S.A., Mannan, M.A., Dawood, E.T. 2019. Workability and compressive strength properties of normal weight concrete using high dosage of fly ash as cement replacement. J Build Rehabil **4**, 26

[2] Balakrishna M and Nataraja M 2013 Proportioning of fly ash concrete mixes a comprehensive approach *Int. J. Emerg. Sci. Eng.* **1** 46-51

[3] Mahajan L and Bhagat S 2020 Fly Ash Effects on Fresh Concrete *International journal of innovative technology and exploring engineering* **9** 1247-1252

[4] Lomesh Mahajan and S.R.Bhagat 2020 The Contribution of Bottom Ash toward Filler Effect with Respect to Mortar *Lecture Notes in Civil Engineering Springer* **72** *145-154*

[5] Subramani T and Ramesh K 2015 Experimental Study On Partial Replacement Of Cement With Fly Ash And Complete Replacement Of Sand With M sand *International Journal of Application or Innovation in Engineering Management* **4** 313-22

[6] Bendapudi S C K and Saha P 2011 Contribution of fly ash to the properties of mortar and concrete *International Journal of Earth Sciences Engineering Structures* **4** 1017-23

[7] Mahajan L and Bhagat S 2020 Hardened behaviour of fly ash incorporated concrete with bottom ash as partial replacement of fine aggregate *International journal of innovative technology and exploring engineering* **9** 2898-2907

[8] Wankhede P and Fulari V 2014 Effect of fly ash on properties of concrete *International journal of emerging technology advanced engineering* **4** 284-9

[9] Hussein A A E, Shafiq N and Nuruddin M F 2013 A comprehensive experimental study on the performance of fly ash concrete *Int. J. Engg. Adv. Technol.* **2** 135-42
[10] Barough A, Shoubi M, Kiani I and Amini Z 2010 Advantages of using fly ash in concrete industry for achieving sustainable development. In: *management in construction research association (MICRA) post graduate conference. UTM, Malaysia,*
[11] Mehta P K 1986 Concrete. Structure, properties and materials
[12] 1959 IS 1199:1959, Methods of Sampling and Analysis of Concrete. Reaffirmed- Dec 2013, BIS, New Delhi, India
[13] 1974 IS 7320:1974 Specifications for Concrete Slump Test Apparatus (4th revision). Reaffirmed- Dec 2013, BIS, New Delhi, India

IOP Conf. Series: Materials Science and Engineering **970** (2020) 012012 doi:10.1088/1757-899X/970/1/012012

Characterization of Alkali Activated Bentonite - Fly Ash as Brick Material

P P Rout, L Garanayak

Department of Civil Engineering, Institute of Technical Education and Research, SOA University, Bhubaneswar, Odisha, India

E-mail: lasyamayeegaranayak@soa.ac.in

Abstract. Brick, one of the most common binding materials which need excess energy during manufacture. Gathering of excess industrial waste creates the environmental pollution as well as a health issue. To developing green environment, industrial waste can be used as an alternation of binder material as a brick. In this present study combination of sodium bentonite and fly ash get activated by sodium hydroxide (NaOH) and calcium oxide (CaO) whose Na_2O/CaO proportion varied as 1 with mole content. Mixing proportion of both solid particles were varied as 40/60 and 60/40 as weight percentages. Casted samples were conducted for the curing period of 0 days, 3 days, 7 days, and 28 days. Mineralogy, morphology, and compressive strength tests are conducted and found that 1.5 moles are the best for developing compressive strength that obtained as 135.8 kPa at 60°C oven curing temperature.

1. Introduction

Bricks are considered as versatile, durable and reliable building material in all over the world. Bricks are manufactured by clay and shale in kiln in large quantities, which needs a huge amount of coal and fuels for developing high temperature i.e., 900 to 1000°C. According to research, Indian brick industry is considered as the second largest producer in all over the world that consume more than 150 million tons of coal annually. Beyond that, diesel for transporting the bricks produce one third of the global industry, which emits around 550million tons of CO_2. For developing eco-friendly building material, less consumption of natural resource (clay and shale) and energy, different type of waste materials is used as an alternation of binding material [1]. New binder material named as geopolymeric brick that more advantageous for its low energy consumption, good mechanical properties, lower pollution impact, good durability, significantly reduce the energy consumption, better resistance to chemical attack, and reduction of greenhouse effect [2, 3]. Although fire bricks are more strength than sun dried brick, but reduction of energy consumption and CO_2 emission motivate to prepared brick using industrial waste products. However, utilization of industrial wastes as brick can be categorized as three production methods i.e., firing, cementing, and geopolymerization [3]. But geopolymer bricks are turns to be most popular which can replace the industrial waste as an alternation of binding material without emission of CO_2 content. Silica and alumina rich material are carried out for a synthesis of geopolymer brick which are found to be most common elements in industrial waste.

Geopolymer also named as “inorganic polymer”, “alkali-activated cements”, “geocements”, “alkali-bonded ceramics”, “hydro-ceramics” etc. It was discovered by Prof Joseph Davidovits in 1972 and considered as alumino-silicate binder chiefly obtained in alkaline environment condition.

Geopolymerization is the technology which is based on the chemical reaction of amorphous silica and alumina rich solids with high alkaline solution at ambient and elevated temperature [3]. It has numerous applications in transportation, industrial, agricultural, residential, mining, oil well cements, high-volume application etc. [4]. Si rich material like fly ash, slag, rice husk and Al rich material like clay which belongs to kaolin, bentonite are the major part of developing geopolymerization [5-10]. Alkali concentration is the most significant factor of geopolymerization. Solubility of alumino-silicate gets increased with increase of hydroxide ion concentration.
Research is still continuing on developing low-cost, light weight and environmentally friendly construction material from industrial waste products as equivalent to traditional brick affirmed in the standard. Geopolymer brick is most popular among firing, cementing brick for less consumption, energy and reduction of global warming effect. Si and Al are the major part of developing the geopolymer product. So, alumina and silica based industrial wastes are considered as – fly ash, blast furnace slag, mine tailing, red mud etc. But very few studies are related to bentonite. Here research is based preparing geopolymer brick using fly ash and bentonite that activated with low concentration of a combining solution of sodium hydroxide (NaOH) and calcium oxide (CaO).

2. Materials & Methods

2.1. Source materials
Fly ash, an industrial waste was collected from NALCO industry situated in Angul of Odisha, India. It is pozzolanic type due to the presence of Al and Si content which helps to provide the geopolymer product. Sodium bentonit (known as a clay colloid material) was collected from Jagatpur of Odisha, India.
Both fly ash and bentonite were carried out for oven dried maintaining temperature as 105-110°C for 24 h. Since fine particles help to improve high compressive strength, hence both fly ash and bentonite were sieved under 150μ sieve to obtain uniform fine particles.

2.2. Chemical activator
Two chemicals, i.e., sodium hydroxide and calcium oxide were used as chemical activator. Sodium hydroxide pellet of Merck Life Science Private Limited were used whose specification varies as: $\geq$ 97%. Calcium oxide powder pure of Sisco Research Laboratories Pvt. Ltd., were used whose specification varies as: min 95%.

2.3. Mixing method
Both solid particles, i.e., fly ash and bentonites were mixed homogeneously with two different proportions i.e., 40/60 and 60/40 as weight percentages. Both were categorized as FB_{40} and FB_{60} respectively. Here F refers to fly ash, B refers to bentonite. Subscription 40 and 60 refer to weight percentages of fly ash.
Chemical solutions were prepared with distilled water one day prior to the casting of the sample variance with mole proportion of Na_2O and CaO as 1:1. Solutions were prepared as two type moles i.e., 1 and 1.5 moles separately. They were notified as NC_1 and $NC_{1.5}$ respectively. Here N refers to Na_2O, C refers to CaO. Subscription 1, 1.5 refers to the mole.
After a homogeneous mix of solid part as said proportion, prepared solution was added whose solution/solid ratio varied as 0.4.

2.4. Curing method
Casted samples were exposed to curing method that categorized as two type i.e., atmosphere, oven drying. Atmosphere samples are provided to no cure in the oven, but only on the ambient condition up to testing ages. Oven drying samples is provided to cure in the oven at 40 and 60°C temperature for 24 h that each sample wrapped with cling film to avoid moisture loss. Then, each oven drying cling free sample was exposed to room temperature.

2.5. Curing ages

Each sample was considered to be cured as 0, 3, 7 and 28 days respectively. Atmosphere curing samples were denoted as A. Oven curing samples are denoted as T. Here T refers to 40°C or 60°C.

3. Result and Discussion

3.1. Physical properties

Oven dried both fly ash and bentonite were tested for specific gravity by following the IS: 2720 (Part III/Sec 1)-1980 [11] and obtained as 2.09 and 2.9 respectively. Here bentonite was found as high specific gravity in comparison to fly ash.

Maximum dry density (MDD) and Optimum moisture content (OMC) of both fly ash and bentonite were calculated by following IS: 2720 (Part VII) -1980 [12, 13] that shown in Figure 1. For fly ash it is observed as 1.378g/cm^3 11.28%; and for bentonite it is observed as 1.358g/cm^3 and 28%, respectively.

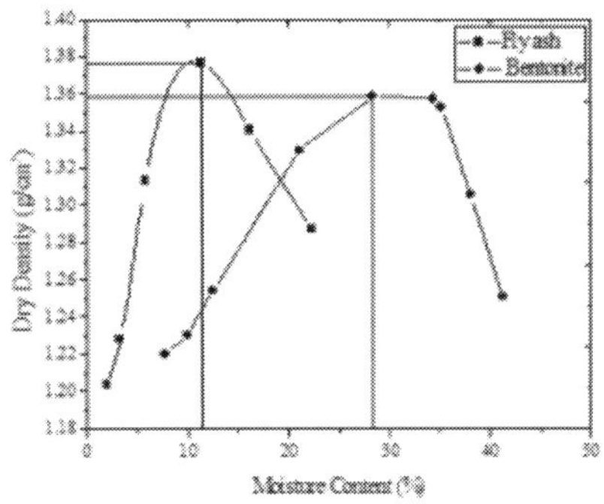

Figure 1. MDD and OMC curve of fly ash and bentonite

3.2. Mineralogy

Figure 2 (a) and (b) shows the XRD image of fly ash and sodium bentonite. From Figure 2 (a), it is observed that, NALCO fly ash is amorphous type whose hump angles (2θ) varies between 10 to 20°. In its quartz and mullite are found as common material that indicate to Al and Si as the major element. From Figure 2 (b), quartz and montomorilonite are found as the major element. Sodium bentonite is found as the amorphous type whose humps angle varies 10 to 30°.

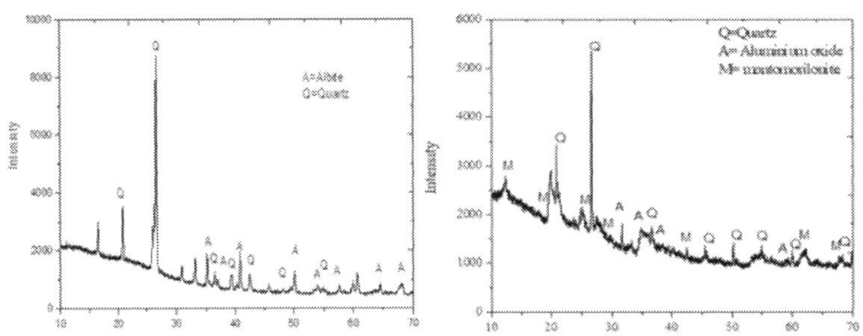

(a) (b)

Figure 2. XRD image of (a) fly ash, and (b) bentonite

3.3. Morphology

Figure 3 (a) shows the SEM image of NALCO fly ash, which is a spherical type with smooth surface whose shape varies from 1 to 100μm. Figure 3 (b) shows the microstructure of sodium bentonite that varies as non-uniform structure and considered as prismatic type.

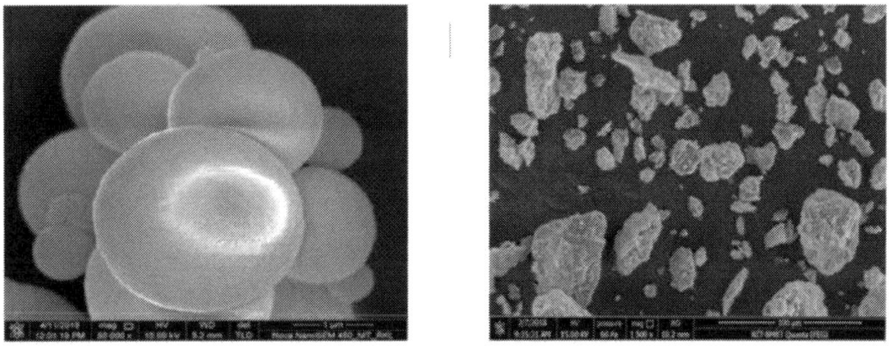

(a) (b)

Figure 3. SEM image of (a) fly ash, and (b) bentonite

3.4. Effect of mole vs solid proportion

Compressive strength was represented as a bar chart that shown in Figure 4 (a) and (b). Two proportions i.e., FB_{40} and FB_{60} get activated with NC_1 and $NC_{1.5}$. The compressive strength of each sample was calculated after curing up to 0, 3, 7, and 28 days.

3.5. Effect of NC_1

NC_1 was the combine solution of NaOH and CaO where the mole content of both Na_2O and CaO were taken as 1. Figure 4 (a) and (b) shows the compressive strength of FB_{40} and FB_{60}. Atmosphere curing and 24h oven curing samples were placed in ambient condition up to different curing aging as mentioned above. It is observed that maximum strength is observed with FB_{40}-NC_1-60°C that obtained as 90.8 KPa after 28 days of curing aging. Similarly, Figure 4 (b) shows 129KPa strength with FB_{60}-NC_1-60°C. This strength was found to be comparatively lower due to low concentration of sodium hydroxide (NaOH) i.e., 1 mole of NaOH was added to 1 mole of calcium oxide (CaO). For low concentration, there might be low reaction formation in between solid and solution which results to no better strength was formed.

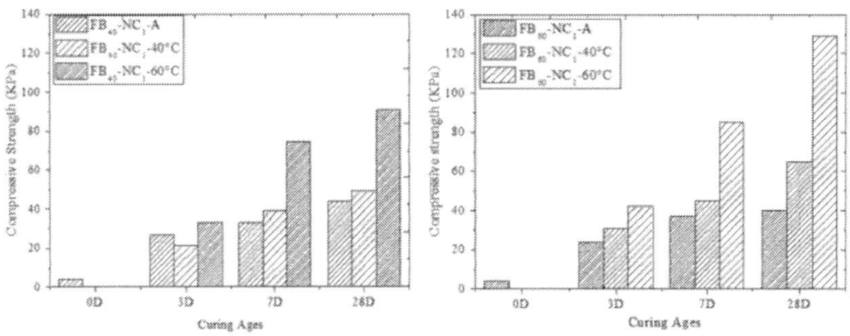

(a) (b)

Figure 4. Compressive strength of (a) FB_{40} (b) FB_{60} activated by NC_1

3.6. Effect of $NC_{1.5}$

$NC_{1.5}$ was the combine solution of NaOH and CaO where mole content is 1.5. Here, maximum strength is observed with FB_{40}-$NC_{1.5}$-60°C that obtained as 102.4 KPa as shown in Figure 5 (a). Similarly, Figure 5 (b) shows the maximum strength with FB_{60}-$NC_{1.5}$-60°C that obtained as 135.8 KPa after 28 days curing aging. In this case NaOH and CaO concentration were increased to 1.5 moles which results to growth of reaction and results to increase strength than previously.

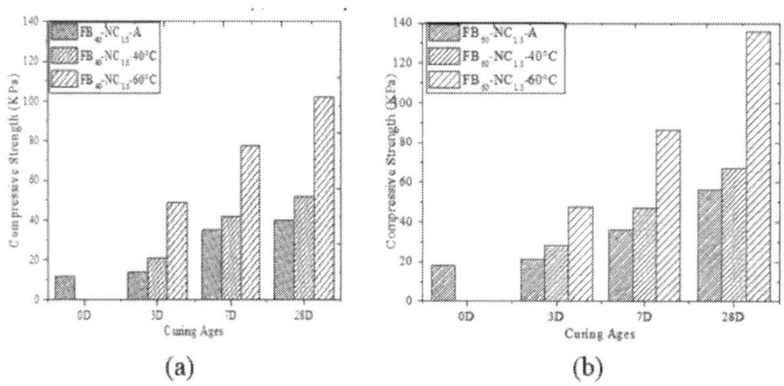

Figure 5. Compressive strength of (a) FB_{40} and (b) FB_{60} activated by $NC_{1.5}$

Although strength in atmosphere curing condition is gradually increased, but better strength is achieved with oven curing conditions. It is conformed that; strength gets increased by increase the mole content of Na_2O and CaO. In alkali activated system, enriched mechanical strength is possible due to Al-O and Si-O bond which developed N-A-S-H gel under highly alkaline solution.

4. Conclusion

Development on greener environment by reducing energy sources, utilizing industrial waste are the major criteria in the construction industry. Now-a-days, alkali activated brick turns to be more popular than traditional brick for less consumption of electric energy, heat energy. Although, researches are still carried on alkali activated brick, but those are related to high concentration of alkali solution. This work is related to developing alkali activated fly ash bentonite brick using low concentrations of chemical activator. Here, sodium hydroxide (NaOH) and calcium oxide (CaO) are used whose preparation method is based on mole content i.e., 1 and 1.5. Combine solution of NaOH and CaO are added as the chemical solution to maintaining a solid solution ratio as 0.4. Atmosphere and oven dry curing (for 24 h) conditions are preferred to compare the unconfined compressive strength of alkali activated fly ash bentonite brick under two different mole proportion (1 and 1.5). It consists with following conclusion:

- Specific gravity of bentonite is maximum as compared to fly ash.
- MDD (maximum dry density) of fly ash and bentonite is 1.378g/cm^3 and 1.358g/cm^3. OMC (optimum moisture content) is 11.28% and 28% respectively.
- NALCO fly ash is major with mullite and quartz which indicates that it is pozzolanic type. Sodium bentonite is major with montomorillonite and quartz. Both solid materials are amorphous type that conforms from hump angles.
- NALCO fly ash is spherical with smooth surface whose shapes are varied as uniformly. Sodium bentonite is varied as a non-uniform shape with prismatic type.
- Oven curing condition is better compared to the atmosphere curing condition.
- 60°C oven curing condition is better for developing strength in alkali activated fly ash-bentonite.
- 1.5 moles of the Na_2O and CaO is better in compared to 1 mole for developing strength.
- Maximum strength is observed as 135.8 KPa from FB_{60}-$NC_{1.5}$-60°C.

References

[1] Ahmari S and Zhang L 2012 Production of eco-friendly bricks from copper mine tailings through geopolymerization *Construction building materials* **29** 323-31

[2] Peng M X, Wang Z H, Xiao Q G, Song F, Xie W, Yu L C, Huang H W and Yi S J 2017 Effects of alkali on one-part alkali-activated cement synthesized by calcining bentonite with dolomite and Na2CO3 *Applied Clay Science* **139** 64-71

[3] Zhang L 2013 Production of bricks from waste materials–A review *Construction building materials* **47** 643-55

[4] Li C, Sun H and Li L 2010 A review: The comparison between alkali-activated slag (Si+ Ca) and metakaolin (Si+ Al) cements *Cement concrete research* **40** 1341-9

[5] Part W K, Ramli M and Cheah C B 2015 An overview on the influence of various factors on the properties of geopolymer concrete derived from industrial by-products *Construction Building Materials* **77** 370-95

[6] Hu M, Zhu X and Long F 2009 Alkali-activated fly ash-based geopolymers with zeolite or bentonite as additives *Cement Concrete Composites* **31** 762-8

[7] Mohammed B S, Liew M S, Alaloul W S, Al-Fakih A, Ibrahim W and Adamu M 2018 Development of rubberized geopolymer interlocking bricks *Case studies in construction materials* **8** 401-8

[8] Murmu A L and Patel A 2018 Towards sustainable bricks production: An overview *Construction building materials* **165** 112-25

[9] Perumal G and Sivakumar A 2014 Performance evaluation of alkali activated fly ash lightweight aggregates *Engineering journal* **18** 77-86

[10] Peng M X, Wang Z H, Shen S H, Xiao Q G, Li L J, Tang Y C and Hu L L 2017 Alkali fusion of bentonite to synthesize one-part geopolymeric cements cured at elevated temperature by comparison with two-part ones *Construction Building Materials* **130** 103-12

[11] 1980 IS code IS 2720 (PartIII/sec 1): 1980 Methods of Test for Soils Determination of Specific Gravity

[12] 1980 IS code IS 2720 (Part VII): 1980 Methods of Test for Soils Determination of Water Content – Dry density Relation using Light Compaction

[13] 1991 IS code IS 2720 (Part 10): 1991 Methods of Test for Soils Determination Unconfined Compressive Strength

Mechanical Properties of Concrete with Sintered Fly Ash Aggregate as Substitute of Natural Fine Aggregate

Bidyadhar Basa[1], Nibedita Pradhan[1], Lipsita Priyadarshini Parhi[2]

[1] Department of Civil Engineering, ITER, SOA Deemed to be University, Bhubaneswar, Odisha, India
[2] Department of Civil Engineering, Hitech Institute of Technology, Bhubaneswar, Odisha, India

E-mail: bidyadharbasa@soa.ac.in

Abstract. Attempts have been made from time long to develop construction materials which should be strong, durable and simultaneously of light weight. In an attempt to reduce the self-weight of the structural concrete, light weight aggregates are being used and implemented. Fly ash is a waste material of coal based thermal power plants. Presently, fly ash is being used in construction industries in different ways. Sintered fly ash aggregates when used as a substitute for natural aggregates benefits the preservation of natural aggregates as well as the utilization of fly ash which poses a great environmental problem for its disposal. Simultaneously, it also reduces the cost of concrete. In the present investigation, sintered fly ash light weight aggregates have been used as a partial replacement of natural sand as fine aggregate in concrete. Varying percentages of natural fine aggregate have been replaced by sintered fly ash light weight aggregates. Experimental investigations have been carried out to study the mechanical properties related to the strength of concrete, namely, compressive strength, split tensile strength and flexural strength with sintered fly ash aggregate as fine aggregate substitute in concrete. In all mixes, silica fume have been used in conjunction with the binder material consisting of ordinary Portland cement.

1. Introduction

Concrete is one of the most resourceful and adaptable construction materials possessing superior qualities in terms of chemical and mechanical characteristics. It has found its application in almost all types of constructions. Presently, in the construction industry throughout the world, most of the concrete prepared is with natural fine aggregate as a major constituent. Its extensive usage results in geological and environmental imbalance and simultaneously, the naturally available sand gets depleted causing a great catastrophe for future generations. Accordingly, use of construction material for eco-friendly and sustainable development has become a great challenge for construction engineers. Keeping in view of the growing need of qualitative construction materials, substantial development has been made in this domain keeping in pace with the growing technology. In this context, there has been a wide range of explorations throughout the globe to come up with superior quality materials for construction which must guarantee in terms of strength and performance level. For sustainable development, it has become mandate to use industrial wastes as alternative construction materials. This not only saves the original materials for future generations but also their successful use eliminates the problem of their disposal. Now in one hand to save the natural materials and use the wastes as their potential substitute and on the

other hand to make the structure light weight, it has become vital for preparing artificial construction materials from industrial and agricultural sources.
In a developing nation like India, construction activities are going on at a very big scale. This has necessitated a large sum of structural materials involving a large sum of construction cost. In this scenario, it is necessary to substitute these commonly used construction materials by appropriate novel materials which should be cost effective and eco-friendly. To the answer for the strain and demand for water-course fine aggregates, researches are being conducted and attempts have been made to utilize quarry dust, polyvinyl waste, plastic waste etc. as substitute for natural river sand as fine aggregate in concrete. The bending performance of concrete is found to be increasing by using polyvinyl wastes as a part substitute of sand [1]. Singh and Siddique [2] studied the effect of iron slag as partial replacement of fine aggregate in concrete and checked its durability and observed that up to 40% replacement of iron slag against NFA gives high strength to concrete. Structural low density concrete necessity can be met by pumice lightweight aggregates in concrete as fine and coarse aggregate substitute, in respect of strength and durability parameters [3]. Siddique [4] investigated the effects of flyash as a fine aggregate substitute in concrete and noteworthy development has been observed in concrete strength. Arumugan et al. [5] used pond ash as an effective alternative material for fine aggregate in concrete. Use of bottm ash as a material in concrete to be used as fine aggregate isubstitute n 10-40% enhances its compression and tension resistance as well as resistance to abrasion. However, the freeze- thaw resistance gets lowered [6]. Singh and Siddique [7] used coal bottom ash as fine aggregate in concrete. Enhanced strength was observed in concretes containing bottom ash.
Fly ash, an industrial by-product from coal firing thermal power plants with current annual generation of approximately 300 million tons can be used in cement-based materials due to its very good pozzolanic properties. There are different ways of using fly ash in concrete industry. It can be used as partial replacement of cement or use of fly ash aggregate in replacement of natural aggregate. Fly ash light weight aggregates produced by sintering at high temperature and by pelletization has proved to be a successful construction material as substitute for natural coarse aggregates [8-10]. From the detailed review of the literature, it is observed that materials like pellets, coconut shell, volcanic pumice light weight aggregate, periwinkle shell etc. can be used as replacement of natural coarse aggregate. In the concrete industry these materials can be used as light weight aggregate. However, it is apparent that no specific study has been reported in the literature regarding utilization of sintered fly ash aggregate for replacement of fine aggregate. This presents wide scope for future investigation in this direction.

2. Experimental Setup

This section provide the details of materials used for the present experimental work, their properties, mixing and casting procedure for assessing different properties of concrete containing various percentage of sintered fly ash aggregates as fine aggregate substitute.

2.1. Materials Used

For the present experimental investigations, following materials are used:

- Ordinary Portland cement (OPC-43 grade)
- Natural fine aggregate
- Sintered fly ash aggregate
- Water
- Natural coarse aggregate (20mm down)
- Silica fume

The properties of the materials are mentioned below.

2.1.1. Cement

OPC 43 grade cement is used having physical characteristics represented in Table 1.

Table 1. Physical characteristics of cement

Properties	Results
Specific gravity	3.15
Initial setting time	40 min
Final setting time	230 min
Normal Consistency	34%

2.1.2. Aggregates

Water course sand available locally and satisfying the criteria of zone-II is used as fine aggregates for making of the concrete specimens. Similarly, Natural Coarse Aggregate (NCA) consists of broken granite rock of particle size ranging between 4.75 mm to 20 mm. Alternatively, Sintered fly ash aggregate used was acquired from Indian Metals and Ferro Alloys (IMFA), situated in Cuttack, Odisha, India. Since the flyash lightweight aggregates are used as fine aggregate substitute in the present study, maximum size of aggregates is limited to 4.75mm.The characteristics of the fine and coarse aggregates used in the present study are mentioned in Table 2.

Table 2. Characteristics of aggregates

Properties	Results		
	Fine Aggregate	Coarse Aggregate	Flyash Aggregate
Bulk density (kg/m^3)	1650	1732	1430
Specific gravity	2.70	2.75	1.58
Fineness modulus	2.84	7.8	-
Water absorption (%)	0.81	0.22	11.5

Figure 1. Sintered fly ash aggregate

2.2. Mix Design and Casting

For the present experimental investigation, volume proportion is considered for mix proportioning. Following IS: 10262-2009 stipulations, concrete mix for M30 grade was designed to achieve a target strength of 38.25N/mm^2. Here OPC 43 grade cement was used as cementitious material. Accordingly, for 1 m^3 of concrete, the ingredient materials are 438 kg of cement, natural fine aggregate 610 kg, natural coarse aggregate 1206 kg and 197 liter of water. Different mixes were subsequently prepared keeping in view of the mix proportions so obtained but natural fine aggregates being replaced by fly ash light weight aggregates in 10, 20, 30, 40, 50 and 100 percent by weight. In all the mixes, silica fume was added as supplementary cementitous material at the rate of 20% by weight of cement. The materials were properly mixed following standard procedure in a pan mixer and were cast into different moulds for further testing as shown in Figure 2. After 24 hours of casting, the specimens were placed inside water for curing. The various mechanical properties were tested after 7, 28 and 56 days of curing. The mix identity corresponding to different mix proportions are given in Table 3.

Figure 2. Casting of Samples

Table 3. Mix identity designation for concrete specimens

Concrete mix Proportion	Mix identity
C80%+NFA100%+SF0%+ SI20%	NA100SF0SI20
C80%+NFA90%+SF10%+SI20%	NA90SF10SI20
C80%+NFA80%+SF20%+SI20%	NA80SF20SI20
C80%+NFA70%+SF30%+SI20%	NA70SF30SI20
C80%+NFA60%+SF40%+SI20%	NA60SF40SI20
C80%+NFA50%+SF50%+SI20%	NA50SF50SI20
C80%+NFA0%+SF100%+SI20%	NA0SF100SI20

C-Cement, NFA -Natural fine aggregate, SF - Sintered fly aggregate, SI - Silica fume

3. Results and Discussion

For the present investigation, to assess the properties of flyash aggregates as a potential material for replacement of natural fine aggregates, tests were carried out to determine their hardened properties after 7, 28 and 56 days of curing. For this, strength of concrete in compression, flexure and splitting tension are evaluated following experimental procedures laid by Indian Standard specifications. The results pertaining to various tests are mentioned in the subsequent sections and discussions are made.

3.1. Compressive strength

Compressive strength of 150 mm cubic samples is found by examining in a compression testing machine after the samples were cured for specified period. The outcomes corresponding to the strength of cubes containing various proportions of fly ash light weight aggregates as a substitute of natural fine aggregates are presented in Figure 3.

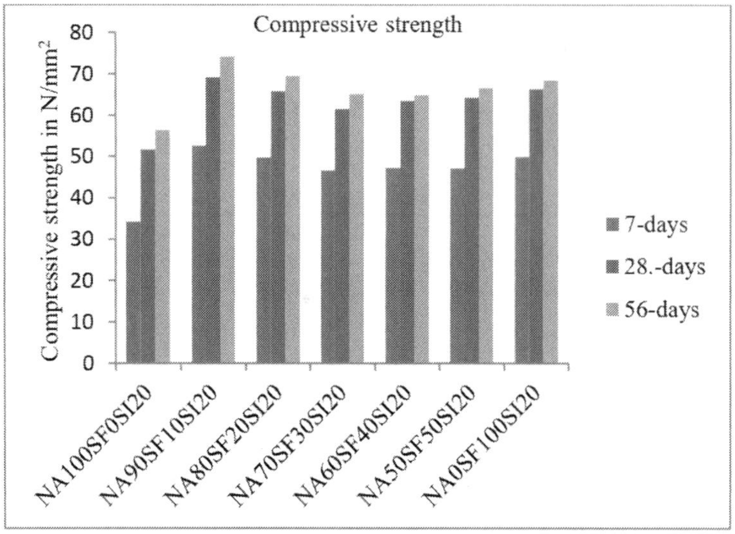

Figure 3. Compressive strength of Cube specimens

From Figure 3 it is obvious that the cube compressive strength of concrete with different percentage of fly ash light weight aggregate are in very good agreement with that of natural fine aggregate. The minimum strength corresponds to 30% replacement. However, this one also is having strength at par with the controlled specimen containing natural fine aggregate. Even at 100% replacement, the strength is found to be higher than that corresponding to NFA concrete. It may be noted here that even though the mix is intended for M30 grade, the strength of concrete is quite higher than the target strength. This may be due to the addition of silica fume which further enhances the formation of C-S-H gel.

3.2. Splitting tensile strength

The tensile strength corresponding to splitting of concrete was determined by examining samples of cylindrical shape of 100 mm diameter and 200 mm high in a compression testing machine following Bureau of Indian Standards guidelines. The consequences are shown in Figure 4.

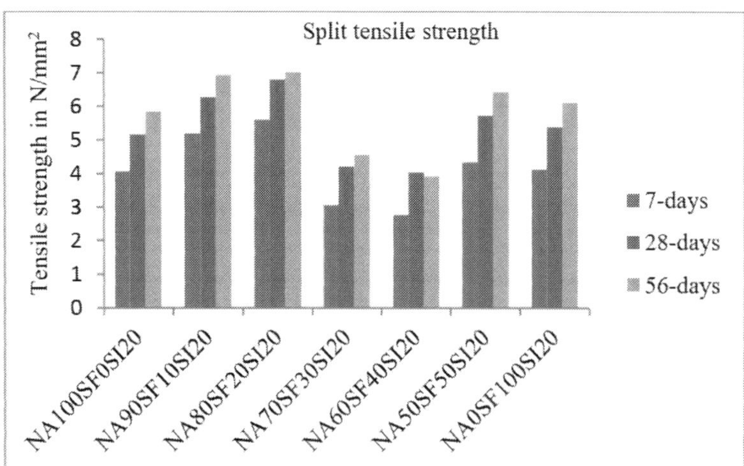

Figure 4. Split tensile strength of cylindrical specimens

The results reveal that the split tensile strength of concrete with different percentage of fly ash light weight aggregate are in very good agreement with that of natural fine aggregate. The maximum strength corresponds to 20 % replacement while, minimum strength corresponds to 40% replacement. Even at 100% replacement, the strength is found to be higher than controlled concrete specimen.

3.3. Strength in flexure

The strength of concrete in flexure was evaluated on prism samples of 100 mm square in cross section and 500 mm long the outcomes are represented in terms of modulus of rupture. The procedure laid down by BIS was followed to derive the flexural strength in a flexure testing machine. The results corresponding to test are presented in Figure 5.

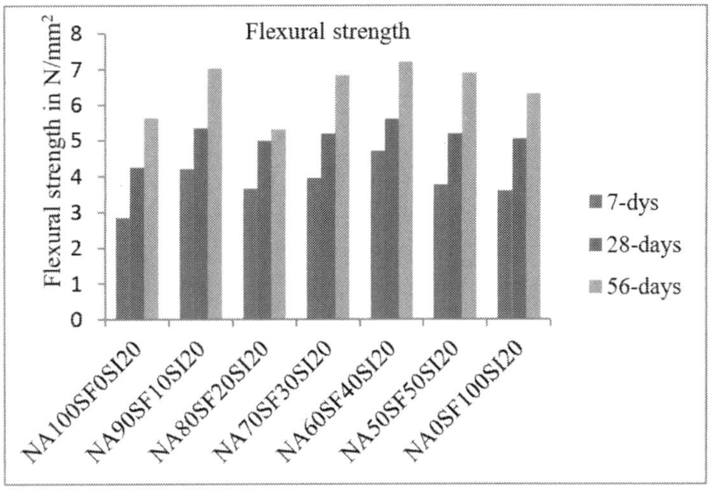

Figure 5. Flexural strength of prism specimens

The graph corresponding to experimental result shows that the strength of concrete in flexure with different percentage of fly ash light weight aggregate also follow the same trend, the strength being in very good agreement with that of controlled specimen with natural fine aggregate. The minimum strength corresponds to 20% replacement.

4. Conclusions

From the above experimental investigations made in the present work, these are the subsequent inferences which can be drawn. Fly ash light weight aggregate concrete shows very good mechanical properties in respect of strength corresponding to cube compression, splitting tension and flexure as compared to natural fine aggregate. At complete replacement, the strength in compression shows an increase of 45.6%, 28.4% and 21% after a curing period of 7, 28 and 56 days. Similarly, the strength corresponding to splitting tension exhibits an enhancement of 1.5%, 4.26% and 4.62% after curing period of 7, 28 and 56 days. The strength in flexure of concrete with full replacement shows respectively 26%, 18.6% and 12.3% increase after 7, 28 and 56 days of curing with respect to controlled specimen. Thus it can be fruitfully used as a prospective substitute for natural fine aggregate. However, since it absorbs more water, it is suitable for interior work rather than exterior structural units. Sintered fly ash aggregates when used as a substitute for natural aggregates benefits the preservation of the natural resources as well as the utilization of fly ash which poses a great environmental problem for its disposal. Simultaneously, it also reduces the cost of concrete. However, micro-structural study like XRD, SEM etc. needs to be carried out to ascertain the actual behaviour.

Acknowledgments
Authors are thankful to the authority of IMFA, Odisha, India for providing the sintered fly ash lightweight aggregates for the present experimental investigations.

References

[1] Ikponmwosa E, Fapohunda C, Kolajo O and Eyo O 2017 Structural behaviour of bamboo-reinforced foamed concrete slab containing polyvinyl wastes (PW) as partial replacement of fine aggregate *Journal of King Saud University-Engineering Sciences* **29** 348-55

[2] Singh M and Siddique R 2013 Effect of coal bottom ash as partial replacement of sand on properties of concrete *Resources, conservation recycling* **72** 20-32

[3] Parhizkar T, Najimi M and Pourkhorshidi A 2012 Application of pumice aggregate in structural lightweight concrete *International Journal of Earth Sciences and Engineering* **8**

[4] Siddique R 2003 Effect of fine aggregate replacement with Class F fly ash on the mechanical properties of concrete *Cement Concrete research* **33** 539-47

[5] Arumugam K, Ilangovan R and Manohar D J 2011 A study on characterization and use of pond ash as fine aggregate in concrete *International Journal of Civil Structural Engineering* **2** 466-74

[6] Bakoshi T, Kohno K, Kawasaki S and Yamaji N 1998 Strength and durability of concrete using bottom ash as replacement for fine aggregate *Special Publication* **179** 159-72

[7] Singh G and Siddique R 2016 Effect of iron slag as partial replacement of fine aggregates on the durability characteristics of self-compacting concrete *Construction Building Materials* **128** 88-95

[8] Basa B and Sethi J 2016 Fresh And Hardened Properties of Self Compacting Concrete using Fly Ash Lightweight Aggregate as Partial Replacement of Natural Coarse Aggregate *International Journal of Civil Engineering Technology* **7** 498-505

[9] Manikandan R and Ramamurthy K 2008 Effect of curing method on characteristics of cold bonded fly ash aggregates *Cement concrete composites* **30** 848-53

[10] Reddy M V S, Nataraja M, Sindhu K, Harani V and Madhuralalasa K 2016 Performance of light weight concrete using fly ash pellets as coarse aggregate replacement *International Journal of Engineering Research Technology* **9** 95-104

IOP Conf. Series: Materials Science and Engineering **970** (2020) 012014 doi:10.1088/1757-899X/970/1/012014

Determination of SSUT of HSC Beams with WASPAS Method

Gopal Charan Behera, Subhasri Panda

Department of Civil Engineering, Government College of Engineering, Kalahandi, Bhawanipatna, Odisha, India

E-mail: beheragb@gmail.com

Abstract. Distressed beams should be retrofitted rather than demolishing. Now a days there are so many retrofitting materials available. Retrofitting is to be done with materials those are easy to apply, economic and less time consuming. Engineers are more interested in ultimate strength of a structure. Strength is related to stiffness. Finding secant stiffness at ultimate torque of a suitable wrapping material for distressed concrete beams that is economic, long lasting, sustainable and appropriate to the common people is the main focus of this investigation. The paper progresses with search of different wrapping strategies and different materials available. Then the paper aims exploring sustainability options for torsional retrofitting with ferrocement. Aim of the paper is to evaluate SSUT of "U" wrapped beam with different methods. This study shows there is reduction of stiffness at ultimate stage of wrapped beams. Experimental, analytical and Soft computing method MARS and WASPAS were discussed here to predict the value of SSUT. Torsionally over reinforced sections are found with more stiffness at ultimate than all other types of beams. The predicted values are well within the acceptable limits.

1. Introduction

Degradation by climate, change in functional requirements or excess load during earthquake, has necessitated reinforced Concrete (RC) structures to be structurally upgraded. Poor maintenance, inferior workmanship or due to codal updates, RC structures may be upgraded rather than demolished [1-2]. Fiber reinforced polymer (FRP) is the commonly used wrapping material for strengthening of structures in present era.

Each Repair technique differs from one to other. Building reoccupancy time limit, cost of repair, amount and type of labour decide type of reapir [3]. Retrofitting of deficient structures by FRP is easy. Retrofitting by FRP against bending and shear has attained a good pace, for torsion it is in back seat [4-5]. Enhancement of strength has been observed with FRP wrapped structures against shear, torsion and flexure [6]. Wrap if it is possible may be in all four sides or it may be on three sides leaving the top face due to existing structures. Many investigation are carried out for full with FRP [7-10] investigated with "U" wrap. Requirement of skilled labour and high cost keeps FRP wrapping familiar in developed countries [11,12], may not be suitable for developing countries. Better torsional capacity is observed in full wrap than "U" wraps [13]. Torsional stiffness is a key parameter. Stiffness of a structure depends on its size, materials used and its properties [14-16]. Stiffness is the other way of measuring ductility. Here determination of the stiffness of beams is the main aim of this investigation. Analytical method is developed to supplement the work. The objective is to strengthen distraught structure to enhance the torsional secant stiffness at ultimate torque (SSUT) by an economical material such as ferrocement. Destruction of prototype structure for experimental work and consumption of time in analytical method push both the methods in back seat. Soft computing procedure like MARS and WASPAS have been employed to evaluate the stiffness of different high strength concrete beams with reinforcement in core and ferrocement "U" wrap on periphery

1.1. Substitute for FRP and Evaluation Processes for Torsional Parameters

Ferrocement is a low cost material and easy to apply. Torsional strength is found to increasing in distressed structures with wrap of ferrocement [17]. Construction on ferrocement is easy. It has low permeability and strength in tension is high [18]. Ferrocement is a combination of welded wire mesh with rich grade of mortar. Ferrocement is a good repairing material, increases ductility and strength of distressed members when wrapped with them. Ferrocement is easy to apply as it has the flowable property along with resistance to sulphate and corrosion. When a distressed structure is wrapped with ferrocement, failure mode shifts to ductile from shear. Studies were performed on "U" jacketing for enhancement of strength on distressed structures. Values obtained from test values shows that there is enough enhancement of strength with ferrocement wrapping. The analytical models were also predicting the enhancement of stiffness and dynamic characteristics [19]. Three methods were adopted to evaluate the torsional parameters. (1)Experimental Evaluation (2) Analytical Method ad (3) Soft computing.

2. Different Evaluation Methods

The different methods adopted here for evaluation of secant stiffness at ultimate are described here.

2.1. Experimental Method

For evaluation of SSUT, high strength beams were prepared and tested respectively. Different quantity of reinforcement was used in beams to make different states of torsion.

Shear stress is induced due torsion which creates shear cracks with 450 angle to longitudinal direction. To accommodate two spirals of crack pattern a length of 2mt. was taken with width 125 mm and depth 250 mm. The concrete strength was kept 60 MPa while mortar was used of 55 MPa. The two end zones with 250 mm each were heavily reinforced to avoid failure due to gripping. To make different states of torsional beams, reinforcement was varied in longitudinal and transverse direction. Reinforcement in both directions is less in under reinforced beam than balance reinforcement. For over reinforced in longitudinal direction, higher quantity of reinforcement than that is required for a balance section in the longitudinal direction. To make over reinforced in transverse direction, less quantity of steel in longitudinal and more steel in transverse direction are kept in high strength concrete. For completely over reinforced sections higher amount reinforcement in longitudinal and transverse directions is kept. Numbers of beams prepared and tested are presented in Table 1. A sketch of section of a beam presented in Figure 1 and details are covered in [20-21].

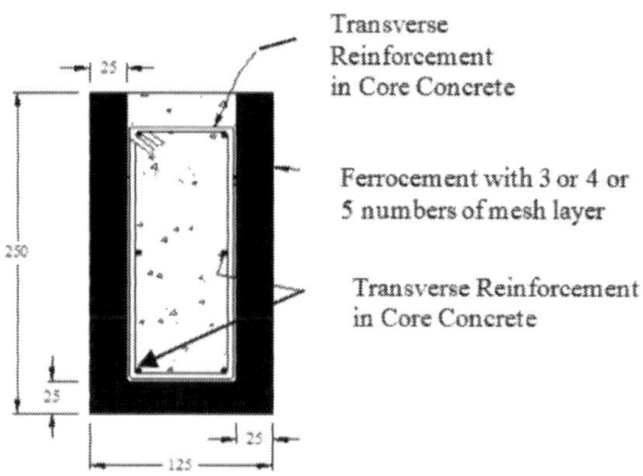

Figure 1. Details of the beams

Table 1. Details of the beam cast

Designation	Amount of steel				
	Reinforcement in core concrete				Ferrocement
	Reinforcement Longitudinal direction		Reinforcement Transverse direction		
	No. of bars, Diameter	Young's modulus (MPa)	Spacing, Diameter	Young's modulus (MPa)	Dia. 0.72 mm No. of mesh layers
BOH					
BO4H					4
L4H	6 nos.,12 mm	440			4
T4H			70 mm c/c, 10mm	445	4
U4H	6 nos.,6 mm	350	70 mm c/c, 6mm	350	4
Lo4H	6 nos.,12 mm	440	70 mm c/c, 6mm	350	4
To4H	6 mm, 6 nos.	350	70 mm c/c, 10 mm	445	4
Co4H	6 nos.,12 mm	440	70 mm c/c, 10 mm	445	4

2.2. Analytical Model

To find out torsional stiffness, experimental method is one. This method is a time consuming process and destruction of specimens makes the process uneconomical as well as wastage of material. From the experimental data, prediction of torsional parameters with change of dimension, material and material properties could not evaluated. To get rid of this situation, analytical method is developed by use of Softened truss model of Hsu [22]. As there is asymmetry in material, soften truss model is taken with some changes. The results predicted are found to be well within acceptable limit. Formulation of analytical model is described in [20—21]. Table 2 shows the values of SSUT.

2.3. Soft Computing Method

The analytical model developed for the evaluation of the torsional parameters, which solves a set of basic equations by iterative process. This does not provide any basic equation to evaluate the parameters taking consideration into the size aspect ratio, material type and their properties. So, soft computing method is adopted to evaluate the parameters.

2.3.1 Multivariate Adaptive Regression Spline (MARS)

In MARS method, some experimental results are taken for fitting and others are used for testing. Equations are developed in simple form to calculate the values for this MARS method. There is no need for any suppositions to be taken for this method. Finally, it provides simple equations to calculate the parameter for which some are calling it white box method. The final equation fetches the relation between various parameters. Friedman (1991) [23] presented this valuable theory. The method is very flexible. MARS predicts the value of secant stiffness at ultimate torque as presented below.

Secant stiffness at Ultimate Torque= maximum of [0, Fly-350] * 4.6414 + maximum of [0, 350- Fly] * 0.9407 +169.9223-maximum of [0, Amount of Longitudinal Steel] * 1.1151 + maximum of [0, Mortar strength-40]* 5.8969 + maximum [0, 40 - Mortar strength] * 14.2836 + maximum of [0, 0.921- Ultimate twist] * 3928.1784+ maximum of [0, 350- Fly] *0.5248

2.3.2. Method of Soft Computing: (WASPAS)

Another method of soft computing such as weighted aggregated sum product assessment (WASPAS) was used to solve multi-criteria decision-making- MCDM problems. Zavadskas [24] proposed this. This is developed by the combination of two methods such weighted product method (WPM) and weighted sum method (WSM). The different steps required to solve the problem by WASPAS method are presented in [25-26].

1st Step: Decision matrix is to be initialized.

2nd Step: Decision matrix is to be normalized by use of the equations given below. Maximize or minimize as per the equation.

$$\bar{x}_{ij} = x_{ij}/max_i x_{ij} \quad (1)$$

$$\bar{x}_{ij} = \quad min_i x_{ij} \quad (2)$$

Where x_{ij} is the assessment value of the ith alternative with respect to the jth criterion.

3rd Step: It involves for calculation of the total relative importance of ith value alternative. It is based on weighted sum method (WSM) with use of this values.

$$Q_i^{(1)} = \sum_{j=1}^{n} \bar{x}_{ij}.w_j \quad (3)$$

4th Step: Again the estimation of the total relative importance value of ith alternative is to be done. It depends upon (WPM) weighted product method. This involves the below mentioned formulae.

$$Q_i^{(2)} = \prod_j^n \bar{x}_{ij}^{w_j} \quad (4)$$

5th Step: In this step, importance is given for accuracy and effectiveness. In this method of WASPAS, decision-making process was given more importance. To determine total relative importance, the following equation was employed.

$$Q_i = \lambda.Q_i^{(1)} + (1-\lambda).Q_i^2 \quad (5)$$

Where λ can take the values either 0 or 1 or it can take the value 0.1. The usual value of λ is 0.5. When the values of λ is changed, there is change in the value of rankings of alternatives and also values of total relative importance of alternatives gets altered.

Table 2. Experimental and Predicted Values of SSUT

Beam	SSUT (kNm^2)			
	WASPAS	MARS	Softened Truss model	Experimental
BH	1362.43	1605.00		1605.00
B4H	1362.43	1111.66	1145.00	1145.00
L4H	558.41	869.40		869.40
T4H	940.41	927.42		915.28
U4H	37.10	69.13	38.00	38.00
Lo4H	65.29	60.93	66.00	68.00
To4H	64.20	69.13	72.75	72.00
Co4H	136.39	130.00	129.00	130.00

3. Discussion on Torsional Parameter SSUT

SSUT values were taken by experimental results. The same has been derived from analytical and soft computing methods. The variation of the SSUT values with states of torsion is presented here. Errors found in the predicted values by these three methods are calculated to find out suitability of these methods.

3.1. SSUT of High Strength Beams

Torsional stiffness of a beam depends on the strength of core material. Stiffness of a high strength beam is not same as normal strength beam. Properties such as softening co-efficient and elastic modulus change with change in strength of concrete. High strength beams with and without reinforcement and "U" wrap were cast and analyzed. All the beams have aspect ratio 2, mortar grade M55 and concrete of grade M60.Here focus is to evaluate SSUT. SSUT depends on the core and outer periphery material and their properties. During loading, structure behaves elastically at initial stage, then micro-cracking stage initiates, beyond that cracking takes place. After cracking, ultimate stage takes place. Beyond cracking, reinforcement plays a vital role. In the tested beams, designation BH refers to a beam in which there is no reinforcement in core concrete and no wrap. Beam B4H refers to a beam without core reinforcement and with "U" wrap. All other six beams are reinforced beams with "U" wrap covering six states of torsion.

The beam BH was found with highest SSUT value of 1605 kNm^2 and U4H was with lowest value of 38 kNm^2. Just after cracking, BH failed without any increase in ultimate torque. This proves unwrapped beam fails after cracking. The unwrapped beams are unable to sustain any load beyond cracking. The other beams B4H, L4H, T4H, Lo4H, To4H and Co4H were found with SSUT values of 1145 kNm^2, 869.40 kNm^2, 915.28 kNm^2, 68 kNm^2, 72 kNm^2 and 130 kNm^2 respectively. Beams when reinforced in both directions of core concrete, difference in SSCT values was very less. SSUT values by experimental and predicted were presented in Table 2. Softened truss model could not predict SSUT of beams of without reinforcement or with singly type of reinforcement. Figure 2 presents the value of SSUT of all beams.

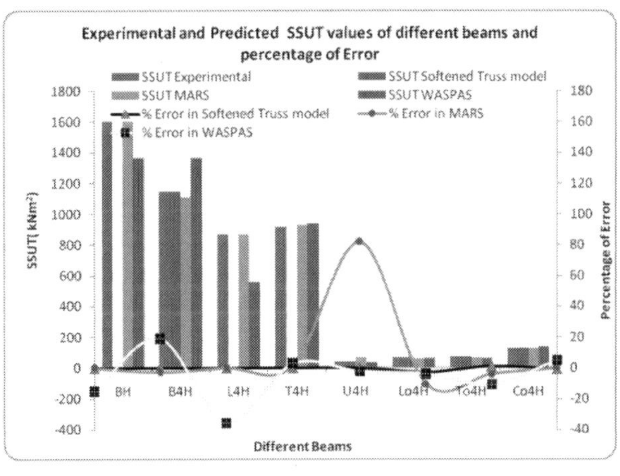

Figure 2. Experimental and predicted values of SSUT and percentage of error in predicted values.

3.2. Percentage of Error in Predicted Values

The percentage of errors in the predicted values was also presented in the same figure. The errors in percentage were found as -2.94, 1.04 and -0.77 respectively for longitudinally over reinforced, transversely over reinforced and completely over reinforced. Other values are found same with experimental results. The percentage of error in MARS was found to be 0.0, -2.91, 0.0, 1.33, 81.92, -10.39, -3.98, 0.0 for plain beam without wrap, B4H with wrap, only longitudinal reinforcement with wrap, T4H, under reinforced with wrap, Lo4H, transversely over reinforced with wrap and Co4H respectively. Only under reinforced beam was found with some error more than 10 % by MARS. The percentage of error in WASPAS was found to be -15.11, 18.99, -35.77, 2.75, -2.37, -3.99,-10.83 and 4.92 for BH, plain beam with wrap, L4H, only transverse reinforcement with wrap, U4H, longitudinally over reinforced with wrap, To4H and completely over reinforced with wrap respectively. The percentage of error is very less except control specimens and beam with only longitudinal reinforcement. So, softened truss model, soft computing models MARS and WASPAS can be employed for prediction of secant stiffness at cracking torque (SSUT).

3.3. Percentage of decrease in SSCT value in comparison to BH and B4H

SSUT values are found to be decreased in comparison to their control specimen BH. It has been noticed that percentage of reduction was found to be 97.63 for U4H, which is the maximum for all beams. The beams with "U" wrap and reinforcement in core concrete in both directions are found to have percentage of reduction more than 90. The percentage decrease of SSUT value of all reinforced beams with respect to control "U" wrap beam B4H was found to be up to 90 and for U4H, percentage decrease is 96.68. This proves under reinforced beams have undergone maximum twist. The same has been plotted in Figure 3. Stiffness at ultimate torque is controlled by the reinforcement present in the core. Completely reinforced beams are provided with maximum reinforcement. So, more SSUT was noticed for this beam.

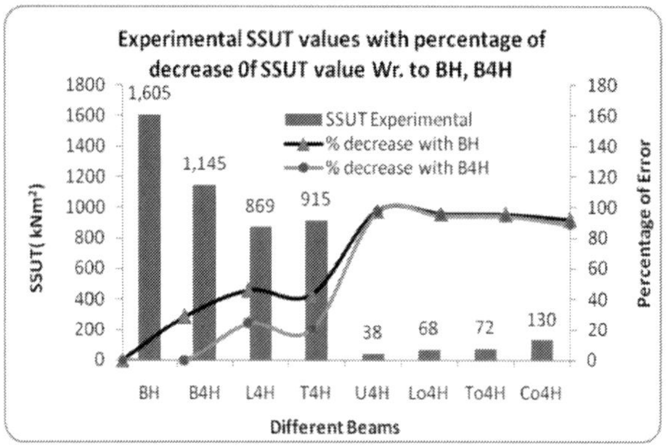

Figure 3. Experimental values of SSUT and percentage of decrease of SSUT with respect to BH and B4H.

4. Conclusion

The experimental study was conducted for calculation of SSCT values of different beams. The secant stiffness at ultimate torque was predicted by other three methods. Main conclusions highlighting the novelty of the work is briefly mentioned in this section.

- The results of SSUT is better predicted by analytical model than soft computing methods such as MARS and WASPAS.
- A significant decrease in SSUT is found with "U" ferrocement wrapped under reinforced high strength concrete beams over the beams without any reinforcement. "U" wrapped beam has 97.63 % reduction of SSCT value over their control specimen.
- SSUT is reliant upon the core concrete reinforcement.
- Under reinforced concrete, beams showed overall decrease in SSUT.
- The experimental values, predicted values from soft computing model and analytical model reveal that the SSCT of a ferrocement "U" wrap beam is more influenced by the state of torsion.

The percentage of error in predicted values for most of the beams are within 10%. So, all these methods; analytical, MARS and WASPAS can be used to predict the secant stiffness at ultimate torque of ferrocement "U" wrap high strength beams.

References

[1] Hii A K and Al-Mahaidi R 2007 Torsional capacity of CFRP strengthened reinforced concrete beams *Journal of Composites for Construction* **11** 71-80

[2] Karayannis C G, Chalioris C E and Sirkelis G M 2008 Local retrofit of exterior RC beam–column joints using thin RC jackets—An experimental study *Earthquake Engineering Structural Dynamics* **37** 727-46
[3] Ghobarah A, Ghorbel M and Chidiac S 2002 Upgrading torsional resistance of reinforced concrete beams using fiber-reinforced polymer *Journal of composites for construction* **6** 257-63
[4] Jing M, Raongjant W and Li Z 2007 Torsional strengthening of reinforced concrete box beams using carbon fiber reinforced polymer *Composite Structures* **78** 264-70
[5] Santhakumar R, Dhanaraj R and Chandrasekaran E 2007 Behaviour of retrofitted reinforced concrete beams under combined bending and torsion: A numerical study *Electronic Journal of Structural Engineering* **7** 1-7
[6] Deifalla A and Ghobarah A 2010 Strengthening RC T-beams subjected to combined torsion and shear using FRP fabrics: Experimental study *Journal of composites for construction* **14** 301-11
[7] Panchacharam S and Belarbi A 2002 Torsional behavior of reinforced concrete beams strengthened with FRP composites. In: *First FIB Congress, Osaka, Japan,* pp 01-110
[8] Salom P R, Gergely J and Young D T 2004 Torsional strengthening of spandrel beams with fiber-reinforced polymer laminates *Journal of Composites for Construction* **8** 157-62
[9] Ameli M and Ronagh H R 2007 Analytical method for evaluating ultimate torque of FRP strengthened reinforced concrete beams *Journal of Composites for construction* **11** 384-90
[10] Chalioris C E 2008 Torsional strengthening of rectangular and flanged beams using carbon fibre-reinforced-polymers–Experimental study *Construction building materials* **22** 21-9
[11] Deifalla A and Ghobarah A 2010 Full torsional behavior of RC beams wrapped with FRP: analytical model *Journal of Composites for Construction* **14** 289-300
[12] Bansal P, Kumar M and Kaushik S 2007 Effect of mesh orientation on strength of beams retrofitted using ferrocement jacket *International Journal of Engineering* **20** 8-19
[13] Charan B G, Gunneswara R T and Rao C 2016 Torsional behaviour of reinforced concrete beams with ferrocement U-jacketing—Experimental study *Case Studies in Construction Materials* **4** 15-31
[14] Salom P R, Gergely J and Young D T 2004 Torsional retrofit of spandrel beams with composite laminates *Proceedings of the Institution of Civil Engineers-Structures Buildings* **157** 69-76
[15] Wong R S and Vecchio F J 2003 Towards modeling of reinforced concrete members with externally bonded fiber-reinforced polymer composites *ACI Structural Journal* **100** 47-55
[16] Jing M and Grünberg J 2006 Mechanical analysis of reinforced concrete box beam strengthened with carbon fiber sheets under combined actions *Composite Structures* **73** 488-94
[17] Li B, Lam E S-s, Wu B and Wang Y-y 2013 Experimental investigation on reinforced concrete interior beam–column joints rehabilitated by ferrocement jackets *Engineering Structures* **56** 897-909
[18] Atcheson M and Alexander D 1979 Development of fibrous ferrocement *Special Publication* **61** 81-102
[19] Chalioris C E, Thermou G E and Pantazopoulou S J 2014 Behaviour of rehabilitated RC beams with self-compacting concrete jacketing–Analytical model and test results *Construction Building Materials* **55** 257-73
[20] Behera G C, Rao T D G and Kameswara C B 2014 Study of Post-Cracking Torsional Behaviour of High-Strength Reinforced Concrete Beams with a Ferrocement Wrap *Slovak Journal of Civil Engineering* **22** 1-12
[21] Behera G C, Gunneswara Rao T and Kameswara C 2014 Analytical model for torsional response of RC beams strengthened with ferrocement U-Wraps *Structural Engineering International* **24** 509-20
[22] Hsu T T 1988 Softened truss model theory for shear and torsion *Structural Journal* **85** 624-35
[23] Rao T G and Seshu D R 2005 Analytical model for the torsional response of steel fiber reinforced concrete members under pure torsion *Cement Concrete Composites* **27** 493-501

[24] Friedman J H 1991 Multivariate adaptive regression splines *The annals of statistics* 1-67
[25] Zavadskas E K, Turskis Z, Antucheviciene J and Zakarevicius A 2012 Optimization of weighted aggregated sum product assessment *Elektronika ir elektrotechnika* **122** 3-6
[26] Madić M, Gecevska V, Radovanović M and Petković D 2014 Multi-criteria economic analysis of machining processes using the WASPAS method *Journal of Production Engineering* **17** 1-6
[27] Madic M, Antucheviciene J, Radovanovic M and Petkovic D 2016 Determination of manufacturing process conditions by using MCDM methods: Application in laser cutting *Engineering Economics* **27** 144-50

Comparative Mechanical Characterization of M30 Concrete Grade by Fractional Replacement of Portland Pozzolana Cement with Industrial Waste Using CoCoSo and CODAS Methods

Dilip Kumar Bagal[1], Bibekananda Naik[2], Biswajit Parida[3], Abhishek Barua[3], Siddharth Jeet[3], Sangam Kumar Singh[4], Ajit Kumar Pattanaik[1]

[1] Department of Civil Engineering, Government College of Engineering, Kalahandi, Bhawanipatna, Odisha, India
[2] Department of Civil Engineering, Bhubaneswar Institute of Technology, Bhubaneswar, Odisha, India
[3] Department of Mechanical Engineering, Centre for Advanced Post Graduate Studies, BPUT, Rourkela, Odisha, India
[4] Department of Mechanical Engineering, Delhi Technical Campus, Bahadurgarh, Haryana, India

E-mail: dilipbagal90@gmail.com

Abstract. Since last few years, marble has been deliberated as some of the utmost significant embellished constructing resources. Marble Powder (MP) and Fly Ash (FA) are nearly those materials, which creates health hazards and pollutes the environment. This present study targets to investigate the effect of marble powder and fly ash as fractional replacement of cement respectively in different concrete mixes. The M30 grade of concrete was chosen with a continual w/c ratio 0.43 where fractional replacement of Portland Pozzolana Cement with Marble Powder and Fly Ash is done in different proportion. After preparing the samples according to Taguchi's method orthogonal array, the mechanical characterizations are performed. Hardened concrete tests such as compressive strength, split tensile strength and flexural strength test have been piloted to appraise the mechanical properties of concrete. Two current multi-response optimization techniques were used i.e. Combinative Distance-based Assessment (CODAS) and Combined Compromise Solution (CoCoSo) method for optimization of process parameters. In order to outline comparative consequence of measured criteria a pairwise comparison matrix was used. The results of all tests were analyzed, comparison has done among concrete mixes and the conclusion is drawn.

1. Introduction

A new revolution took place in the construction industry when cement was invented in the 19th century. Concrete manufacturing is one of the technique related to building and construction industry have been developed after the invention of cement. Marble industry was one of the pristine industry related to building and construction. Nowadays deployment of marble remaining in many sectors advantages in regulating environmental hazards. Similarly, Fly Ash is used extensively as a fractional replacement of cement; however, its inclusion in concrete provides various aids, such important lessening in early strength due to moderately relaxed hydration of Fly Ash [1-13].

The present investigation work is done to find the optimal percentage of replacement of cement with combined application of fly ash and marble powder in M30 grade of concrete. Multi-response optimization was done by Combinative Distance-based Assessment (CODAS) and Combined Compromise Solution (CoCoSo) method. Hardened concrete tests such as compressive strength, split tensile strength and flexural strength test has been piloted to evaluate the mechanical properties of concrete with the addition of various proportions of fly ash and marble powder.

2. Experimental Analysis and Methodology

Materials used for fabrication of test specimen are as follows:

1. Portland Pozzolana Cement (PPC)
2. Marble Powder (MP)
3. Fly Ash (FA)
4. Coarse Aggregate
5. Fine Aggregate
6. Water

1. Portland Pozzolana Cement (PPC): It makes concrete extra resistant, thicker as related to Ordinary Portland Cement. PPC produces a smaller amount of warmness of hydration and deals better resistance to outbreak of antagonistic water.
2. Marble Powder (MP): It is formed from handling plants where marble blocks sawing and polishing is done.
3. Fly Ash (FA): It is formed from burning of anthracite coal and bituminous coal, which has roughly 7% lime.
4. Natural Fine Aggregate (NFA): Regionally accessible sand confirming to IS specifications turned into used as the satisfactory combination in the concrete coaching. The specific gravity of Natural Fine Aggregate is 2.66. The bulk density of Natural Fine Aggregate is 1415 Kg/m^3. Water absorption of sand is 13.89%.
5. Natural Coarse Aggregate (NCA): The sieve analysis test has been carried out for the NCA and the procedures are followed as per the IS code. As a result, the aggregate has taken between sizes of 10 mm to 20 mm. The specific gravity of aggregate has found out which further used in the mix design procedure. Fineness modulus range of 10 mm to 20 mm size coarse aggregate is 6.0 to 6.9.
6. Water: 15% tap water was used for preparing the concrete mix.

Table 1. Test Specimen size details

Tests	Sample size	Nos.
For cube's compressive strength	150mm X 150mm X 150mm	9
Cylinder's split tensile strength	100mm diameter X 200mm height	9
Beam's flexural strength	10mm X 100mm X 50mm	9

2.1. Mixing and Casting

M30 grade of concrete was designed as per the Indian Standard code of practice (IS 10262 (2009)). All the materials such as PPC, NCA, NFA, MP and FA of particular quantity are added in the concrete mixture machine. All the materials mixed properly in dry condition. Required amount of water has been added slowly it forms a homogenous mixture. Specimens were casted for 90 days for each concrete mix based on L9 orthogonal array. The size details of test specimen are shown in Table 1. Table 2 shows the input parameters.

Table 2. Input parameters

Factors	Symbol	Level 1	Level 2	Level 3
cement	A	320 kg/m^3	340 kg/m^3	360 kg/m^3
Fly ash	B	5 %	15 %	20 %
Marble powder	C	5 %	15 %	20 %

2.2. Curing

Mould has been removed after 24 hours of casting period. Specimens are marked clearly after removed from the mould. All specimens are kept in curing tank and place there safely for 90 days, which provide adequate moisture and temperature to cement hydration for adequate period.

2.3. Hardened Concrete Test

Prepared concrete specimens were tested using 600 KN universal testing machine and with a crosshead speed of 10mm/min.

1. Compressive Strength Test: Structural design codes are based on compressive strength. Tests have been conducted of the test specimen for 90 days. Cube specimens (150 X 150 X 150) mm cured in water has taken out. It has been tested immediately after drying. Compressive strength (MPa) was found out using the expression:

$$F_{compressive} = P/A \tag{1}$$

Where, P = Load in kN and A = Surface area of the cube

2. Split Tensile Strength Test: Wet cylinder specimen (200 mm height X 100 mm diameter) has been taken from water after 90 days of curing. It has been tested immediately after drying. Split tensile strength (MPa) was found out using the expression:

$$F_{Split\ tensile} = 2P/\pi DL \tag{2}$$

Where, P = Load in kN, D = diameter in mm and L = length in mm.

3. Flexural Strength Test: Wet beam specimen (10 X 100 X 50) mm has been taken from water after 90 days of curing. It has been tested immediately after drying. The flexural stress in three-point bending test was found using formula:

$$\sigma_f = 3FL/2bd^2 \tag{3}$$

Where F = maximum load, L = space amongst the supports, b and d are breadth and wideness of specimen respectively.

2.4. Weight Calculation between Criterions by means of Pair-Wise Comparison

For formulating the pair-wise comparison matrix, Saaty's nine-point preference scale was used. Output response weights were premeditated by following equations [14]-

$$GM_i = \left(\prod_{j=1}^{n} b_{ij}\right)^{1/n} \tag{4}$$

$$w_j = GM_i/\sum_{j=1}^{n} GM_i \tag{5}$$

Output response weights were attained as 0.60, 0.13, 0.27, which will be used in CoCoSo method and CODAS method for optimization.

2.5. Combined Compromise Solution Method

The following steps are used to crack CoCoSo problem [15, 16]:

1. Determination of initial decision-making matrix using equation (6):

$$X_{ij} = \begin{bmatrix} x_{11} & x_{12} & \cdots & x_{1n} \\ x_{21} & x_{22} & \cdots & x_{2n} \\ \cdots & \cdots & \cdots & \cdots \\ x_{m1} & x_{m2} & \cdots & x_{mn} \end{bmatrix}; \text{i=1, 2,.., m; j=1, 2,.., n} \quad (6)$$

2. Via compromise normalisation equation, normalisation of norms values is prepared:

$$r_{ij} = \frac{x_{ij}-minx_{ij}}{maxx_{ij}-minx_{ij}} \quad \text{; for benefit} \quad (7)$$

$$r_{ij} = \frac{maxx_{ij}-x_{ij}}{maxx_{ij}-minx_{ij}} \quad \text{; for cost} \quad (8)$$

3. Determination of S_i and P_i :

$$S_i = \sum_{j=1}^{n}(w_j r_{ij}) \quad (9)$$

$$P_i = \sum_{j=1}^{n}(r_{ij})^{w_j} \quad (10)$$

4. Three appraisal score are used to generate relative weights of other options derived using equation (11), (12), (13):

$$k_{ia} = \frac{P_i+S_i}{\sum_{i=1}^{m}(P_i+S_i)} \quad (11)$$

$$k_{ib} = \frac{S_i}{minS_i} + \frac{P_i}{minP_i} \quad (12)$$

$$k_{ic} = \frac{\lambda(S_i)+(1-\lambda)(P_i)}{(\lambda maxS_i+(1-\lambda)maxP_i)} \quad (13)$$

5. The ranking of all the alternatives is found based on k_i values from higher to lower:

$$k_i = (k_{ia}k_{ib}k_{ic})^{\frac{1}{3}} + (k_{ia} + k_{ib} + k_{ic}) \quad (14)$$

2.6. Combinative Distance-based Assessment (CODAS) Method

The steps for CODAS method are as follows [17]:

1. Determination of the initial decision-making matrix using equation (15):

$$X = [x_{ij}]_{nXm} = \begin{bmatrix} x_{11} & x_{12} & \cdots & x_{1m} \\ x_{21} & x_{22} & \cdots & x_{2m} \\ \cdots & \cdots & \cdots & \cdots \\ x_{n1} & x_{n2} & \cdots & x_{nm} \end{bmatrix} \quad (15)$$

Where x_{ij} ($x_{ij} \geq 0$) is performance value of ith alternative on j^{th} criterion.

2. The normalisation of criteria values using equation (16):

$$n_{ij} = \begin{cases} \frac{x_{ij}}{maxx_{ij}} & if\ j \in N_b \\ \frac{minx_{ij}}{x_{ij}} & if\ j \in N_b \end{cases} \quad (16)$$

Where N_b and N_c represent the benefit and cost criteria respectively.

3. Weighted normalized matrix is premeditated using the following equation:

$$r_{ij} = w_j n_{ij} \quad (17)$$

Where wj ($0<wj<1$) is weight of jth criterion, and $\sum_{j=1}^{m} w_j = 1$.

4. Determination of the negative-ideal solution (point) using equation (18) and (19):

$$ns = [ns_j]_{1Xm} \quad (18)$$

$$ns_j = min\, r_{ij} \tag{19}$$

5. Calculation of Euclidean distances and Taxicab distances for options using negative-ideal solution using equation (20) and (21) respectively:

$$E_i = \sqrt{\sum_{j=1}^{m}(r_{ij} - ns_j)^2} \tag{20}$$

$$T_i = \sum_{j=1}^{m}|r_{ij} - ns_j| \tag{21}$$

6. Construction of relative assessment matrix using equation (22):

$$R_a = [h_{ik}]_{nXn} \tag{22}$$

$$h_{ik} = (E_i + E_k) + (\Psi(E_i - E_k)X(T_i - T_k)) \tag{23}$$

Where $k \in \{1,2,\ldots,n\}$ and ψ denotes threshold function defined as follows:

$$\Psi(x) = \begin{cases} 1 & if\ |x| \geq \tau \\ 0 & if\ |x| \leq \tau \end{cases} \tag{24}$$

7. Assessment score for respective alternative is calculated using equation (xxv) and their ranking is done according to the decreasing values of H_i and the alternative with highest value of H_i is best choice among all alternatives:

$$H_i = \sum_{k=1}^{n} h_{ik} \tag{25}$$

3. Results and Discussion

Samples were made using L9 orthogonal array of Taguchi's design shown in Table 3. The experimental results for compressive strength, flexural strength and split tensile strength are listed in Table 3.

Table 3. L9 orthogonal array design for experimental runs and results

Run No.	A	B	C	Compressive strength, MPa	Split tensile strength, MPa	Flexural Strength, MPa
1	320	5	5	49.24	4.83	4.700
2	320	15	15	49.03	4.30	4.732
3	320	20	20	39.55	4.25	3.904
4	340	5	15	40.36	3.80	4.442
5	340	15	20	39.51	4.50	4.290
6	340	20	5	40.21	3.88	4.108
7	360	5	20	45.68	4.68	4.382
8	360	15	5	45.27	4.12	4.394
9	360	20	15	48.96	3.82	4.026

3.1. Optimization using CoCoSo Method

Normalised decision-making matrix was formed shown in Table 4. The comparability sequence matrix was formed using formulas (9) and (10), respectively. The values of k_a, k_b, and k_c are calculated using Equations (11), (12) and (13). Equation (14) produces the ranking score by k_i for finding final ranks for the options.

Table 4. Computational data of CoCoSo Method

Expt. No.	Compressive strength	Split tensile strength	Flexural Strength	k_{ia}	k_{ib}	k_{ic}	k_i	Rank of k_i
1	1.000	1.000	0.964	0.173	12.836	1.000	5.974	1
2	0.978	0.485	1.000	0.159	11.160	0.917	5.253	2
3	0.004	0.437	0.005	0.063	2.387	0.365	1.319	8
4	0.087	0.000	0.653	0.057	2.420	0.330	1.292	9
5	0.000	0.680	0.470	0.089	4.064	0.514	2.126	6
6	0.072	0.078	0.251	0.071	2.310	0.411	1.338	7
7	0.634	0.854	0.581	0.145	9.352	0.836	4.486	3
8	0.592	0.311	0.595	0.126	7.374	0.729	3.622	5
9	0.971	0.019	0.152	0.0927	7.5130	0.5546	3.1501	4

From k_i values of alternatives, it was detected that investigational results obtained in experiment no. 1 is the best result according to the ranking.

3.2. Optimization using CODAS Method

The normalized matrix was calculated using equation (16) as listed in Table 5. Weighted normalized performance values was evaluated using weights of different criteria, and then the negative-ideal solution is determined. Euclidean distances and Taxicab distances of every alternative were computed. The relative assessment matrix and assessment scores (H_i) of every alternative were calculated using Table 6 and equations (22) to (24). The results are presented in Table 6.

Table 5. The normalized decision-making matrix

Expt. No.	Compressive strength	Split tensile strength	Flexural Strength
1	0.8024	0.7867	0.8298
2	0.8058	0.8837	0.8242
3	0.9990	0.8941	0.9990
4	0.9789	1.0000	0.8780
5	1.0000	0.8444	0.9091
6	0.9826	0.9794	0.9494
7	0.8649	0.8120	0.8900
8	0.8728	0.9223	0.8876
9	0.8070	0.9948	0.9687

Table 6. Computational results using CODAS Method

Expt. No.	E_i	T_i	H_i	Rank
1	0.0015	0.0016	0.6158	1
2	0.0128	0.0147	0.5145	2
3	0.1279	0.1792	-0.5213	9
4	0.1105	0.1482	-0.3655	6
5	0.1210	0.1490	-0.4603	8
6	0.1161	0.1670	-0.4153	7
7	0.0417	0.0586	0.2535	3
8	0.0489	0.0770	0.1886	5
9	0.0476	0.0689	0.2005	4

According to assessment scores, the ranking of alternatives is E1>E2>E7>E9>E8>E4>E6>E5>E9. Therefore, E1, i.e. concrete specimen no. 1 is the best concrete mixture with respect to the valuation of the CODAS method.

After applying two different optimization technique, Cement with 320 kg/m^3, Marble Powder with 5% and Fly Ash with 5% content is the optimal parameter setting for fabrication of M30 concrete. Figure 1 shows the split tensile strength, compressive strength and flexural strength of all the experimental run where the values obtained in 1st experiment is optimal values attained using optimal parameter settings obtained using CoCoSo and CODAS method.

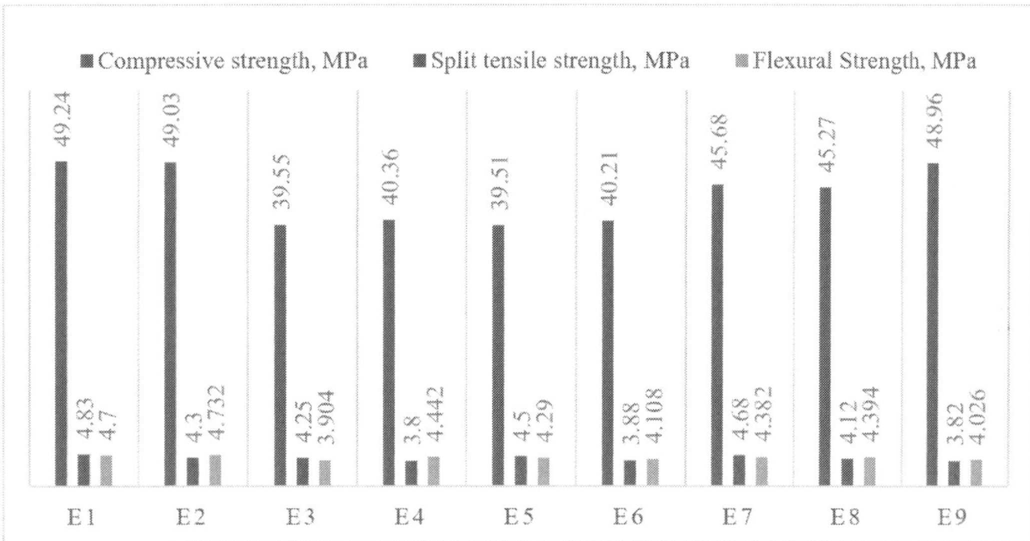

Figure 1. Output responses obtained in L9 experimental run

4. Conclusions

The present work explores the development of concrete by replacing of PPC with marble powder and fly ash. To find an optimum concrete mixture, recent multi-criteria decision-making optimization methods have been employed. Split tensile strength, compressive strength and flexural strength were evaluated for different replacement of marble powder and fly ash based concrete. The results of all tests were optimized using CoCoSo and CODAS method, compared and the following conclusions are made from this study.

From the split tensile strength, compressive strength and flexural strength results of 90 days, it was found that with 5% replacement of marble powder and fly ash respectively with PPC getting great strength as comparing to control specimen.

Table 7. Optimal parameters using two optimization approach

Algorithm	Optimal Setting	Cement	Marble Powder	Fly Ash
CoCoSo Method	A1 B1 C1	320 kg/m^3	5 %	5 %
CODAS Method	A1 B1 C1	320 kg/m^3	5 %	5 %

The predictive values found using CoCoSo and CODAS method is similar to each other. Optimal factor setting for concrete specimen fabrication by replacing of PPC with marble powder and fly ash using two different optimization methods are stated in Table 7. The optimum values will be convenient for industrial assortment of amalgamation of PPC, marble powder and fly ash for construction.

References

[1] Sahoo K K, Naik B, Paul S, Dhangdamajhi G, Bagal D K and Barua A 2019 Comparative Analysis Based on Partial Replacement of Portland Cement with Fly ash and Iron slag in M25 and M40 Grade Concrete *International Journal of Management Technology Engineering Economics* **9** 971-82

[2] Kumar M, Saxena S K and Singh N B 2019 Influence of some additives on the properties of fly ash based geopolymer cement mortars *SN Applied Sciences* **1** 481

[3] Naik B, Pradhan S S and Bagal D K 2019 Mechanical Characterization Based on Partial Replacement Analysis of Portland Pozzolana Cement with Industrial Waste in M30 Grade Concrete *International Journal of Applied Engineering Research*

[4] Prusty R, Mukharjee B and Barai S 2013 Mechanical Properties of Recycled Aggregate Concrete Containing Colloidal Nano-Silica Designed by Taguchi Method *International Journal of Civil Engineering Applications* **3** 5-8

[5] Khodabakhshian A, De Brito J, Ghalehnovi M and Shamsabadi E A 2018 Mechanical, environmental and economic performance of structural concrete containing silica fume and marble industry waste powder *Construction Building Materials* **169** 237-51

[6] Prusty R, Mukharjee B B and Barai S V 2015 Nano-engineered concrete using recycled aggregates and nano-silica: Taguchi approach *Advances in concrete construction* **3** 253

[7] Mishra M, Agarwal A and Maity D 2019 Neural-network-based approach to predict the deflection of plain, steel-reinforced, and bamboo-reinforced concrete beams from experimental data *SN Applied Sciences* **1** 584

[8] Keshavarz Ghorabaee M, Zavadskas E K, Turskis Z and Antucheviciene J 2016 A New Combinative Distance-based Assessment (CODAS) Method for Multi-Criteria Decision-Making *Economic Computation Economic Cybernetics Studies Research* **50**

[9] Naik B, Paul S, Mishra S P, Rout S P, Barua A and Bagal D K 2019 Performance Analysis of M40 Grade Concrete by Partial Replacement of Portland Pozzolana Cement with Marble Powder and Fly Ash Using Taguchi-EDAS Method *JASC: Journal of Applied Science Computations* **6** 733-43

[10] Rao B K 2016 Study on marble powder as partial replacement of cement in normal compacting concrete *IOSR Journal of Mechanical Civil Engineering e-ISSN: -, p-ISSN: -334X* **13** 01-5

[11] Prusty R, Mukharjee B and Barai S 2014 A Taguchi Approach for Investigation of Mechanical Properties of Recycled Aggregate ConCrete Containing *International Journal of SEWC*

[12] Mahapatra C K and Barai S V 2019 Temperature impact on residual properties of self-compacting based hybrid fiber reinforced concrete with fly ash and colloidal nano silica *Construction Building Materials* **198** 120-32

[13] Uysal M 2018 The Use of Waste Maroon Marble Powder and Iron Oxide Pigment in the Production of Coloured Self-Compacting Concrete *Advances in Civil Engineering* **2018**

[14] Lohe M, Makarande S and Shende N 2015 Utility of Marble Dust & Fly Ash in Different Grades of Concrete *International Journal Of Engineering Research* **3** 270-3

[15] Naik B, Paul S, Barua A, Jeet S and Bagal D K 2019 Fabrication and Strength Analysis of Hybrid Jute-Glass-Silk Fiber Polymer Composites Based on Hybrid Taguchi-WASPAS Method *International Journal of Management, Technology Engineering Economics* **9** 3472-9

[16] Yazdani M, Zarate P, Zavadskas E K and Turskis Z 2019 A Combined Compromise Solution (CoCoSo) method for multi-criteria decision-making problems *Management Decision*

[17] Barua A, Jeet S, Bagal D K, Satapathy P and Agrawal P K 2019 Evaluation of Mechanical Behavior of Hybrid Natural Fiber Reinforced Nano SiC Particles Composite Using Hybrid Taguchi- CoCoSo Method *International Journal of Innovative Technology Exploring Engineering* **8** 3341-5

Impact of Ultrafine Ground Granulated Blast Furnace Slag on the Properties of High Strength Durable Concrete

Kaliprasanna Sethy[1], S Rojalin Nanda[2]

Department of Civil Engineering, Government College of Engineering, Kalahandi, Bhawanipatna, Odisha, India

E-mail: kaliprasanna87@gmail.com

Abstract. Ultrafine ground granulated blast furnace slag (UFGGBFS), because of its pozzolanic nature, could be an extraordinary resource for the advanced development needs, since slag cements concrete can be of elite, if properly planned. Improvement of the durability properties of concrete to support a more extended life expectancy and creating a climate amicable greener concrete are turning out to be significant standards in getting high quality concrete. Fusing Ultrafine Ground Granulated Blast Furnace Slag (UFGGBFS) as a mineral admixture enhances the workability of fresh concrete and decreased the link between pores; in this way, decreasing the permeability and improving the obstruction of the solid against chloride infiltration. The measure of ozone depleting substance delivered in making the concrete and the energy needed to create the concrete are significantly decreased with the utilization of UFGGBFS. Ultrafine GGBFS (UFGGBFS) with a normal molecule size less than 10m and a Blaine surface area more than 600m^2/kg can significantly improve the properties of the concrete as far as diffusion and chemical reactivity impacts. Compared with GGBFS, the UFGGBFS increases the rate of hydration and pozzolanic responses and has a superior filling impact. In this study, the early development of mechanical strength and permeability properties of high strength concrete with UFGGBFS is examined. Total five mixes with 180kg/m^3 each of both Ordinary Portland Cement (OPC) concrete and Ultrafine Ground Granulated Blast furnace Slag (UFGGBFS) concrete were designed. The mixes starts from 0% replacement and gradually increases up to 20% (i.e 0, 5, 10, 15 and 20%) of identical all total cementitious materials with UFGGBFS substitution were planned and a sum of 145 samples from the five blends were projected. Compressive strength parameters, flexural strength parameter, modulus of elasticity parameter and water permeability test outcomes are introduced.

1. Introduction
Establishing quality concrete in the current atmosphere doesn't rely exclusively upon accomplishing a high quality property. Improving the durable quality of the concrete to support a more drawn out life expectancy and delivering greener concrete are getting one of the principle rules in acquiring quality concrete. By utilizing modern byproducts, for example, Ground Granulated Blast furnace Slag (GGBFS) from industries as mineral admixture halfway supplanting Ordinary Portland Cement (OPC) in the concrete, the measure of ozone depleting substance delivered in making the concrete and the energy needed to create the concrete are diminished. GGBFS is an industrial byproduct produced when liquid iron blast furnace slag is quickly chilled by drenching it in water. To create environmental friendly concrete the best way is to replace the OPC content with some good mineral admixtures in the concrete. It has been all around archived that GGBFS is an awesome mineral admixture to be utilized in improving the properties of the concrete because of its beneficial outcomes on its feasible turn of events and the

climate. Significant level of strength value and durability of concrete are a need in diminishing the pace of decay of concrete. In mixing GGBFS with OPC, an improved flowable and reduced bleeding concrete can be accomplished.

It is fine reported that with the additional mix of GGBFS, the early strength of the concrete is influenced; anyway as the concrete curing age increases, the mechanical parameters of the concrete improves and matches that of the normal concrete at 56 days. Conversely, with the diminished fineness of the UFGGBFS, rising the Blaine surface area and henceforth escalating the speed of hydration and pozzolanic response, the early strength of the designed concrete will improve altogether. The properties of GGBFS help the concrete in opposing chloride induced corrosion and the designed concrete will have a diminished pore connection which helps in preventing chloride entrance. The complete pore volume of concrete additionally diminishes as the slag fineness gets higher.

In this report, the impact of UFGGBFS substitution on the durability and strength properties of high quality concrete is considered. Five blends in with various measure of UFGGBFS substitution were considered. The measure of UFGGBFS substitution was set from 0% to 20%. Compressive strength property, flexural strength property, and modulus of elasticity tests were performed to examine the impact of UFGGBFS on the fresh and hardened properties of the concrete. The UFGGBFS substitution brought about higher early strength, lower permeability, and better durable property.

2. Experimental investigations

A whole of 145 samples comprising of 100mm and 150mm dia. cylindrical specimens, 100mm cubical specimens, and 150mm cube specimens were examined. Compressive strength parameter, flexural strength parameter, and modulus of elasticity give the stage to focus on the mechanical viewpoint. The quality of UFGGBFS utilized in this examination was inspected utilizing particle size analyzer and chemical study. The physical properties and oxide compositions obtained were recorded in Table 1.

Table 1: Chemical and Physical Composition of Ultrafine slag

Chemical constituent	% by mass	Physical constituent	Range
CaO	32-35	Bulk density	600-700kg/m^3
Al_2O_3	18-20	Surface area	12000cm^2/gm
Fe_2O_3	1.8-2	Particle shape	Irregular
SiO_2	28-32	Particle size, d_{10}	< 2u
MgO	8-10	d_{50}	< 5u
SO_3	0.3-0.7	d_{90}	< 9u

To get a minimum slump value of 150 mm, a polycarboxylate based high range water reducing agent (HRWRA) was utilized.

Table 2: Mix Proportion per cubic meter (For SSD Condition)

Constituents	Mixes				
	M-300	M-305	M-310	M-315	M-320
Cement (OPC)	533.33	506.67	480	453.33	426.67
UFGGBFSS	0	26.67	53.33	80	106.67
Water	160	160	160	160	160
Aggregate (20mm)	605.03	604.48	603.94	603.39	602.85
Aggregate (10mm)	600.94	600.4	599.85	599.32	598.77
Natural Sand	677.46	676.85	676.24	675.63	675.10
Density	2576.76	2575.06	2573.36	2571.67	2569.97

In this present study, a whole of five blends were examined. The amount of UFGGBFS replacement was such an order that 0%, 5%, 10%, 15%, 20%. Mix with 0% replacement of UFGGBFS is abbreviated as Mix 300 and have a binder content of 533.33 kg/m^3. Mix 305 maintains the similar binder content to

Mix 300 with the 5 % UFGGBFS replacement. Similarly M-310, M-315, M-320 were obtained. The particulars of the blends are introduced in Table 2. The samples were cured in water in lab temperature of near 25 °C for 3, 7, 28 and 56 days. Five varieties of samples were casted in two batches as indicated by each blend plan.

3. Results and discussion

3.1. Workability of the designed concrete

The workability of designed concrete blends was obtained by conducting slump test and the outputs are reported in Table 3. The slump value of 180 mm was obtained in Mix-320 and only 140 & 120 mm was obtained in Mix-310 and Mix-315 respectively.

Table 3: Slump test value (mm)

Mix abbreviation	% of Admixture Dosage	Slump value (mm)
M-300	0.40	130
M-305	0.30	140
M-310	0.25	140
M-315	0.20	120
M-320	0.15	180

Figure 1. Freshly mixed and casted concretes

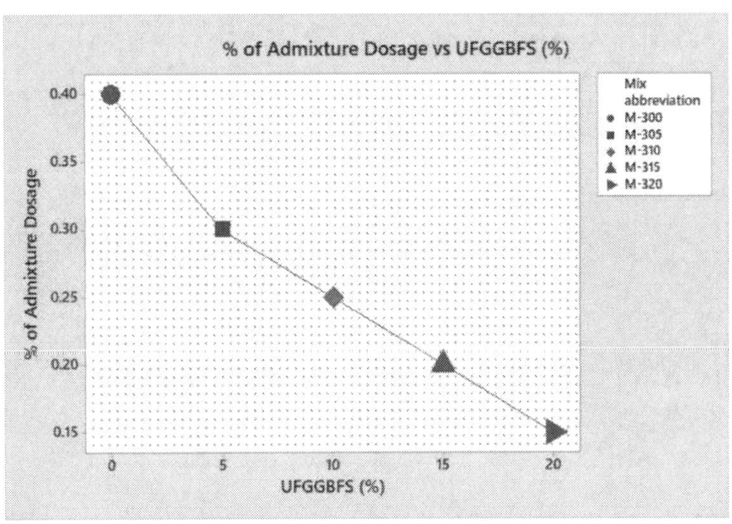

Figure 2. Relation between UFGGBFS with Admixture dosage in (%)

3.2. Hardened properties (Compressive strength)

Compressive strengths of the cubical specimens inspected are assembled and recorded in Table 4. A research of the impact on the compressive strength properties due to increase volume of UFGGBFS in concrete was observed. As presented in Table 4, each mix achieved higher compressive strength at 28days curing, whereas the 56days curing strength remains almost same as the 28days compressive strength. The UFGGBFS prompts a higher rate of hydration and pozzolanic response compared with the normal GGBFS. Likewise, the UFGGBFS can top off the pores in the Interfacial Transition Zone (ITZ) better contrasted with GGBFS.

Figure 3. Compressive strength test set up

Table 4: Compressive strength results (kg/m^3)

Mixes	3 Days	7 Days	28 Days	56 Days
M-300	67.45	83.51	91.33	93.46
M-305	66.51	76.55	91.80	93.92
M-310	74.67	75.43	88.29	96.78
M-315	75.96	83.61	93.40	98.53
M-320	68.57	85.76	93.55	105.78

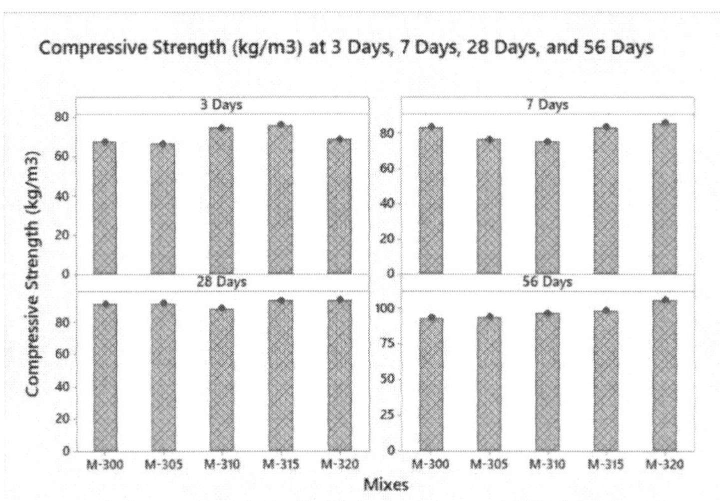

Figure 4. Compressive strength values at different ages

3.3. Hardened properties (Split tensile strength)

Figure 6 is the graphical representation of the outcomes of the Split tensile strength tests got from this investigation. The standard deviation of the strength values is relatively small, representing to a reliable blend. From the outcomes, it tends to be seen that the blends in with UFGGBFS accomplished higher flexural strength compared with the normal OPC concrete. Addition of UFGGBFS, Mix-320 acquired a split tensile strength of 5.37 MPa after 56 day compared with 3.92 MPa for the normal OPC concrete, i.e Mix-300.

Figure 5. Split tensile strength set up

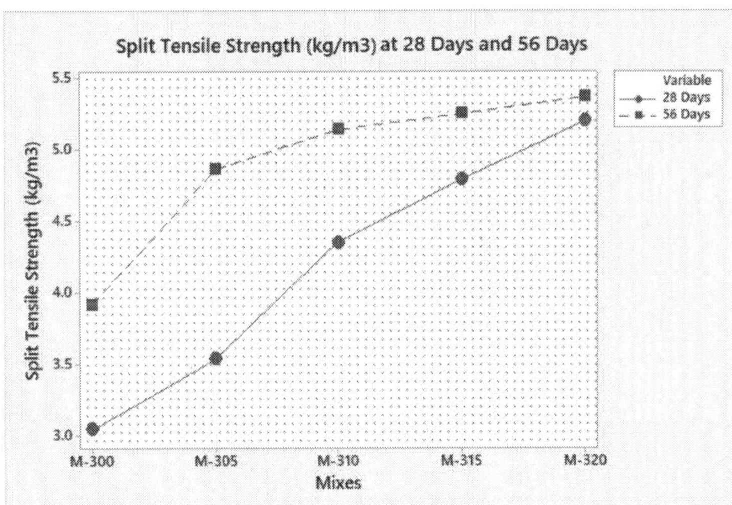

Figure 6. Relation between concrete mixes with split tensile strength

3.4. Hardened properties (Elastic modulus)

The static modulus of elasticity estimations of different concrete blends is introduced. Each value is the average of three estimations of 56 days.

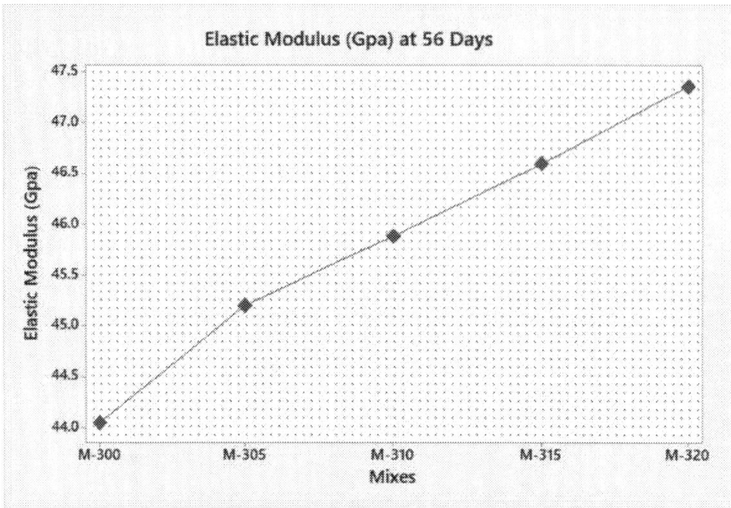

Figure 7. Elastic modulus of the designed concretes

4. Water permeability

Water permeability depth was measured using the permeability testing apparatus. The 28 days and 56 days water permeability test outcomes are presented in the Figure 9. A notable difference between 28 days and 56 days was obtained where all the 56 days permeability depth were quite less than 28 days permeability results.

Figure 8. Water penetration depth shown using marker

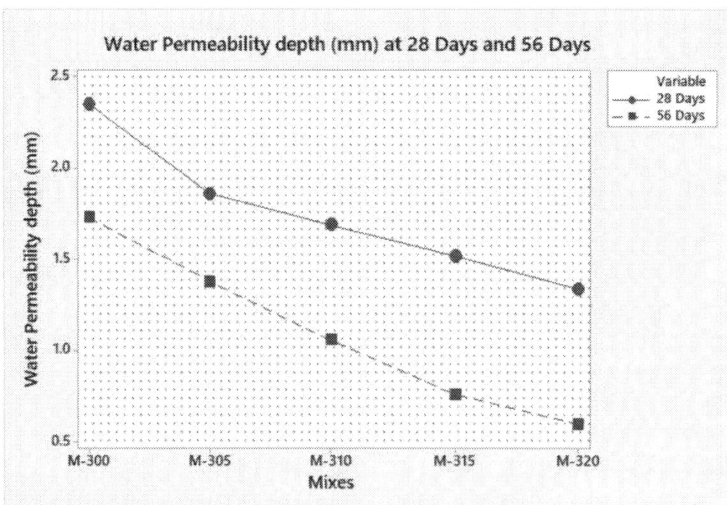

Figure 9. Relation between design mixes with permeability depth

5. Conclusions

The accompanying ends can be made dependent on the aftereffects of this investigation:

- UFGGBFS has a bigger whole surface area, and subsequently, a greater amount of it is accessible for hydration and pozzolanic response, compared with ordinary GGBFS. Likewise, improved workability and high consistency were accomplished by using UFGGBFS. This will prompt an early strength improvement regarding compressive and flexural strength, as contemplated.
- A continuously increased compressive and flexural strength was accomplished with the step by step increased substitution of UFGGBFS at each age.
- With the inclusion of UFGGBFS into the concrete, it is probable to get a reliable blend, as the high surface area of UFGGBFS enhances the rheology of freshly mixed concrete.
- With the inclusion of UFGGBFS, there is a noteworthy improvement in the mechanical properties of the concrete. The improvement is more evident for higher grade concrete.
- With the inclusion of UFGGBFS, the permeability of concrete is notably decreased. Because of the diminished permeability, chloride infiltration into the hardened concrete is decreased. This denoted a critical improvement in the durability part of the concrete.

References

[1] Fong, W. F. K, and Jaime, S. K. Y (2002).“Production and application of recycled aggregates.” Proc., Int. Conf. on Innovation and Sustainable Development of Civil Engineering in the 21st century, China Civil Engineering Society, Hong Kong.

[2] Gonzalez, B., and Martínez, F., "Concretes with aggregates from demolition waste and silica fume". Materials and Mechanical Properties Building and Environment, Vol. 43, 2008, pp 429–437.

[3] IS 516 -1959 Indian Standard methods of tests for strength of concrete, Bureau of Indian Standards, New Delhi.

[4] IS 2386 (Part III) – 1963 “Indian Standard Methods of Test for Aggregate for concrete. Bureau of Indian Standards, New Delhi.

[5] Indian standard Code of Practice for Ordinary Portland Cement – 43 Grade IS 8112 -1989 Bureau of Indian Standards, New Delhi.

[6] IS: 10262- 2009 "Guidelines for concrete mix design proportioning"Bureau of Indian Standards, New Delhi.

[7] Katz, A Treatments for the Improvement of Recycled Aggregate ASCE Journal of Materials in Civil Engineering, Vol. 16, No. 6, December 1, 2004.

[8] Kou SC, Poon CS, Chan D. Influence of fly ash as cement replacement on the properties of recycled aggregate concrete. J Mater Civil Eng 2007; 19(9):709–18.

[9] Kou,S S and Poon CS “Properties of concrete prepared with PVA-impregnated recycled concrete aggregates” Cement and Concrete Composites. 32 (2010)649–654.

[10] Kwan, A. K. H., Wang, Z. M. and Chan, H. C. Mesoscopic study of concrete II: nonlinear finite element analysis. Computers and Structures 1999;70(5): 545-556.

[11] Limbachiya M. C., T. Leelawat and R. K. Dhir, “Use of recycled concrete aggregate in high-strength concrete” RILEM Journal of Materials and Structures. Vol. 33, November 2000, pp 574-580.

[12] Mehta, P. K. and Aitcin, P. C. Microstructural basis of selection of materials and mix proportions for high strength concrete. High Strength Concrete, Second International Symposium, ACI SP121 1990: 265-279.

[13] Parekh,D. N., and Modhera, C. D., “Assessment Of Recycled Aggregate Concrete" Journal of Engineering Research and Studies, Vol.II, Issue I/1-9, January-March 2011.

[14] Prokopski G, Halbiniak J. Interfacial transition zone in cementitious materials. CemConcr Res 2000;30:579–83.

[15] Saravana Kumar P and Dhinakaran G, “Effect of Admixed Recycled Aggregate Concrete on Properties of Fresh and Hardened Concrete”, ASCE Journal of Materials in Civil Engineering, Vol. 24. (4), April 2012.

[16] Tam VWY, Gao XF, Tam CM. Microstructural analysis of recycled aggregate concrete produced

from two-stage mixing approach. CemConcr Res 2005; 35:1195–203.
[17] Tam VWY, Tam CM and Le KN. "Removal of Cement Mortar remains from Recycled Aggregate using Pre-Soaking Approaches", Resources Conservation and Recycling Vol. 50(1) 2007, pp. 82-101.
[18] Tsujino M et al (2008). Completely recyclable concrete of aggregate recovery type by using microwave heating technology.Proceedings of the third ACF international Conference, Ho Chi Minh, Vietnam.

Mechanical Characterization of Hybrid Polymer SiC Nano Composite Using Hybrid RSM-MOORA-Whale Optimization Algorithm

Dilip Kumar Bagal[1], Biswajit Parida[2], Abhishek Barua[2], Bibekananda Naik[3], Siddharth Jeet[2], Sangam Kumar Singh[4], Ajit Kumar Pattanaik[1]

[1] Department of Mechanical Engineering, Government College of Engineering, Kalahandi, Bhawanipatna, Odisha, India
[2] Department of Mechanical Engineering, Centre for Advanced Post Graduate Studies, BPUT, Rourkela, Odisha, India
[3] Department of Civil Engineering, Bhubaneswar Institute of Technology, Bhubaneswar, Odisha, India
[4] Department of Mechanical Engineering, Delhi Technical Campus, Bahadurgarh, Haryana, India

E-mail: dilipbagal90@gmail.com

Abstract. Colossal materials do not provide a wide range of applications due to their mechanical property limitations. Fiber reinforced polymer composites are now considered as an important class of engineering materials. The processing and mechanical characterization of a new class of multi-phase composites consisting of polyester resin reinforced with jute-glass-silk fiber. This research investigation deals with mechanical properties of hybrid polymer composite. Woven palm fiber, teak wood dust and SiC Nano particles have been used for fabrication of the composites. Experiments have been planned as per Response Surface Method. After preparing the composite material by hand layup technique and then the mechanical characterizations are performed. Multi-response optimization has been carried out using Multi objective optimization on the basis of the ratio analysis method (MOORA) and Whale Optimization Algorithm. Optimal results have been verified through confirmatory experiments. Based on the experimental observations density, flexural and Ultimate Tensile Strength, it is concluded that teak wood powder influences the mechanical properties more than that of other two reinforcements but Nano particle filled composites shows better properties.

1. Introduction

In material science, there are numerous sorts of composite material planned so as to improve and amplify the likely quality of the last items. Most basic kind of fiber fortified composite utilized in weighty industry are glass and carbon filaments strengthened thermoset composites because of their boss properties. Consequently, there is a need to supplant engineered filaments with regular strands that are natural benevolent without yielding the quality and solidness offered by these composites. The utilization of characteristic plant filaments as support in fiber-strengthened plastics is accepting more consideration as of late, in view of their points of interest, for example, inexhaustibility, low thickness, and high explicit quality. Characteristic fiber fortified composite can be generally solid, and Light weight, liberated from wellbeing perils and biodegradable. It can possibly be utilized as building materials. Because of their numerous preferences they are generally utilized in the aeronautic trade, in countless business mechanical designing applications, for example, machine parts; inner ignition motors; vehicles; warm control and electronic bundling; railroad mentors and airplane structures; mechanical segments, for example, drive shafts, tanks, brakes, pressure vessels and flywheels; measure enterprises gear expecting protection from high-temperature erosion, oxidation, and wear; dimensionally stable segments; sports and relaxation hardware, marine structures, and biomedical gadgets [1-5].

As of late, miniature, sub-miniature and Nano-scale particles have been considered as filler materials for epoxy to create elite composites with upgraded properties. Numerous analysts have discovered that an extraordinary assortment of Nano-and miniature inorganic fillers, for example, Nano- Si_3N_4, ZnO, SiC, Al_2O_3, SiO_2, TiO_2 and MnO_2 can largely improve the mechanical and tribological properties of the polymer composites.

In this study, Response surface method is used for Experimental design, and multi objective optimization procedures has been employed i.e. Multi objective optimization on the basis of the ratio analysis method (MOORA) and Whale Optimization Algorithm are used to find optimum results. The present work is expected to analyze the effect of Nano particle with natural fiber reinforcement.

2. Experimental Procedure

Unsaturated polyester resin Ecmalon 4413 matrix material, methyl ethyl ketone peroxide (MEKP) as hardener material and cobalt octoate as accelerator are used for composite specimen fabrication. Different types of fiber materials have also been employed for fabrication of composite like woven palm fiber matt along with SiC Nano particle and teak wood dust. Composites were fabricated using mold of dimension 20 cm x 10 cm [3-5]. Different % of reinforcement were used as shown in Table 1.

Table 1. Factors and Their Levels

Factor	Symbol	Level 1	Level 2	Level 3
Palm	A	5%	10%	15%
Wood	B	2%	4%	6%
Nano	C	0.5%	1%	1.5%

The composite was kept for curing for around 24 hours. fifteen samples were prepared according to Response Surface Methodology [6-9]. The specimens were cut into the required size by using electric cutter which has been shown in Figure 1.

Prepared composites are tested by using a capacity of 600 KN universal testing machine (UTM BSUT 60JD) and with a cross head speed of 10mm/min. Ultimate tensile strength (MPa) was found out using the expression:

$$\sigma_u = F/A \quad (1)$$

Where σ_u the ultimate tensile strength (MPa), F is the maximum load (kN) applied and A is the cross-sectional area (m^2) of the composite. The flexural stress in a three-point bending test is found out by using formula

$$\sigma_f = 3FL/2bd^2 \quad (2)$$

Where σ_f is the flexural strength, F is the maximum load applied, L is the distance between the supports, and b and d are breadth and thickness of the specimen respectively.

Figure 1. Prepared material, before and after testing of composite sample

3. Multi objective optimization on the basis of the ratio analysis method (MOORA)

The MOORA method (Multi objective optimization on the basis of the ratio analysis) has been used to disregard unsuitable substitutions by selecting the most appropriates an also by collation the selection parameter. The steps are [10-14].

- Decision matrix formation

$$X = \begin{bmatrix} x_{11} & x_{12} & .. & .. & x_{1n} \\ x_{21} & x_{22} & .. & .. & x_{2n} \\ :: & :: & :: & :: & :: \\ x_{m1} & x_{m1} & .. & .. & x_{mn} \end{bmatrix} \quad (3)$$

- Ratio system formation by decision matrix normalization using the equation (4).

$$x_{ij}^* = x_{ij}/\left[\sum_{i=1}^{m} x_{ij}^2\right]^{\frac{1}{2}} \qquad (j = 1, 2, \ldots\ldots n) \quad (4)$$

- Calculation of overall assessment value and ranking of y_i is done from highest to lowest value to know the best alternate among the entire attributes:

$$y_i = \sum_{j=1}^{g} w_j x_{ij}^* - \sum_{j=g+1}^{g} w_j x_{ij}^* \quad (5)$$

4. Whale Optimization (WO)

It is a nature galvanized metaheuristic optimization process and which is derived from hunting behaviour humpback whales. They search and hunt for food by the extraordinary method entitled bubble-net feeding method where they create bubbles via enclosing or over a '9'- formed route. This performance of probing is exhibited arithmetically as dual levels [15].

4.1. Probing and enclosing target

Probing for the target can be demonstrated using Eqs. (6) and (7):

$$D = |C.X_{rand} - X| \quad (6)$$

$$X(t+1) = X_{rand} - A.D \quad (7)$$

Here, coefficient vectors A and C are

$$A = 2.a.r - a \quad (8)$$

$$C = 2.r \quad (9)$$

Here, 'a' = linearly subsiding commencing 2 to 0 & 'r' = arbitrary number amongst 0 and 1.

$$D = |C.X^*(t) - X(t)| \quad (10)$$

$$X(t+1) = X^*(t) - A.D \quad (11)$$

4.2. Spirally apprising locus

Point apprising is signified by Eq. (12):

$$X(t+1) = \begin{cases} X^*(t) - A.D, & if\ p < 0.5 \\ D.e^{bl}.\cos(2\pi l) + X^*(t), & if\ p \geq 0.5 \end{cases} \quad (12)$$

IOP Conf. Series: Materials Science and Engineering **970** (2020) 012017 doi:10.1088/1757-899X/970/1/012017

where 'l' = amongst -1 and 1; 'p' = arbitrary numeral amongst 0 and 1; 'b' = continual for recounting the coiled form.

5. Results and Discussion

Table 2 shows the RSM experimental design using which samples were prepared. The experimental results were optimized and analyzed by MINITAB 18 software.

Table 2. Experimental Results

Sl. No.	Palm	Wood	SiC	Density (g/cc)	UTS (MPa)	Flexural Strength (MPa)
1	5	2	1.0	1.54	84.69	1773.76
2	15	2	1.0	1.38	65.33	1674.05
3	5	6	1.0	1.68	104.44	2577.58
4	15	6	1.0	1.52	85.08	2477.87
5	5	4	0.5	1.48	114.90	2157.24
6	15	4	0.5	1.32	95.54	2057.53
7	5	4	1.5	1.45	109.09	2460.10
8	15	4	1.5	1.29	89.72	2360.39
9	10	2	0.5	1.29	77.66	1533.15
10	10	6	0.5	1.43	97.41	2336.97
11	10	2	1.5	1.26	71.84	1836.01
12	10	6	1.5	1.40	91.59	2639.83
13	10	4	1.0	1.20	111.01	2487.44
14	10	4	1.0	1.35	111.18	2485.11
15	10	4	1.0	1.30	110.13	2466.36

MOORA optimization method was applied to find out the optimal parameters for fabrication of composite. The MOORA overall assessment value is designed using equation (5) and ranked according to the highest value of the overall assessment value shown in Table 3.

Table 3. Computational values using MOORA

Sl. No.	Density	UTS	Flexural strength	Y_i	Rank
1	0.094	0.075	0.067	-0.048	3
2	0.084	0.058	0.063	-0.037	1
3	0.103	0.093	0.098	-0.088	7
4	0.092	0.076	0.094	-0.077	5
5	0.090	0.102	0.082	-0.093	10
6	0.080	0.085	0.078	-0.082	6
7	0.089	0.097	0.093	-0.102	12
8	0.078	0.080	0.089	-0.091	9
9	0.079	0.069	0.058	-0.048	3
10	0.087	0.087	0.089	-0.088	8
11	0.077	0.064	0.070	-0.057	4
12	0.085	0.081	0.100	-0.096	11
13	0.073	0.099	0.094	-0.120	15
14	0.082	0.099	0.094	-0.111	13
15	0.079	0.098	0.093	-0.112	14

In the above table, the overall assessment value is highest in experiment no. 2. Hence, experiment number 2 which is having the factor setting A3 B1 C2 is the optimal parameter combination for fabrication of composite. From Figure 2, A3 B1 C1 (larger the better in this case) is the optimal parameter combination for fabrication of composite

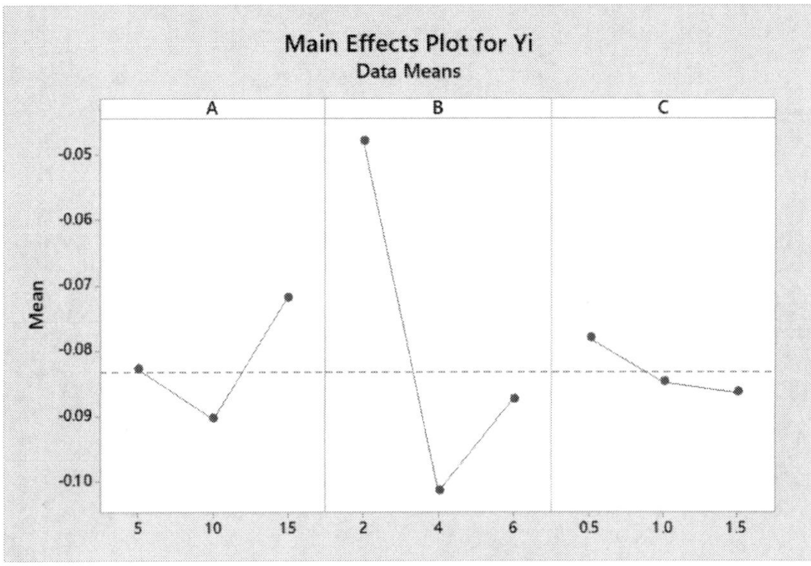

Figure 2. Main-effect plot for Yi

After performing MOORA method optimization, a regression equation was generated which will serve as a fitness function in Whale Optimization Algorithm techniques which will be used in the later stage. Employing the MOORA method the regression equation is developed as:

$$Yi = 0.1538 - 0.01166a - 0.08112b - 0.0581c + 0.000637a^2 + 0.008901b^2 + 0.02489c^2 - 0.000000ab + 0.000000ac + 0.00000bc \quad (13)$$

According to previous investigations by respective researchers, it has been reported that meta-heuristic optimization techniques results are more precise then statistical one, hence in this study also metaheuristic optimization techniques is used for finding a more precise composite fabrication constraint. The algorithm was coded in MATLAB R2018a software and ran on SONY VAIO VPCEH3AEN Notebook Computer with the configuration of the Intel Core i5 2nd Gen processor with a speed of 3.1 GHz, 4GB of RAM and Microsoft Windows 7 Ultimate OS.

For achieving a compound score of the responses, Whale Optimization was formulated to achieve the best fitness. In the Whale Optimization technique, the pod of the whale was set to 100 with iteration perimeter to 50. Figure 3 shows the convergence curve plotted by Whale Optimization. Figure 4 shows the best fitness value spawned -0.1182 along with parameter configuration with 10% Palm fiber, 6% teak wood powder and 1.5% SiC Nano-particle which can be reported as optimum configuration.

The analysis of variance (ANOVA) was done by using the values of overall assessment value calculated using the MOORA method for finding the most influential fabrication parameter and goodness of the relationship between them. From Table 4, teak wood powder with 86.63 % is the most influential parameter among the 3 parameters.

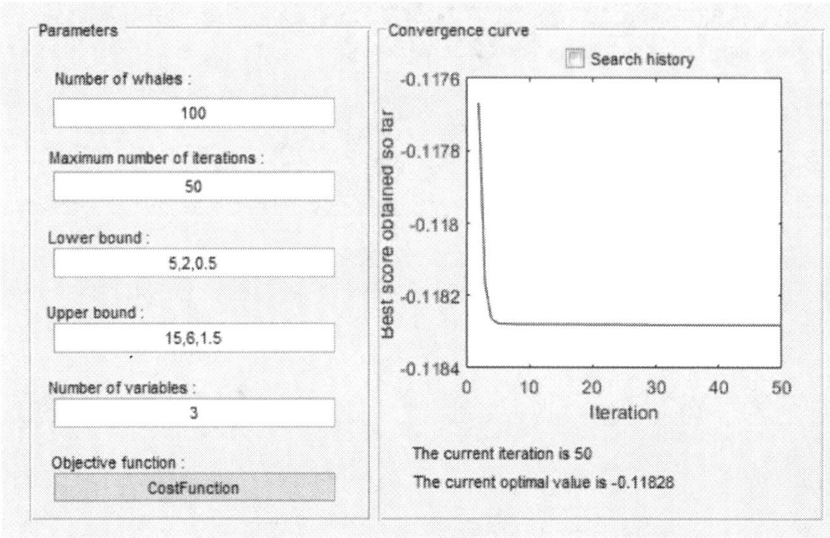

Figure 3. Convergence Curvature of WO

Table 4. ANOVA of Y_i

Source	DF	Contribution	Adj SS	Adj MS	F-Value	P-Value
A	2	9.70%	0.001177	0.000588	101.42	0.000
B	2	86.63%	0.007822	0.003911	674.16	0.000
C	2	3.14%	0.000280	0.000140	24.16	0.000
Error	8	0.52%	0.000046	0.000006		
Lack-of-Fit	6	0.00%	0.000000	0.000000	0.00	1.000
Pure Error	2	0.52%	0.000046	0.000023		
Total	14	100.00%				

To check the enrichment of output quality features after outcome of best factor setting, a confirmatory experiment is done.

Table 5. Initial and optimal level performance

Algorithm	Optimum setting	Predicted value	Expt. value
RSM-MOORA	A3 B1 C1	-0.0374	-0.0371
MOORA-WO	A2 B3 C3	-0.1182	-0.1101

6. Conclusions

This research presents forecast and optimization of parameters prominent to maximization of density, flexural strength and ultimate tensile strength thru static loading of fabricated composite. The analytical values determined using hybrid MOORA method coupled with Whale optimization is more precise as compared to value determined by RSM based MOORA method approach.

According to ANOVA, teak wood powder is most influencing parameter for the composite fabrication than other two factors if the maximization of density, flexural strength and ultimate tensile strength are concurrently considered. Also, the confirmatory test results are found exceedingly good contract with those forecasted which can be beneficial for fabrication of desired composite.

References

[1] Ahmed K S, Vijayarangan S and Naidu A 2007 Elastic properties, notched strength and fracture criterion in untreated woven jute–glass fabric reinforced polyester hybrid composites *Materials Design* **28** 2287-94

[2] Ahmed K S and Vijayarangan S 2008 Tensile, flexural and interlaminar shear properties of woven jute and jute-glass fabric reinforced polyester composites *Journal of Materials Processing Technology* **207** 330-5

[3] Barua A, Jeet S, Bagal D K, Satapathy P and Agrawal P K 2019 Evaluation of Mechanical Behavior of Hybrid Natural Fiber Reinforced Nano SiC Particles Composite Using Hybrid Taguchi-COCOSO Method *International Journal of Innovative Technology Exploring Engineering* **8** 3341-5

[4] Parida B, Barua A, Jeet S and Bagal D 2018 Fabrication and mechanical characterization of jute-glass-silk fiber polymer composites based on hybrid RSM-GRA-FIS and RSM-TOPSIS approach *International Journal for Research in Engineering Application Management Decision* **4** 25-33

[5] Naik B, Paul S, Barua A, Jeet S and Bagal D K 2019 Fabrication and Strength Analysis of Hybrid Jute-Glass-Silk Fiber Polymer Composites Based on Hybrid Taguchi-WASPAS Method *International Journal of Management, Technology Engineering Economics* **9** 3472-9

[6] Bagal D K, Barua A, Jeet S, Satapathy P and Patnaik D 2019 MCDM Optimization of Parameters for Wire-EDM Machined Stainless Steel using Hybrid RSM-TOPSIS, Genetic Algorithm and Simulated Annealing *International Journal of Engineering and Advanced Technology (IJEAT)* **9** 366-71

[7] Jeet S, Barua A, Bagal D K, Pattanaik A K, Agrawal P K and Panda S N 2019 Multi-Parametric Optimization During Drilling of Aerospace Alloy (UNS A97068) Using Hybrid RSM-GRA, GA and SA *International Journal of Management, Technology And Engineering* **9**

[8] Sahoo B, Barua A, Jeet S and Bagal D K 2018 Multi Objective Optimization of WEDM Process Parameters Using Hybrid RSM-GRA-FIS, GA and SA Approach *International Journal of Research in Advent Technology* **6** 1752-61

[9] Jeet S, Barua A, Parida B, Sahoo B B and Bagal D K 2018 Multi-objective optimization of welding parameters in GMAW for stainless steel and low carbon steel using hybrid RSM-TOPSIS-GA-SA approach *I Technical Innovation in Modern Engineering Science* **4** 683-92

[10] Barua A, Jeet S, Bagal D, Bagal P and Pattanaik A 2019 Comparative Analysis Based on MCDM Optimization of Printing Parameters Affecting Compressive and Tensile Strength of Fused Deposition Modelling Processed Parts *International Journal of Technical Innovation in Modern Engineering Science* **5** 383-92

[11] Barua A, Jeet S, Cherkia H, Bagal D K and Sahoo B B 2019 Parametric Optimization of FDM Processed Part for Improving Surface Finish Using MOORA Technique and Desirability Function Analysis *International Journal of Applied Engineering Research* **14**

[12] Barua A, Jeet S, Parida B, Samantray A and Bagal D 2018 Comparative evaluation and optimization of 4-cylinder CI engine camshaft material using finite element analysis: a hybrid MOORA technique and Taguchi based desirability function analysis approach *International Journal of Technical Innovation in Modern Engineering Science* **4** 105-14

[13] Jeet S, Barua A, Cherkia H and Bagal D K 2019 Comparative Investigation Based on MOORA, GRA and TOPSIS Method of Turning of Nickel-Chromium-Molybdenum Steel under the influence of Low Cost Oil Mist Lubrication System *International Journal of Applied Engineering Research* **14** 8-20

[14] Panda S N, Bagal D K, kumar Pattanaik A, Patnaik D, Barua A, Jeet S, Parida B and Naik B 2020 Comparative Evaluation for Studying the Parametric Influences on Quality of Electrode Using Taguchi Method Coupled with MOORA, DFA, and TOPSIS Method for Electrochemical Machining *Recent Advances in Mechanical Infrastructure*: Springer pp 115-29

[15] Tanvir M H, Hussain A, Rahman M, Ishraq S, Zishan K, Rahul S and Habib M A 2020 Multi-Objective Optimization of Turning Operation of Stainless Steel Using a Hybrid Whale Optimization Algorithm *Journal of Manufacturing Materials Processing* **4** 64

IOP Conf. Series: Materials Science and Engineering **970** (2020) 012018 doi:10.1088/1757-899X/970/1/012018

Effect of Ground Granulated Blast Furnace Slag on the Properties of Sea Shell Concrete

Kishor Chandra Panda, Bharat Bhusan Gouda, Priyanka Mohapasayat

Department of Civil Engineering, Government College of Engineering, Kalahandi, Bhawanipatna, Odisha, India

E-mail: kishoriit@gmail.com

Abstract. This paper opens up how the properties of seashell mix concrete influenced by the effect of ground granulated blast furnace slag (GGBS). This experimental program consists of thirteen concrete mixes. One is conventional concrete & other concrete mixes are formed by the partial replacement of fine aggregate with crushed seashell and cement with GGBS. The proportion of crushed seashell was added 0, 10 %, 20 % & 30 % to the conventional concrete. Then seashell mix concrete of 10 % was blended with 10 %, 20 % & 30 % of GGBS mix concrete and like this 20 %, 30 % of seashell mix concrete was intermixed with 10 %, 20 % & 30 % of GGBS mix concrete respectively. For this experiment M30 grade concrete mix was prepared. Tests like slump test for workability of concrete were carried out on different mixes of fresh concrete. Also, on hardened concrete samples of different mix, test like compressive, split-tensile & flexural strength test were allotted after 7 days and 28 days of curing under water. From the experimental observations show that use of seashell in concrete decreases the strength gradually but in case of GGBS, the strength increases gradually in all type of concrete mixes. The result declares higher compressive, flexural and split-tensile strength at 10 % partial replacement of fine aggregate with crushed seashell and 20 % partial replacement of cement with GGBS.

1. Introduction

Food, Cloth and Shelter are basic need of human. We use varieties of materials for our shelter. Concrete mainly contains water, fine and coarse aggregates which are granular in nature and cement. We have been used these ingredients from way back. Since ancient time we use lime, stone, sand and aggregate for creating our homes, monuments, some huge structures etc. Now a days concrete is employed as an artifact for any form of construction work. We can directly prepare the concrete on the site. It is very easy to prepare. The demand for concrete as construction material is increasing constantly. Even though it is economical, it is not very eco-friendly. To deal with the negative impacts on environment, sustainable construction materials are being researched in huge amount. On the other hand, industrial byproducts have also become headache lately. Estimations revealed that the cement's production rose from about 1.5 billion tonnes in 1995 to 3.2 billion tons in 2016. The demand of key ingredients of concrete are increased also. So there are chances of shortage of these materials in future because it is available from nature and it takes long time to refill the erode space. So in this experiment our main moto is to use the waste material and form a new generation concrete which is not only eco-friendly but also has the equal and more load bearing capacity or strength.

In this experiment we use the seashell which is the dead shell of sea animals. Our India is surrounded by sea around its three sides. So there are a lot of places where we can get these seashells. The seashell has the most amazing properties of binding because it consists of lime. We can use this seashell as a replacing material of cement, fine aggregate, coarse aggregate. In this project we use the seashell for partially replacing fine aggregate. For this effective use of seashell, the natural source like sand which has eroded from sea bank or river bank will somehow reduce. So, overall concrete plays key role in infrastructural development. But there are downsides of using concrete as well. One of which is its bad effects on environment (as it produces heavy amount of carbon dioxide gas). One practical approach for solving this problem while expecting optimal results is using supplementary material which are also cementitious in nature i.e. fly ash, slag etc. A recycled aggregate can successfully replace natural aggregate in concrete. Recycled aggregates are composed of substances that have previously been used in construction and later collected after demolition of the constructed infrastructure. GGBS is produced as a by-product in the construction of pig iron. It is slightly alkaline in nature and does not pose any threat to reinforcement of concrete. So the second waste which has used in this project is ground granulated blast furnace slag (GGBS). Steel furnace slag (SFS) along with Blast furnace slag (BFS) go back a long way to be used as industrial byproducts. GGBS is partially substituted cement in this experiment. We should take care of our environment as well as give our best to the nature and for the development of our construction practices [1-10].

In this project our main purpose is to make an economical, eco-friendly, sustainable concrete and study the effect, and the properties of ground granulated blast furnace slag on seashell mix concrete. Some of the researchers have investigated the effect of cementitious material over crushed seashell concrete Adewuyi and Adegoke [1] used periwinkle shell as a coarse aggregate substitution with a percentage of 35.4% and 42.5 % and getting a 28 days cube strength value of 21 N/mm2 and 15 N/mm2 for concrete mix 1:2:4 and 1:3:6 respectively. Chakravarthy and Mutusva [2] analyzed that in early stages it was found that partially adding and increasing the seashells as coarse aggregate reduces the workability of concrete. Mohanlaxmi et al. [3] researched about how replacement of sea shell as fine aggregate could increase overall strength of concrete. Panda et al. [4] analysed that with increase in seashell amount, workability gradually lessened. When 10 % crushed seashell is employed in place of fine aggregate as well as 20 % RHA is employed in place of cement, it provides best results for concrete mechanical properties. Padhi and Panda [5] partially replaced the fine aggregate with fine rubber (FR) and cement with Silpozz in a proportion of 5 %, 10 % and 15 %. The result shows that, as quantity of FR rises, workability of both conventional and self compacting concrete lessens. In case of FR replacement, the compressive strength decreases and the flexural strength increases up to 5 % replacement of FR. Silpozz enhances the strength of SCRC.

Panda and Pradhan [6] partially replaced cement by Silpozz and lime stone powder. The result shows concrete specimens having lime stone powder result in increased workability. Jena and Panda [7] replaced sand with crusher dust in cement concrete and finally found out a suitable mixture proportion that also helped in retaining quality. They used seven number of mix sample for this experiment. The result was obtained that at 28 days curing, with 100 % replacement of sand with crusher dust compressive, flexural and split tensile strength rises. Lenka and Panda [8] used the supplementary cementitious material as Metakaolin (MK). The authors used MK as the replacing material of cement from 5 – 20 % for both SCC and CC. The outcome reveals that concrete having MK had great impact on strength. This paper focuses on the properties of seashell mix concrete influenced by the effect of ground granulated blast furnace slag (GGBS).

2. Material Used

2.1 Materials

Portland Pozolana Cement (PPC), Natural fine aggregate (fineness modulus 2.30, water absorption 0.70 and specific gravity 2.62) and natural coarse aggregate (fineness modulus 6.08, water absorption 0.82 and specific gravity 2.80). the results obtained from the testing of the materials as per the code IS:

383:1970 [9]. Sea shell and GGBS are used in this experimental study. Seashells' passing through 4.75mm IS sieve, fineness modulus 2.28 and specific gravity 2.50 are collected from Balasore. The uncrushed seashell and crushed seashell are displayed in Figure 1 and the particle size distribution curve of sea shell is displayed in Figure 2.

Figure 1. Uncrushed and crushed seashell below 4.75 mm

Table 1. Chemical constitution and physical features of GGBS

Chemical constitution	RA (%)	Physical features	
Silica (SiO_2)	31-37	Colour	off-white
Alumina (Al_2O_3)	22-27	Specific gravity	2.9
Magnesia Oxide (MgO)	4-10	Maximum dry density	2.56gm/cc
Calcium Oxide (CaO)	30-50	Optimum moisture content	20 %

GGBS is used as partial replacement of cement which is supplied by Rourkela Steel Plant, Rourkela. GGBS is processed under a temperature of 800°C with containing a higher percentage of CaO (30-50 %) and having some additional mineral in it i.e. SiO_2, Al_2O_3 and MgO. The chemical constitution and physical features of GGBS are presented in Table 1.

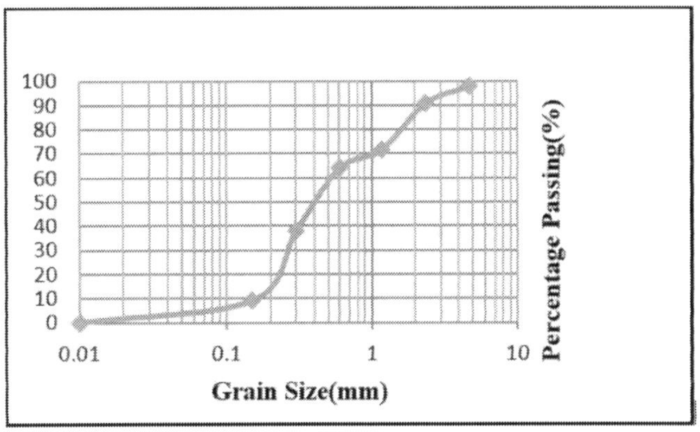

Figure 2. Particle size distribution curve of sea shell

2.2 Mix Proportion

For the concrete, the material which is used in this experiment were PPC, sand (NFA), natural coarse aggregate (NCA), Sea shell (C) & ground granulated blast furnish slag (GGBS). The grade of concrete was M30 which was designed as per IS: 10262-2009 [10]. Here the mix proportions 1:1.6:2.8 were taken in this experiment. The three different mix of concrete design were 10 %, 20 %, and 30 % for sea shell and GGBS replaced with natural fine aggregate and cement respectively. The same mixes have been tested in a constant water cement ratio 0.375 for conventional concrete. In this study total thirteen concrete mix proportions are used, first mix i.e. MC0G0 which stands for conventional concrete. In next three concrete mixes seashell is replacing NFA by 10, 20 & 30 %. Second three mixes were made by replacing cement with GGBS by 10, 20 & 30 % with a constant percentage of seashells (10 %). Third three mixes were made by replacing GGBS with cement by 10, 20 & 30 % with a constant percentage of seashells (20 %). The last three mixes were made by replacing GGBS with cement by 10, 20 & 30 % with a constant percentage of seashells (30 %).

2.3 Mixing and casting

For the concrete NCA, NFA, GGBS and seashell were weighted and placed in the concrete mixer and thoroughly mixed until the mixture become equivalent. The required sum of water was mixed to the respective mix during the mixing. The above procedure is done for casting & curing. Then after the discharge of the concrete mixture from mixture machine, immediately workability test was done such as slump test. From the mix design calculation, the slump value must come within 25 – 50 mm. The test specimens were cast in steel mould of cube (150 x 150 x 150 mm), cylinder (100 mm diameter x 200 mm height) and prism (100 x 100 x 500 mm). Then it compacted by using of table vibrator and demoulded after 24 hours. There after the curing was done for period of 7 and 28 days. For the slump test with different percentage of replacement of sea shell with NFA and GGBS with cement, degree of workability is described in Table 2.

Table 2. The details of mix proportion along with identification

Mix Identity	Water/Cement Ratio	Slump Value (mm)	% of Change of Slump value w.r.t Control Specimen	Degree of Workability
MC0G0	0.375	48 mm	0	Medium for all concrete mixes
MC10G0	0.375	43 mm	-10.41	
MC20G0	0.375	37 mm	-22.91	
MC30G0	0.375	34 mm	-29.16	
MC10G10	0.375	45 mm	-6.25	
MC10G20	0.375	52 mm	8.33	
MC10G30	0.375	48 mm	0	
MC20G10	0.375	39 mm	-18.75	
MC20G20	0.375	42 mm	-12.5	
MC20G30	0.375	46 mm	-4.16	
MC30G10	0.375	37 mm	-22.91	
MC30G20	0.375	43 mm	-10.41	
MC30G30	0.375	48 mm	0	

3 Discussion

To analyse concrete's workability slump test was conducted. It was noticed that increase in quantity of crushed seashell in place of fine aggregate causes decrement in the value of slump. Also with both increment in replacement level of cement with GGBS and increment in crushed seashell, again results in the workability of concrete regaining its value. Angular shape and asymmetrical texture of seashell are the reason behind reduced workability.

3.1. Compressive strength

Figure 3 represents the test result of compressive strength. The graph shows 7 days and 28 days compressive strength of different concrete mixes. Compressive strength of MC0G0 after 7 and 28 days curing are 37.70 MPa and 51.25 MPa respectively. The graph shows that with increment in the partial substitution of crushed seashell in concrete from 10 % to 30 %, the compressive strength starts to decrease gradually. The compressive strength after span of 7 and 28 days of curing for specimen MC10G0 is 33.10 MPa and 47.25 MPa respectively, for specimen MC20G0 30.25 MPa and 43.70 MPa and for specimen MC30G0 25.60 MPa and 37.70 MPa in 7 days and 28 days respectively. The decrease in strength is due to adding of more percentage of seashells in conventional concrete. While increasing the percentage of seashells, the binding between crushed seashell and cement become loose due to the more water absorption of seashell that of fine aggregate. As the cement is partially substituted by GGBS and mixed with the different percentage of seashell concrete, the compressive strength starts increasing. The compressive strength is high in 10 % substitution of fine aggregate with crushed seashell and 20 % substitution of cement with GGBS. The compressive strength which is maximum in MC10G20 concrete mix is 41.5 MPa and 55.75 MPa in 7 days and 28 days respectively.

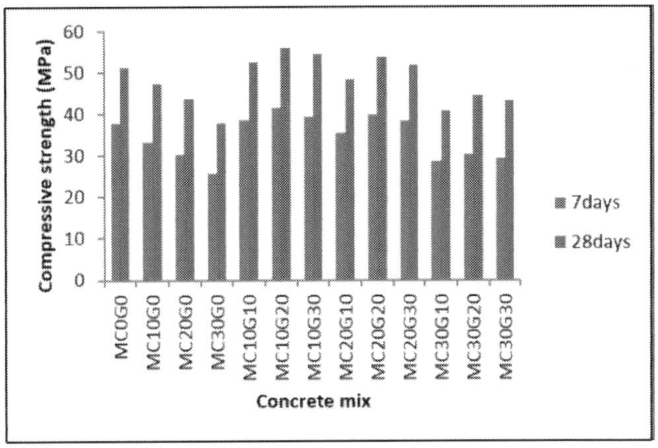

Figure 3. Compressive strentgth vs concrete mix.

3.2. Flexural strength

Figure 4 represents the test result of flexural strength. The graph shows 7 and 28 days flexural strength of different concrete mixes.

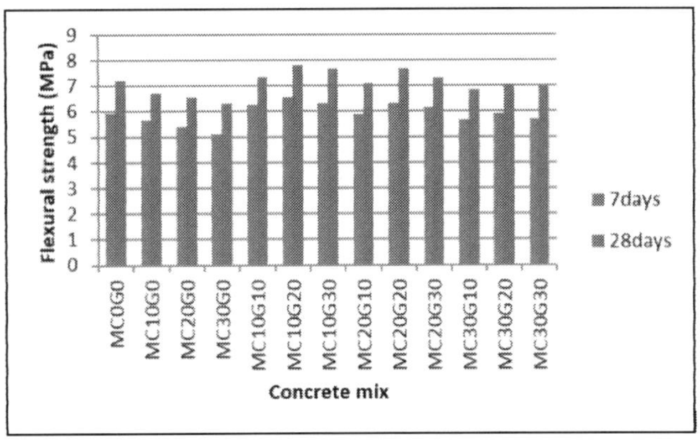

Figure 4. Flexural strength vs concrete mix.

IOP Conf. Series: Materials Science and Engineering **970** (2020) 012018 doi:10.1088/1757-899X/970/1/012018

From Figure 4 flexural strength of MC0G0 after 7, 28 days curing are 5.95 MPa and 7.20 MPa respectively. The decrease in strength is due to adding of more percentage of seashells in conventional concrete. While increasing the percentage, the binding between crushed seashell and cement become loose due to the more water absorption of seashell that of fine aggregate. When cement is partially replaced with GGBS and mixed with the different percentage of seashell concrete, there is increment in flexural strength. The flexural strength is high when the percentage of partial substitution of fine aggregate with crushed seashell is 10 % and that of cement with GGBS is 20 %.

3.3. Split tensile strength

Figure 5 signifies the test result of 7 days and 28 days split tensile strength of different concrete mix. As the replacement level of crushed seashell in concrete is increased from 10 % to 30 %, the split tensile strength starts to decrease gradually. Strength reduction is due to addition of more percentage of seashells in conventional concrete. While increasing the percentage the binding between crushed seashell and cement become loose due to the more water absorption of seashell that of fine aggregate. The split tensile strength which is maximum in MC10G20 concrete mix is 3.65 MPa and 4.75 MPa in 7 and 28 days respectively.

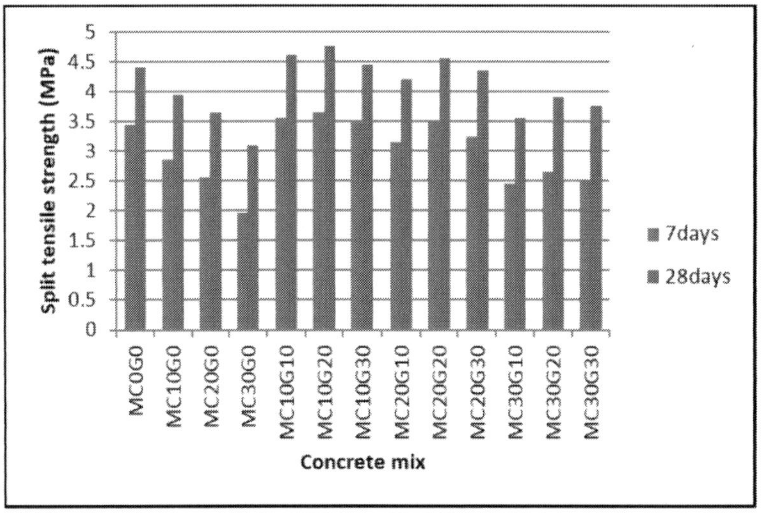

Figure 5. Split tensile strength vs concrete mix.

4 Conclusion

In this paper, the effect of normal concrete and seashell concrete with their mechanical properties is investigated. The result of different concrete mix is evaluated to find out the best composition which gives excellent mechanical properties. GGBS is replaced with cement and seashell is used on behalf of natural fine aggregate in a proportion of 10 %, 20 % and 30 %. It is noticed that with increasing the percentage of replacement of NFA by seashell, concrete's workability is lessening. Likewise, with increasing seashell quantity, concrete's strength is lessening. Concrete's strength and workability increase as the GGBS is added to the seashell mix concrete. Utmost strength is attained with 10 % seashell and 20 % of GGBS. There is growth in compressive strength of concrete upto 10 % substitution of seashell and 20 % substitution of GGBS, there after there is lessening in compressive strength. Rise in split tensile strength is perceived up to 10 % substitution of seashell and 20 % replacement of GGBS, there after it decreases. Flexural strength of concrete increases up to 10 % changeover of seashell and 20 % substitution of GGBS, there after it decreases. Water absorption yielded higher value for all mixes containing seashell and GGBS. It is found that with increase in days, all the three test strength increases. But when the percentage of seashell increases as NFA, the strength decreases.

References

[1] Adewuyi A and Adegoke T 2008 Exploratory study of periwinkle shells as coarse aggregates in concrete works *ARPN Journal of Engineering and Applied Sciences* **3** 1-5

[2] HG N C and Mutusva T 2015 Investigation of Properties of Concrete with Seashells as a Coarse Aggregate Replacement in Concrete *MATTER: International Journal of Science and Technology* **1**

[3] Mohanalakshmi M V, Indhu M S, Hema M P and Prabha M V 2017 Developing concrete using sea shell as a fine aggregate *International Journal for Innovative Research in Science and Technology (IJIRST)* **3** 282-6

[4] Panda K C, Behera S and Jena S 2020 Effect of rice husk ash on mechanical properties of concrete containing crushed seashell as fine aggregate *Materials Today: Proceedings*

[5] Padhi S and Panda K 2016 Fresh and hardened properties of rubberized concrete using fine rubber and silpozz *Advances in Concrete Construction* **4** 049

[6] Pradhan S and Panda K 2017 Assessment of fresh and hardened properties of concrete using supplementary cementitious materials *Materials Today: Proceedings* **4** 9837-41

[7] Jena S and Panda K 2017 Influence of crusher dust on the properties of concrete *Indian Journal of Civil Engineering and Technology* **8** 93-603

[8] Lenka S and Panda K 2017 Effect of metakaolin on the properties of conventional and self compacting concrete *Advances in Concrete Construction* **5** 031

[9] 1970 IS 383:1970: Specifications for Coarse and Fine Aggregates from Natural Sources for Concrete (Second Revision), Bureau of Indian Standard, ManakBhavan, Bahadurshah Zafar Marg, New Delhi, India

[10] 2009 IS: 10262:2009, Guide lines for concrete mix design proportioning, Bureau of Indian Standards, New Delhi, India

IOP Conf. Series: Materials Science and Engineering **970** (2020) 012019 doi:10.1088/1757-899X/970/1/012019

Enabling Technologies of IoT and Challenges in Various Field Of Construction Industry in the 5G Era: A Review

Sriparna Paul[1], Bibekananda Naik[1], Dilip Kumar Bagal[2]

[1] Department of Civil Engineering, Bhubaneswar Institute of Technology, Bhubaneswar, Odisha, India
[2] Department of Mechanical Engineering, Government College of Engineering, Kalahandi, Bhawanipatna, Odisha, India

E-mail: bibekcivil10@gmail.com

Abstract. In this new age, the Internet and its usefulness became the backbone of people in our everyday lives and developed a corresponding wave of technological developments. One of the latest technological upgrade developments is the Internet of Things (IoT), which attracted researchers with great interest. Using this upgraded technology, we can interact with devices via. Internet Sensors, actuators, or computers. The Internet of Things (IoT) is not a new concept in my research. We have been introducing it for decades, but incorporating IoT in the construction industry is a new exciting subject that leads us to the next Construction Industry 4.0 revolution. This paper offers an overview of IoT technology, tracks and classifies IoT Technology's position in various fields with exclusive studies and statistics. This analysis paper also focuses on various IoT applications in contemporary -day production strategies and addresses issues related to IoT implementation in production. This study paper also identifies key points to recognise the use of virtual technologies for building operations and lays a good basis for future studies for smart construction works. Researchers are also doing their best to introduce IoT technology in the field of civil engineering in order to provide trouble-free work in construction industries and to improve productivity, as the construction industry is least digitised.

1. Introduction

The Internet of Things (IoT) is now being considered as the next stage of the internet revolution and has become one of the trend-setting research for academic and commercial organisations Singh et. al [1] claimed that, after collaborating with IoT technologies, it is primarily possible to communicate face-to-face interface interactions or vehicle-to- vehicle interactions through the Internet by authorising innovative software programmes to assist sensors, actuators or devices according to their vision and needs. Condoluci et. al [2], Gupta, et. al [3] and Al-Fuqaha et. al [4] addressed the instability of expectations and fluctuations in launched technologies, but presents considerable problems in accessing large communication links This causes complex difficulties [5] as it is one of the key challenges. Several IoT creation attempts have been made, but something is still missing in one IoT vision conveyed by Sarkar et. al [6] and Kim et. al [7].

2. IoT Overview

Initially, Kevin Ashton, the founder of the MIT auto-identification centre in 1999, presented the idea of IoT [8, 9]. Ashton said, "The Internet of Things, like the internet, has the power to change the world or maybe even more" [8]. In 2005, the International Telecommunications Union (ITU) officially presented the IoT (ITU, "The Internet of Things" (2005)). Many organisations and researchers suggested many definitions of IoT.

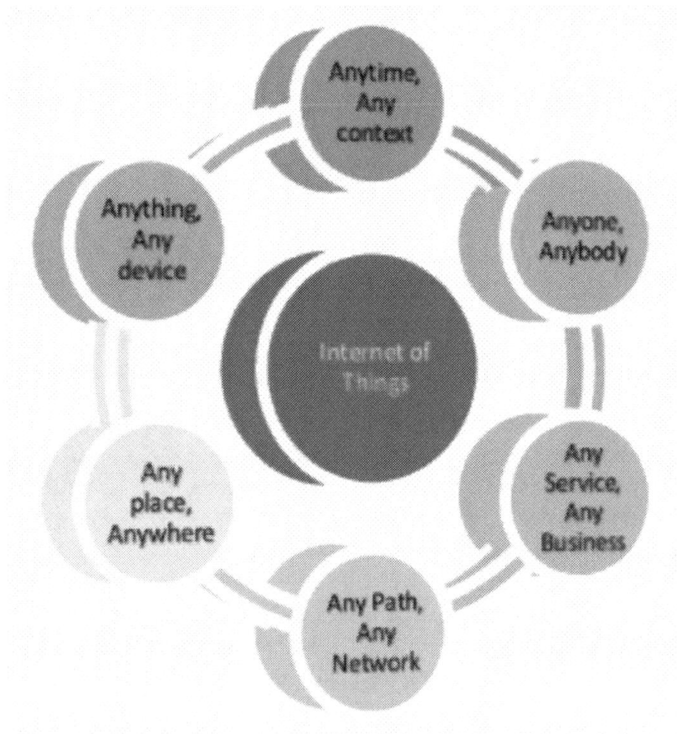

Figure 1. The IoT can connect anything in anywhere using any path

2.1. Features of IoT

1. Ubiquitous Connectivity: IoT has become very famous due to its ubiquitous connectivity which is one of the important features of IoT because of this feature, it is being widely used. For example, using the two mobile phones, we can transfer the data, which was once upon a time a science fiction. We have not even thought that we could have smart phones today in our hands specially when the initial sets which we call GSM handsets or CDMS sets. We have not even thought about these things, but later on, Bluetooth was invented which enables us to communicate with another mobile phone.
2. Internet Protocol (IP)-based networking: IP based networking has helped in connecting and communication between the two electronic components using internet, the IP-based networking and it plays a very significant and important role.
3. Computing Economics: IoT is driven by industry investment in Research, Development and Manufacturing. There is something called Moore's Law. It is one of the important feature that we need to consider here. Moore's Law states that in a dense integrated circuit continues to deliver greater computing power at lower price and lower power consumption and doubles the number of transistors about every two years.

Many researchers continued their studies and resulted that by 2020, over 20 billion devices use IoT allows the device to control in lower expenses on the radio. With the help of IoT services, all the things can be connected in order to gather information and enable the devices to interact with each other over standard and different protocol domains and applications. IoT Technology establishes advanced connectivity with the help of the internet in order to make the surrounding areas automated little by little among the devices or systems or devices, which are increasing day by day.

2.2. IoT Application

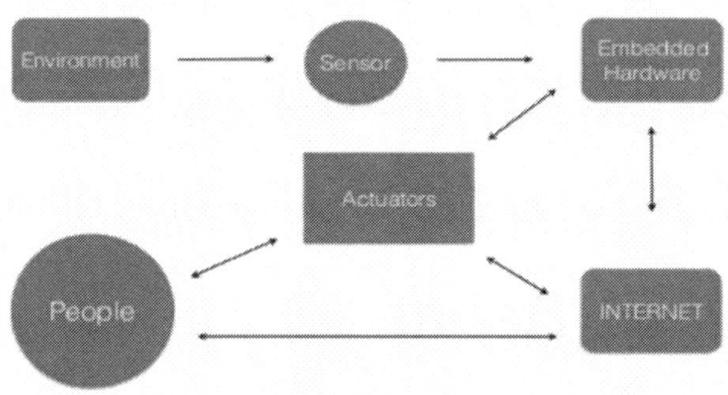

Figure 2. IoT system

2.2.1. Environmental Monitoring

Nowadays, most of the cities and university campuses are implementing IoT technologies in order to exchange the data and thus helps in emerging IoT network advances on computing and wireless communication technologies. The main objectives of advancements on emerging IoT networks are to simplify the administration process, manage real-time access control, monitoring the safety of the environment, and so on.

Marstijepović and Williams [10] described that in the surrounding area, any activities can be monitored with the help of IoT implementation that leads to environmental monitoring i.e. it monitors the quality of air, water, and soil and thus, creates a smart environment. In order to monitor the quality of the environment, applications of sensors are typically utilized so that a risk-free environment can be assessed to the individuals.

Palma et. al [11] studied that the Internet of Things connects all object to the web and make it accessible at any time. This paper also stated that most of the environmental monitoring system is using a distributed framework based on wireless sensor networks (WSN) in order to implement IoT Technology. For example: When we surf the net, we can find that on the right section, similar content page appears which are near to your favourite which is possible due to the embedded computing system in order to improve our life style.

2.2.2. Infrastructure Management

With the help of IoT devices, critical infrastructures can be controlled and also can schedule repair and maintenance activities by co-ordinating between different service providers and users of the facilities [12]. Meola, A. [13] stated that the application of IoT devices for infrastructure monitoring is likely to improve the quality of service and reduce the cost of infrastructure in emergency conditions. Zhang, Q. [14] stated that IoT Technology can be implemented in the field of waste management automatically and can be optimized as required.

2.3. Industrial Applications

In Industrial Applications also, the advanced IoT technology can be implemented in order to investigate the worth of products in real-time and boosts up the marketing. With the help of smart technology, the record will be automatically updated on asset placement in industrial storage and saves the labour time and money both at the same time within a short period.

2.4. Medical and Healthcare Systems
Da Costa et. al. [15], Sternberg et. al [16], Kricka, [17], Gatouillat et al. [18] stated that the IoT Technology also plays an important role for health and medical related purposed, data collection and analysis for research and monitoring. Other consumer devices such as wearable heart monitors or connected scales in order to encourage healthy living can encourage healthy living, which is only possible after the implementation of IoT Technology. IoT platforms can also monitor end-to-end health for chronic patients and can also help in managing medication requirements.

2.5. Building and Home Automation
Iot allows connect everything in a home that have the potential to monitor and remote control thus ensuring comfort, security and low energy consumption. The integration of smart devices in the built environment with IoT has helped the life easier for all.

2.6. Large Scale Deployments
Example of a large deployment is the one completed by New York Waterways in New York City to connect all the city's vessels and be able to monitor them live 24/7. The network was designed and engineered by Fluid mesh Networks, a Chicago-based company developing wireless networks for critical applications. The NYWW network is currently providing coverage on the Hudson River, East River, and Upper New York Bay. With the wireless network in place, NY Waterway is able to take control of its fleet and passengers in a way that was not previously possible. New applications can include security, energy and fleet management, digital signage, public Wi-Fi, paperless ticketing and others.

3. Application of IoT in Civil Engineering

3.1. Ready Mix Concrete

With the help of IoT Technology, the sensors can monitor concrete temperature, its maturity, relative humidity and communicate to our cell phones and computers wirelessly through the cloud while the concrete is being placed at the construction site, the concrete producer, contractors and engineers can monitor the concrete performance and make decision in real-time throughout the entire job.

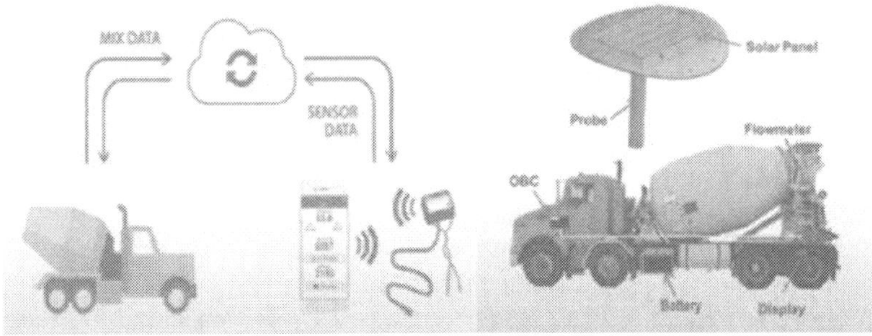

Figure 3. IoT application in concrete mixer

Companies have developed the IoT technology in order to track the dispatched trucks to the delivery location and communicate real-time data on anything that changes to the concrete mix before delivery. The trucks are complete with a stainless steel probe and sensors including thermometer and accelerometer in order to transmit data to the on-board computer (OBC). The computer can then calculate slump and concrete volume for concrete monitoring. Sensors are available to monitor rebar corrosion rate, half-cell corrosion potential, electrical resistivity of concrete and many more, example – Smart concrete.

3.2. Structural Health Monitoring

Its leads to continuous monitoring of buildings for human safety, furnishes information about the alteration in a single part or in the whole structure caused by action of the environment, materials aging or accidental events. Typically, SHM systems monitors humidity, accelerations, temperature, compressive stress, tensile stress and building materials degradation. It leads to deployment of sensors (smart objects) in checkpoints. SHM System includes sensing and data acquisition subsystem, data management system and data access and retrieval system. IoT enabled-SHM System offers advantages in reduction of monitoring costs thus maintaining structural integrity for longer periods and reducing costs related to demolition and rebuilding.

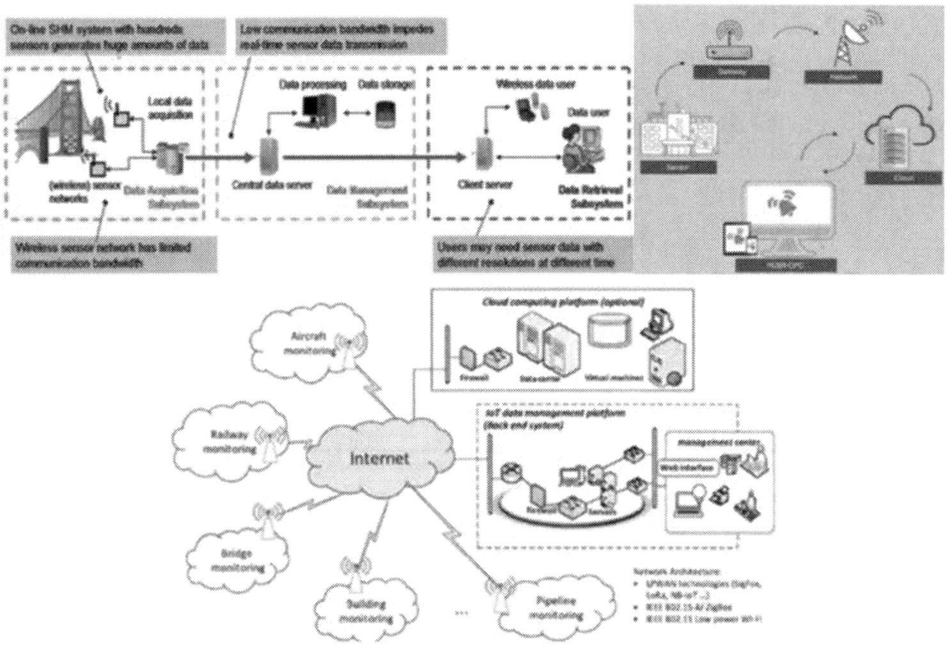

Figure 4. Structural Health Monitoring System

4. Conclusions

Various studies have been conducted using IoT and its feasibility have been investigated in the field of civil construction. With the help of IoT Technology, Ready mix concrete temperature, its maturity, relative humidity can be monitored remotely which will help the concrete producer, contractors and engineers to monitor the concrete performance and make decision in real-time throughout the entire job. Structural health monitoring can be done for continuous monitoring of buildings for human safety and maintenance cost reduction. IoT enabled drone cameras can also help in mapping and also surveillances the construction site for better building of the structures. IoT comes up as a promising technology, which can help construction sector to work more efficiently.

References:

[1] Singh D, Tripathi G and Jara A J 2014 A survey of Internet-of-Things: Future vision, architecture, challenges and services. In: *2014 IEEE world forum on Internet of Things (WF-IoT)*: IEEE) pp 287-92

[2] Condoluci M, Araniti G, Mahmoodi T and Dohler M 2016 Enabling the IoT machine age with 5G: Machine-type multicast services for innovative real-time applications *IEEE Access* **4** 5555-69
[3] Gupta A and Jha R K 2015 A survey of 5G network: Architecture and emerging technologies *IEEE access* **3** 1206-32
[4] Al-Fuqaha A, Guizani M, Mohammadi M, Aledhari M and Ayyash M 2015 Internet of things: A survey on enabling technologies, protocols, and applications *IEEE communications surveys and tutorials* **17** 2347-76
[5] Palattella M R, Dohler M, Grieco A, Rizzo G, Torsner J, Engel T and Ladid L 2016 Internet of things in the 5G era: Enablers, architecture, and business models *IEEE Journal on Selected Areas in Communications* **34** 510-27
[6] Sarkar C, SN A U N, Prasad R V, Rahim A, Neisse R and Baldini G 2014 DIAT: A scalable distributed architecture for IoT *IEEE Internet of Things journal* **2** 230-9
[7] Kim M, Ahn H and Kim K P 2016 Process-Aware Internet of Things: A Conceptual Extension of the Internet of Things Framework and Architecture *KSII Transactions on Internet and Information Systems* **10** 4008-22
[8] Ashton K 2009 That 'internet of things' thing *RFID journal* **22** 97-114
[9] Joshi G P and Kim S W 2008 Survey, nomenclature and comparison of reader anti-collision protocols in RFID *IETE Technical Review* **25** 234-43
[10] Marstijepovic S and Williams S 2016 Environmental monitoring and field surveillance reference guide. pdf Available at http://www.undp.org/content/dam/montenegro/docs/publications/ee/WBEPandENVSEC/
[11] Palma D, Agudo J E, Sánchez H and Macías M M 2014 An internet of things example: Classrooms access control over near field communication *Sensors* **14** 6998-7012
[12] Istepanian R S, Hu S, Philip N Y and Sungoor A 2011 The potential of Internet of m-health Things "m-IoT" for non-invasive glucose level sensing. In: *2011 Annual International Conference of the IEEE Engineering in Medicine and Biology Society*: IEEE) pp 5264-6
[13] Meola A 2016 Why IoT, big data & smart farming are the future of agriculture *Business Insider* **20**
[14] Zhang Q 2016 *Precision agriculture technology for crop farming*: Taylor & Francis)
[15] da Costa C A, Pasluosta C F, Eskofier B, da Silva D B and da Rosa Righi R 2018 Internet of Health Things: Toward intelligent vital signs monitoring in hospital wards *Artificial intelligence in medicine* **89** 61-9
[16] Engineer A, Sternberg E M and Najafi B 2018 Designing interiors to mitigate physical and cognitive deficits related to aging and to promote longevity in older adults: a review *Gerontology* **64** 612-22
[17] Kricka L J 2019 History of disruptions in laboratory medicine: what have we learned from predictions? *Clinical Chemistry and Laboratory Medicine* **57** 308-11
[18] Gatouillat A, Badr Y, Massot B and Sejdić E 2018 Internet of medical things: A review of recent contributions dealing with cyber-physical systems in medicine *IEEE internet of things journal* **5** 3810-22

IOP Conf. Series: Materials Science and Engineering **970** (2020) 012020 doi:10.1088/1757-899X/970/1/012020

Effect of Length of Fibers on Mechanical Properties of Normal Strength Concrete

Gopal Charan Behera, Subhasri Panda, Priyanka Kanda

[1]Department of Civil Engineering, Government College of Engineering, Kalahandi, Bhawanipatna, Odisha, India

E-mail: kanda.priyanka01@gmail.com

Abstract. Concrete as the most common building material possess very less tensile strength. To overcome this problem, different types of fibers are added with concrete matrix from from the ancient era. Investigation of fiber reinforced concrete reveals that short length fibers help in bridging micro cracks while long fibers are able to arrest macro cracks and also enhance toughness of the member. The combinations of different lengths of fibers are found to be more effective than mono length fibers for a fixed volume fraction of fiber. Combination of different lengths of fibers is termed as hybrid fiber and when different lengths of fibers are mixed in concrete it is termed as graded fiber. This investigation aims in finding out the enhancement of mechanical properties of a particular grade of concrete with addition of different lengths of fibers and with graded fiber with a fixed amount of fiber.M30 grade concrete with 1.5% weight of fiber is considered in this present investigation. The results shows that the mechanical strength of concrete increases with addition of fibers, decreases with long length of fibers with same volume fraction of fiber and increases with graded fiber. Workability decreases with addition of fiber, with higher length of fiber. There is enhancement of workability with graded fiber than same volume fraction of long fibers.

1. Introduction

During stacking of structure, the miniature airs out may up and proliferate after versatile zone. Short length strands or miniature filaments arbitrarily scattered in solid assistance to capture the launch of miniature breaks and upgrade the pre-splitting quality. Mechanical quality of solid ascents with increment in volume of filaments. The expansion of higher volume of strands prompts functional issues, for example, packaging, balling and decrease in usefulness, thus, explores have watched decrease in quality and sturdiness [1-3]. In a given volume if the shorter strands are utilized, closer will be their separating and simpler will be capturing opening of miniature breaks. These strands may at first add to postpone the augmenting the splits yet might be pulled out after the miniature breaks will be changed into full scale breaks. Long strands connect the breaks and improve the post-top misshapening of cement. Combination of short and long fibers provides better result than mono length fibers. The crux of the problem is borrowed from the idea of use of well graded aggregate rather than single graded coarse aggregate [1-13].

The aim of the present investigation is to study the effect of graded fiber on properties of normal strength concrete. M 30 grade conventional concrete is considered for this investigation. Mono glass fibers with length of 6 mm, 12 mm and 20 mm are considered for this investigation. Graded glass fiber was

IOP Conf. Series: Materials Science and Engineering **970** (2020) 012020 doi:10.1088/1757-899X/970/1/012020

produced from these three lengths of fibers. The opportunity of the exertion is restricted to fiber dosage of 1.5 % weight of cement. The experimental results were compared with theoretical results.

2. Experimental Program

In this phase materials and martial properties, casting procedure, testing and experimental results are presented.

2.1. Material and Material Properties

In this section, materials used for casting of cubes (150 x 150 x 150) mm^3, cylinders (150 mm dia. 300 mm height) and prisms (150 x 150 x 150) mm^3 are described. Properties of materials such as cement, gravel, sand, water and glass fiber are presented in the following sections.

2.1.1. Cement

53 grade Ordinary Portland cement (OPC) is used in the entire research program. Normal consistency was determined as 36 % while specific gravity of the 53 grade cement was 3.12. Initial setting and final setting time from the test as per [8] was found 92 minutes and 350 minutes respectively. The 28 days compressive strength of cement was 56.2 N/mm^2.

2.1.2. Sand

River sand is used as fine aggregate. Sand properties confirms to [9]. The properties are mentioned in Table 1 along with coarse aggregate.

2.1.3. Gravel

Coarse aggregate such as gravel for the present investigation are taken from local area. Gravel is washed and dried before it is used for casting. Max. size of the coarse aggregate was 20 mm.

2.1.4. Water

Portable water freshly collected is used. It is colourless and odourless as [10].

Table 1. Properties of Fine and Coarse Aggregate

Properties	Fine Aggregate	Coarse aggregate
Specific gravity	2.61	2.7
Bulk density	1600 kg/m^3	1450 kg/m^3
Fineness modulus	2.70	--
Zone	Zone-III	--
Impact Value	--	6.82%
Crushing Value		53.5%
Water absorption	--	0.5%

Table 2. Properties of Alkali resistant glass fiber

Properties of Glass fiber	Values
Tensile strength	1700 N/mm^2
Young's modulus	73 GPa
specific gravity	2.6
Specific surface Area	105 m^2/kg
Diameter	14 m
Dosage of fiber	1.5% weight of cement

2.1.5. Glass fiber

Single type of glass fiber was used in this study with different lengths. Glass fiber was used in this program are having properties as mentioned below. Figure 1 shows the glass fiber.

Figure 1. Alkali Resistant Glass fiber

2.2. Mix Design

Normal Conventional Concrete (NCC) of M30 grade was designed by using IS code - 10262:2019 [11] method and adding different sorts of fibre such as 6 mm, 12 mm, 20 mm with a fixed percentage i,e. 1.5 % weight of cement , the optimum percentage of glass fiber at which previous researchers have been found more strength as compared to normal concrete. The slump test and compaction factor test is carried out on the fresh concrete properties. The strength properties such as compressive strengths of normal conventional concrete and addition of 1.5 % of glass fiber with different lengths such as 6mm, 12mm and 20mm of the total weight of material of concrete were studied. The mix was prepared with ratio of 1: 1.1: 2.5. For this water cement ratio was taken 0.40. The control mix was 100 % concrete and 1.5 % glass fiber and variation of glass fiber length 6 mm, 12 mm and 20 mm.Here 0G represents glass fiber is nil, 12G represents only 12 mm glass fiber while GG represents graded glass fiber which is combination of 6 mm, 12 mm and 20 mm glass fiber.

Table 3. Different Specimens cast for investigation

Mix Designation	0G	6G	12G	20G	GG
Cement (%)	100	100	100	100	100
Glass fiber (%) (with weight of cement)	0	1.5	1.5	1.5	1.5
Glass fiber length (mm)	0	6	12	20	6 mm-0.5% 12 mm-0.5% 20 mm-0.5%
Aspect Ratio (L/d ratio)		428.57	857.14	1428.57	428.57-(wight-0.5% of cement) 857.14-(wight-0.5% of cement) 1428.57-(wight-0.5% of cement)

2.3. Experimental Program

For finding out the compressive, flexural and split tensile strength of glass fiber concrete, six cubes, six cylinders and six number of prisms respectively of size mentioned earlier were prepared. Here concrete was first prepared and taken in a tray where the fibers were added uniformly. Workability test was conducted and then specimens were cast. All specimens were cured in water at room temperature till taken for testing.

Table 4. Experimental Values

Test Values	Designation				
	0G	6G	12G	20G	GG
Slump (mm)	85	55	50	40	50
7 Days Comp. Strength (N/mm^2)	28.17	35.18	34.62	31.6	34.2
28 Days Comp. Strength (N/mm^2)	40.10	50.12	49.32	45.00	48.75
7 Days Split Tensile. Strength (N/mm^2)	3.56	4.03	3.91	3.56	3.9
28 Days Split Tensile. Strength (N/mm^2)	4.39	5.16	4.98	4.71	5.00
7 Days Flexural Strength (N/mm^2)	3.76	4.94	4.78	4.33	4.78
28 Days Flexural Strength (N/mm^2)	5.17	6.50	6.36	5.76	6.27

3. Interpretation of Test Results

This section deals with interpretation of results obtained on fresh and hardened concrete. On fresh properties workability test was conducted. In hardened state compressive, split tensile and flexural test was conducted. The test results presented in Table 4 was compared with the values of different codes.

3.1. Workability

In terms of slump, all mixtures exhibited satisfactory slump flows in the range of 85-40 mm, which is an indication of a good deformability. There was decrease in slump value with addition of fibers and there was reduction in slump value with long fibers tan short fibers with same fiber content. The reason may be that the long fibers might be distributed uniformly in concrete matrix and arresting concrete matrix against flowing. Figure 2 shows the slump value of different mixes and percentage of reduction of slump with respect to control specimen G0. The graded fiber (combination of 6 mm, 12 mm and 20 mm fiber) showed better slump value. 0G, 6G, 12G, 20G and GG showed 35.29 %, 41.18 %, 52.94 % and 41.18 % decrease in slump value than the control specimen G0.

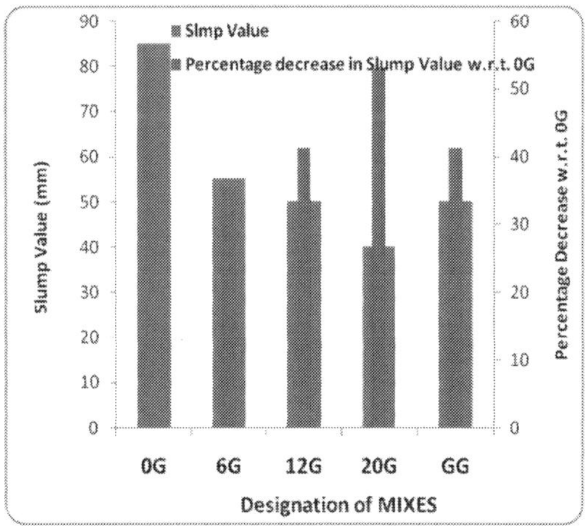

Figure 2. Slump Value of different mixes and percentage decrease in slump value w.r.t. 0G

3.2. Compressive strength

Compressive strength of concrete specimen prepared with 0G, 6G, 12G, 20G and GG was found to be 28.17 MPa, 35.18 MPa, 34.62 MPa, 31.6 MPa and 34.2 MPa in 7 days and 40.10 MPa, 50.12 MPa,

49.32 MPa, 45.00 and 48.75 MPa in 28 days respectively. The same is plotted in Figure 3. Figure 4 also represents the compressive value captured in computerized UTM during testing.

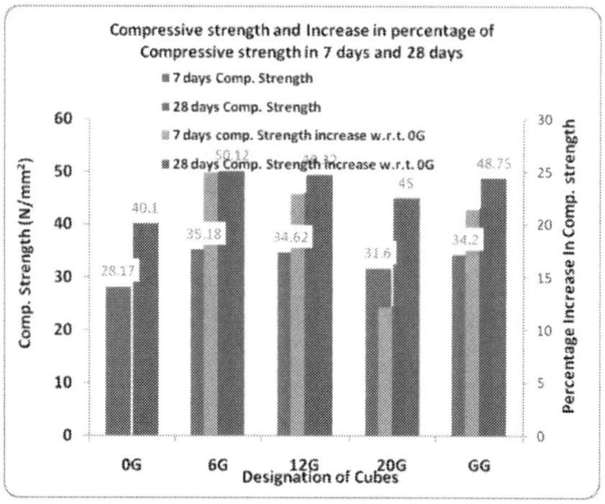

Figure 3. Compressive strength Value of different mixes and percentage increase in compressive strength value w.r.t. 0G

It is found that there was enhancement of compressive strength with smaller aspect ratio and strength decreased with higher aspect ratio with same weight of fiber content. The reason may be large amount of fibers in low aspect ratio available in arresting micro cracks than less number fibers in higher aspect ratio in same volume fraction of fiber content. 6G, 12G, 20G and GG with a volume fraction of 1.5 % weight of cement exhibited an enhancement of 24.88 %, 22.90 %, 12.18 % and 21.41 % compressive strength over the control specimen 0G in 7 days respectively. The same was found to be increased 24.99 %, 22.99 %, 12.22 % and 21.57 % over the control specimen in 28 days respectively for 6mm, 12 mm, 20 mm and graded fiber of combination of above three fibers with a volume fraction of 1.5 %. Graded fibers exhibit better compressive strength than long length fibers of same fiber content.

Test Results

Peak Load(kN)	633.8
Compressive Strength (N / Sq-mm)	27.556
Disp at Max Load(mm)	28.17
E-Mod (N/Sq mm)	0
Poission Ratio	0

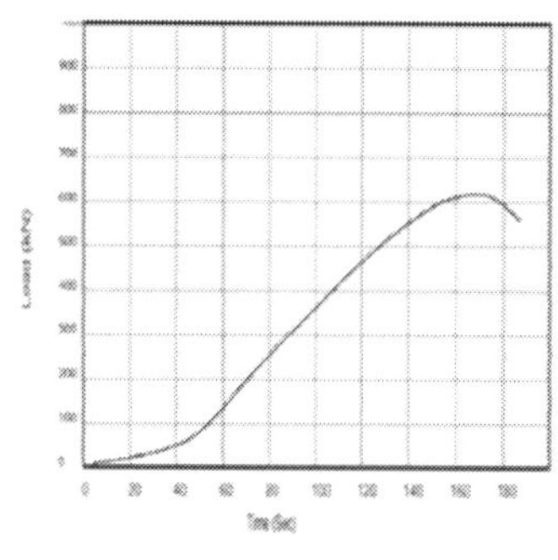

Figure 4. Cube test of specimen 0G

3.3. Split Tensile Strength

The split tensile strength of concrete specimen without addition of fiber and with fiber of 1.5% of 6 mm, 12 mm ,20 mm and a combination of all fibers 0G, 6G, 12G, 20G and GG was found to be 3.56 MPa and 4.03 MPa, 3.91 MPa, 3.82 MPa and 3.90 MPa in 7 days respectively. The same for all these specimens were found to be 4.39 MPa, 5.16 MPa, 4.98 MPa, 4.71 MPa and 5.00 MPa in 28 days respectively.

The split tensile strength of specimen 0G, 6G, 12G, 20G and GG in 7days and 28 days was presented in Table 5. The 7 days and 28 days test value along with different theoretical values calculated by codes is plotted against the percentage of slag in the Figure 5.
Here it is found that Huste et al. predicts closely to the test values while the others overestimate. The increase in the spilt tensile strength of 6G, 12G, 20G and GG was found to be 13.24%, 10.05 %, 7.40 % and 9.73 % over control specimen 0G respectively for 7 days split tensile strength. The increase in the spilt tensile strength of 6G, 12G, 20G and GG was found to be 17.48 %, 13.39 %, 7.31 % and 13.76 % over control specimen 0G respectively for 28 days split tensile strength and Plotted in Figure 6.

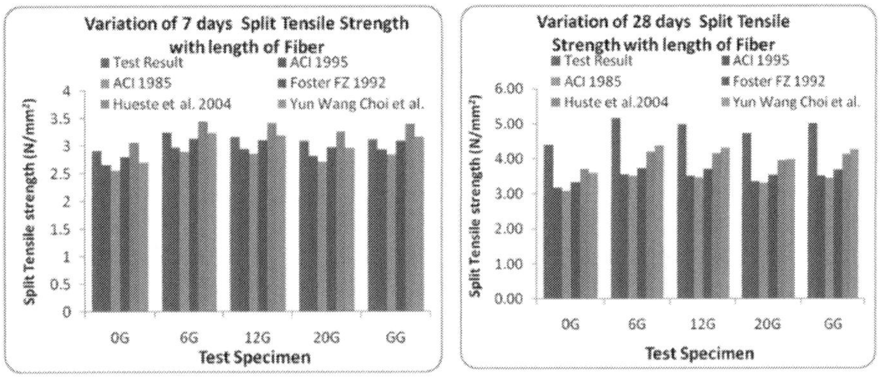

Figure 5. 7 days and 28 days Split Tensile strength vs different length of Fibers

The maximum rise in strength was found to be in 6mm fiber. This may be due to the fact that more numbers of fibers are found in concrete in smaller length of fibers than long fibers in a particular volume fraction of fiber. Again long fibers may be bundled during mixing. Graded fiber exhibit more strength than long fibers.

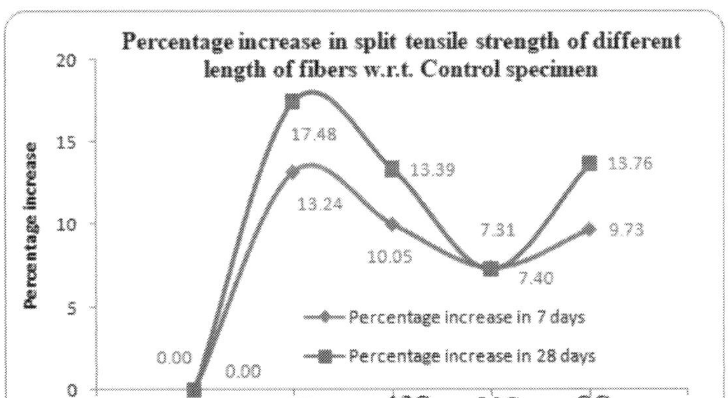

Figure 6. Percentage increase in Split Tensile strength in 7days and 28 days vs different length of Fibers w.r.t. control specimen

3.4. Flexural Strength

7 days Flexural strength of concrete specimens of 0G, 6G, 12G, 20G and GG was found to be 3.76 MPa, 4.94 MPa, 4.78 MPa, 4.33 MPa and 4.78 MPa respectively. The same for 28 days was found to be 5.17 MPa, 6.50 MPa, 6.36 MPa, 5.76 MPa and 6.27 MPa respectively for specimens of 0G, 6G, 12G, 20G and GG as shown in Figure 7.

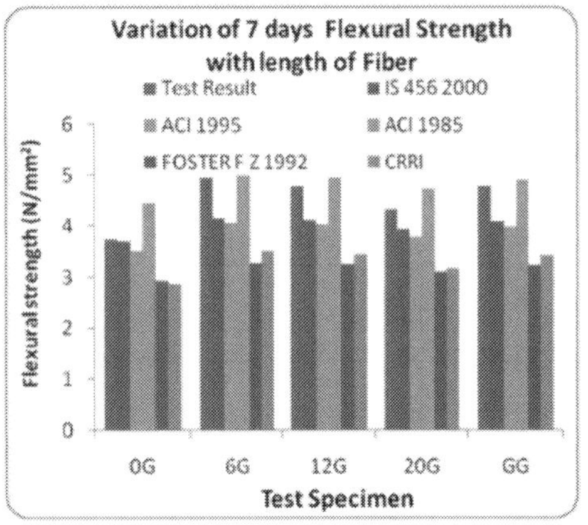

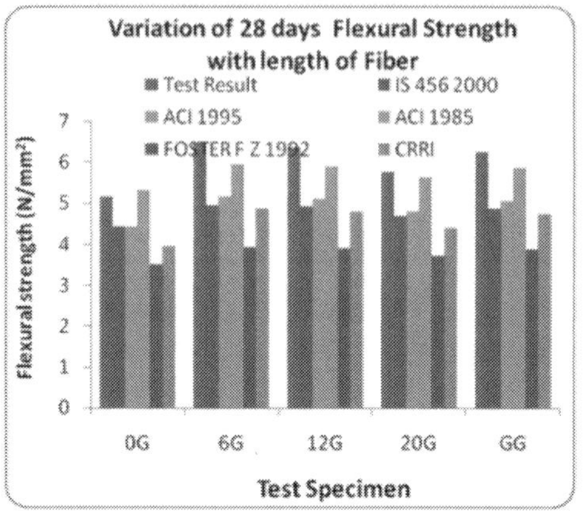

Figure 7. 7 days and 28 days Flexural strength of different length of Fibers

The figures show that ACI-1985 approximately predicts the flexural strength. The increase in flexural strength of 6 mm fiber 6G with respect to control specimen 0G was found to be 31.55 % in 7 days strength and the same was found to be 25.64 % for 28 days strength. For 12 mm fiber and 20 mm fiber the value was found to be reduced. The reason has already been described in previous split tensile strength. The percentage of increase in 7 days and 28 days flexural strength of 6G, 12G, 20G and GG over 0G were presented in Figure 8.

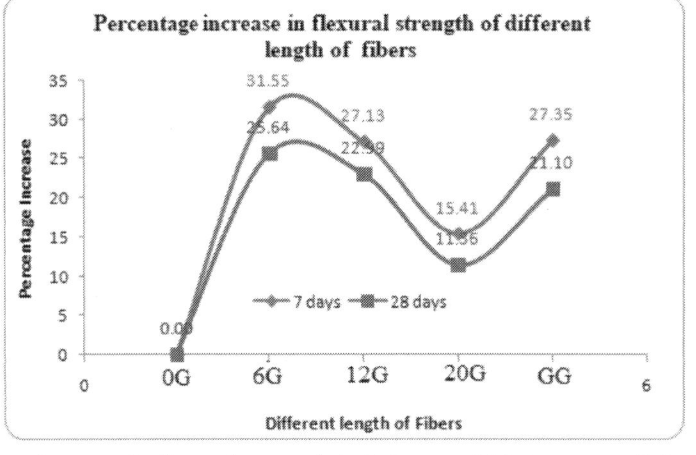

Figure 8. Percentage increase in flexural strength in 7 days and 28 days over different length of Fibers

4. Conclusions

The tests were performed to determine the fresh and mechanical properties of fiber added concrete. From the test results and theoretical values following conclusions were drawn.

- There is decrease in workability of concrete with mixing of glass fiber. Higher aspect ratio glass fiber added concrete has lower workability value than lower aspect ratio glass fiber for same amount of fiber.
- There is enhancement of workability for graded fiber.
- Slump value of glass fiber concrete with fiber aspect ratio 1428.57 decreases 52.94 % than that of concrete without fiber. The same was found to be 41.18 % for graded fiber.
- Mechanical properties of concrete increases with addition of glass fiber.

- Lower aspect ratio fibers are able to increase strength more than higher aspect ratio fibers for same amount of fiber in concrete. Increase in strength for graded fiber is more than fibers with more length.
- Graded fiber has an enhancement of compressive strength 21.5 7 % over zero fiber content.
- Maximum increase in split tensile strength is noticed as 17.48 % with aspect ratio 428.57 over concrete without fiber.
- ACI-1985 closely predicts flexural strength.

References

[1] Alhozaimy A, Soroushian P and Mirza F 1996 Mechanical properties of polypropylene fiber reinforced concrete and the effects of pozzolanic materials *Cement and Concrete Composites* **18** 85-92

[2] Khaloo A, Raisi E M, Hosseini P and Tahsiri H 2014 Mechanical performance of self-compacting concrete reinforced with steel fibers *Construction and Building Materials* **51** 179-86

[3] Kasagani H and Rao C 2018 Effect of graded fibers on stress strain behaviour of Glass Fiber Reinforced Concrete in tension *Construction and Building Materials* **183** 592-604

[4] Kasagani H and Rao C 2016 The influence of hybrid glass fibres addition on stress–Strain behaviour of concrete *Cement, Wapno, Beton* **2016** 361-72

[5] Gupta S, Aakash A R, Arora J and Saini A 2017 Partial Replacement of Cement by Flyash & Glass Fibre in Light Weight Fibre Reinforced Concrete *International Journal Of Engineering Research and Technology* **6** 24-7

[6] Kumar D N and Ramujee K 2017 Strength Properties of Binary Blended Geopolymer Concrete Incorporated with Polypropylene and Basalt Fibers *i-Manager's Journal on Civil Engineering* **8** 15

[7] Chandramouli K, Srinivasa R, Pannirselvam N, Seshadri S and Sravana P 2010 Strength properties of glass fiber concrete *ARPN journal of Engineering and Applied sciences* **5** 1-6

[8] 1987 IS 12269:1987: Specification for 53 Grade ordinary Portland cement'. Bureau of Indian Standard, ManakBhavan, Bahadurshah Zafar Marg, New Delhi, India

[9] 1970 IS 383:1970: Specifications for Coarse and Fine Aggregates from Natural Sources for Concrete (Second Revision), Bureau of Indian Standard, ManakBhavan, Bahadurshah Zafar Marg, New Delhi, India

[10] 2000 IS 456:2000: Code of Practice for Plain and Reinforced Concrete Structures, Bureau of Indian Standard, ManakBhavan, Bahadurshah Zafar Marg, New Delhi, India

[11] 2019 IS 10262:2019: Concrete Mix Proportioning — Guidelines, 2nd Revision, Bureau of Indian Standard, ManakBhavan, Bahadurshah Zafar Marg, New Delhi, India

[12] Mansur M and Islam M 2002 Interpretation of concrete strength for nonstandard specimens *Journal of Materials in Civil Engineering* **14** 151-5

[13] Rashid M, Mansur M and Paramasivam P 2002 Correlations between mechanical properties of high-strength concrete *Journal of Materials in Civil Engineering* **14** 230-8

IOP Conf. Series: Materials Science and Engineering **970** (2020) 012021 doi:10.1088/1757-899X/970/1/012021

Influence of Natural Fibers as Additive on Characteristics of Stone Mastic Asphalt

Pragnya Parimita

Department of Civil Engineering, Silicon Institute of Technology, Sambalpur, Odisha, India

E-mail: payalpragnya@gmail.com

Abstract. The purpose of this article is to emphasize on the development of stabilized Stone Mastic Asphalt Mix for laudable and tangible pavement performance. Attempts have been made to stabilize SMA Mix using synthetic fibers which is too costly and not readily available. The concept of using natural fibers and waste materials to replace these energy intensive synthetic fibers or polymer additives is a recent development in this field. India, being an agricultural economy, produces fairly huge quantity of natural fibers such as coconut, sisal, banana, sugar cane, jute etc. The addition of these natural fibers as additives to SMA has really thrust a revolutionary change to the transport system. This research is an attempt to propose an ideal surface course for the pavement of India using natural fibers stabilized SMA that can face rough weather conditions and more loads. Analyzing the illustrated results of the comparative study on the performance of the stabilized mixtures it is inferred that additive plays a prominent role in the volumetric, mechanical and drain down characteristics of SMA. Improved Marshall Quotient and Marshall Stability value show higher rutting resistance and improved tensile strength shows better Cracking Resistance. This study brings out the importance of additives in SMA and proposes an eco-friendly alternative to Synthetic fibers.

1. Introduction

The development of economic scenario of a nation relies on the strength of transportation, hence strength of roads gains paramount importance. The ruts and its repeated mending weaken the infrastructure of national economy. In spite of multifarious steps to strengthen pavements, ruts continue, so also national loss and this gives birth to innovative thoughts to check this decay. SMA is one of the acceptable and successful options for the durability and load bearing capacity, for its rut resistance and water resistance. It flourished in Germany during 1960s but it gained acceptance in all countries for its effectiveness. SMA Mix faces drain down problem for laxity of binder and matrix; hence, innovative efforts continue to check the drain down by the use of additives as stabilizing agents. The addition of fibres in the required proportion works as better interlock between aggregates. The flexible pavement committee has announced the authenticity of SMA in Indian conditions, but it is realized that the imported fiber used as stabilizer to the aggregates leads to an additional cost of construction. For sometimes, it was thought of using waste tyres, carpet, waste plastic bags and bottles as stabilizing agent but its processing cost appeared to be high, economic burden in acquisition of fiber continues. Therefore it is necessary to think of a fibre that can be obtained in plenty but at low cost. [7-12].

In the context of our country, it is marked that natural fibres like jute, sisal, banana and coconut are accumulated from varied sources .It is rightly felt that the cost of construction can be minimized by the application of such fibres. However, out of these, banana and coconut fibres are of better choice for their plenty production in one hand and easy collection from the waste in the other. The huge production of coconut in the coastal area and the large-scale banana plantation in all parts of our country show an easy way of collecting fibres from the waste i.e. from coconut coir and banana stem [4-7].

2. Materials

Proper material, suitable mixing and right proportion play prominent role in the SMA. Properly graded aggregates of good quality sized between 0.075mm and 19mm mixed with sufficient bitumen provide an elastic quality. The right proportion of bitumen, coarse aggregates, fillers and fine aggregates makes the bituminous mix strong.

The function of bituminous surfacing is largely determined by the quality and proportion of materials. SMA mixtures are prepared by using aggregates of good quality, bituminous binder, stabilizer and mineral fillers. In SMA stone contact is important as it creates a gap graded skeleton like stone structure. The voids of the structural matrix are filled by high viscosity bituminous mastic. 70% stone on stone contact is ensured after compaction of the mix. Mastic stiffness of the required proportion is obtained for the use of crushed sand in the mix, Along with indigenous materials of western Odisha such as aggregates, bituminous binder, banana fibres and coir are used as additives to attain stabilization.

2.1. Aggregates

The strength and stiffness of SMA along with rut resistance largely depend on the size and quality of aggregates. The ministry of road transport and highways (MORTH) [9], National Cooperative Highway Research Project (NCHRP) and National Asphalt Pavement Association (NAPA) have determined the nominal maximum aggregate size (NMAS) to be 19mm, 13mm and 10 mm respectively for the better results of SMA mixture. For the purpose of study, coarse aggregates of 13mm nominal size have been used. In Table 1, the gradation of the aggregates required as per specification in MORTH for stone mastic asphalt has been reflected. Qualities of the aggregates were checked by performing various tests as per IRC: SP: 79 - 2008 specifications shown in Table 2.

Table 1. Aggregates Gradation Used

Size of sieve (mm)	Cumulative percentage passing	Gradation Used	percentage retained	Weight per sample (gm)
19	100	100	0	0
13.2	90-100	95	5	60
9.5	50-75	65	30	360
4.75	20-28	25	40	480
2.36	16-24	20	5	60
1.18	13-21	17	3	36
0.6	12-18	15	3	36
0.3	10-20	14	1	12
0.075	8-12	10	4	48
<0.075	--	--	9	108

Table 2. Physical Characteristics of Aggregates used

Property	Test Results	Standard Range
Impact value (%)	13.22%	27% maximum
Crushing Value (%)	17.65%	30% maximum
Los Angeles Abrasion Value (%)	16.24%	35% maximum
Water Absorption (%)	0.8%	1.3% maximum
Combined elongation and flakiness index (%)	32.29%	35% maximum
Specific Gravity	2.67	3 maximum

2.2. Bitumen

Bitumen has its special place and role in the SMA Mix as it binds polymers, aggregates, fillers and additives. A slightly higher amount of bitumen is used as compared to conventional mixes for better performance. As bitumen is viscoelastic in nature climatic condition, time of loading and maximum temperature of that locality ascertain the grade of bitumen used for construction and keeping this in view VG30 grade of bitumen has been used for the SMA samples.

Table 3. Physical Characteristics of bitumen used

Properties	Test Results	Standard Range
Penetration value ($1/10^{th}$ of mm)	65.2	50-70
Ductility value(cm)	89.8	Minimum 40
Softening Point (℃)	51.5	Minimum 47
Specific gravity	1.01	0.97-1.02

2.3. Filler

In SMA Mix, fillers are used to improve the strength and quality of the mixture. Rock dust, slag dust, hydraulic cement, fly ash and hydraulic lime are added as fillers. Besides providing stiffness to the mix, the fillers help to check rutting. Further, the longevity of the mix is increased as fillers lower the drain down of binders. For adorable performance proportion of filler material passing through 0.075mm sieve should be greater than 95%. Increasing filler amount more than required limit leads to crack formation in the mix. On the other hand if filler amount is reduced it decreases the resistance against rutting. In the preparation of samples, the fillers having specific gravity of 2.87 have been used.

2.4. Fibers

For confirmation of stability in the SMA Mix, addition of fibre as an additive is essential. Fibres collected from manmade materials like waste tyre and carpet are found less efficacious, less eco-friendly than that of fibres collected from natural sources. Collection of natural fibres is also less expensive as it is obtained from the waste. Cellulose fibres are directly collected from jute, cotton, flex and ramie, but manmade fibres are collected and used as consumer or industrial products like rayon and acetate etc. Natural fibres collected from banana stem and waste part of ripe coconut have been used for study purpose for their strong intermolecular forces, high linearity of cellulose molecules that attach crystalline nature to fibres collected from rich natural sources. Coconut fiber is more effective in the SMA Mix for its easy availability, higher cellulose and lignin content that provide extra water resistance as well as stability. Coir fiber with specific gravity value of 1.47 has been used for sample preparation.

Banana fiber as an additive in the SMA Mix has its own merit for its profusion, high tensile strength, and high resistance to wear and tear. Its specific gravity is 1.33 which is used for sample preparation.

2.5. Optimum Bitumen Content

It is marked that without fibre the optimum bitumen content for SMA is 6.36%, almost 5.95% of the total mix by weight. But after adding 0.3% of coconut fibre the optimum bitumen content goes up to 6.46% which is 6.06% by weight of the total mix. But in case of banana fibre the estimate becomes different. With adding of 0.3% of banana fibre the optimum bitumen content of the mix becomes 6.36% which is 6.03% by weight of the total mix. All these values have been obtained by calculating the average value of bitumen content corresponding to 4% air voids, maximum density and maximum stability.

3. Experimental Procedure

3.1. Marshall Test

Marshall Stability Test was conducted as per ASTM D 6927 - 06 specifications and procedure. For the measurement of plastic deformation resistance of the cylindrical specimen after compaction, loading rate of 51 mm per minute has been applied to a specimen along its diameter. In the Marshall Method Density-void analysis and Stability-flow tests are two major features. Marshall Stability value is the maximum load sustained at a standard test temperature of 60^0 C by the test specimen. In terms of one fourth of mm unit, the deformation of specimen is evaluated due to maximum loading which is termed as flow value. This test helps to know the optimum bitumen content required for the gradation of aggregates used and desired traffic.

3.2. Test Specimen preparation

- To achieve specified gradation for SMA as per MORTH required amount of aggregates of different sizes were accumulated and mixed.
- With weight of 1200 gm. 3 samples of SMA were made ready for different cases by increasing the bitumen content by 0.5% from 5 to 7 %. with and without fibre.
- To achieve sufficient fluidity for mixing both aggregates and bitumen were heated with proper observation so as to avoid overheating.
- All materials were mixed at stipulated temperature to get homogeneous SMA Mix.
- Using a standard hammer 50 blows were applied to the prepared mix so placed on specified mould.
- After removing the samples from the mould and bringing the mix to surrounding temperature, the samples were analyzed.

Parameters calculated were Specific Gravity of the mix *(Gt)*, Bulk specific gravity *(Gm)* of the mix, Percentage of Air voids *(Vv)* in the mix, Percentage volume of bitumen *(Vb)* in the mix, Voids in mineral aggregate *(VMA)* and Voids filled with bitumen *(VFB)*.

4. Results and Discussions

Table 4. Marshall Test Values for samples without fiber

Sample No.	1	2	3	4	5
Binder content (%)	5	5.5	6	6.5	7
Bulk density (gm./cc)	2.330	2.332	2.333	2.335	2.331
% Air voids	5.3	4.71	.04	3.37	3.03
% VFB	67.5	71.76	76.27	80.56	83.01
Stability (kg)	790.26	849.91	973.179	1064.01	856.81
Flow(mm)	2.31	2.53	2.92	.23	3.81

IOP Conf. Series: Materials Science and Engineering **970** (2020) 012021 doi:10.1088/1757-899X/970/1/012021

Table 5. Marshall Parameters of Samples with 0.3% Of Coconut fiber

Sample no	1	2	3	4	5
Binder content (%)	5	5.5	6	6.5	7
Fiber content(%)	0.3	0.3	0.3	0.3	0.3
Stability (kg)	934.25	1012.1	1141.86	1291.43	1105.16
Flow (mm)	2.16	2.38	2.8	3.07	3.63
Marshall Quotient	432.52	425.25	407.8	420.66	304.45
Bulk Density(gm./cc)	2.319	2.321	2.323	2.324	2.320
Air voids (%)	5.63	4.91	4.25	3.58	3.07
Vb (%)	10.91	11.95	12.89	13.938	14.96
VMA(%)	16.54	16.86	17.14	17.518	18.03
VFB(%)	65.96	70.87	75.2	79.56	82.97

Table 6. Marshall Parameters of Samples with 0.3% Of Banana fibers

Sample no	1	2	3	4	5
Binder content (%)	5	5.5	6	6.5	7
Fiber content(%)	0.3	0.3	0.3	0.3	0.3
Stability (kg)	901.81	976.6	1115.91	1229.34	1080.33
Flow (mm)	2.2	2.43	2.84	3.14	3.68
Marshall Quotient	409.91	401.89	392.92	391.5	293.56
Bulk Density(gm./cc)	2.317	2.319	2.322	2.323	2.319
Air voids (%)	5.49	4.88	4.24	3.48	3.01
Vb (%)	11.04	12.0	12.89	14.07	15.09
VMA(%)	16.53	16.88	17.13	17.55	18.1
VFB(%)	66.78	71.0	75.24	80.17	83.37

4.1 Graphs

Graphs were plotted based on the experimental observations and referring those, the Optimum Binder Content (OBC) of the stabilized Marshall Mix for 0.3% of fibers (OFC) were obtained for each of the case.

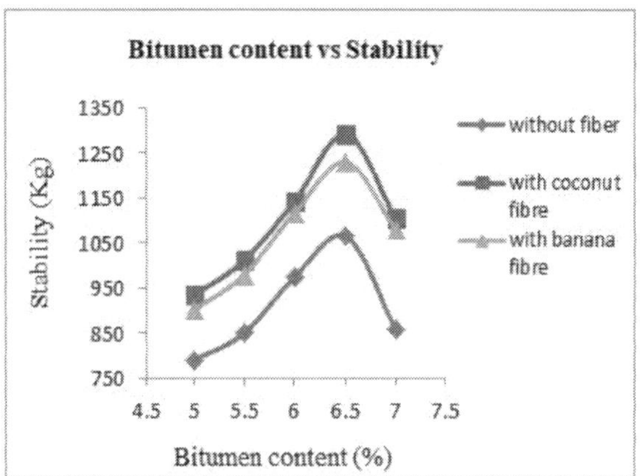

Figure 1. Percentage bitumen vs. Stability.

Observing all the above test results, Figure 1 and recorded tables for samples with or without fibres, it is marked that with rise in bitumen content the stability value goes on increasing to a certain level, and then falls for all the cases. By using 0.3% coconut fibre and 6.5% bitumen content, the maximum stability value found was 1291.43 kg and using 0.3% banana fiber and 6.5% bitumen content maximum stability obtained was 1229.34 kg. On analysis it is found that mix with fiber possess more stability value than mix without fiber and sample containing coconut fibre is the most stable one.

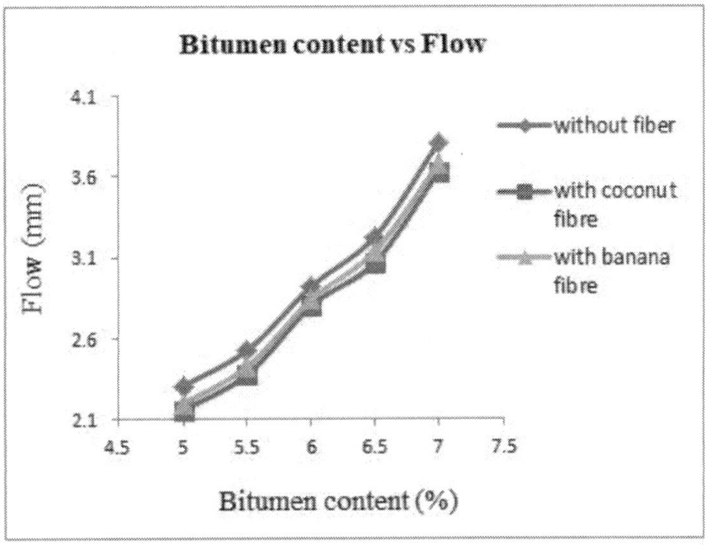

Figure 2. Percentage bitumen vs Flow value.

Figure 2 shows that increasing bitumen content leads to increase in flow value in all cases but the rate of increase is higher at higher bitumen content. Samples with fiber possess lower flow value than samples without fiber.

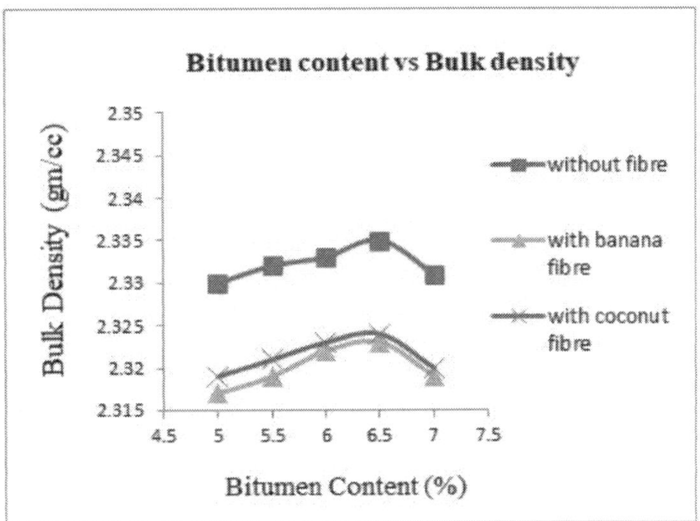

Figure 3. Percentage bitumen vs. Bulk specific gravity.

Figure 3 shows that the Bulk Density gradually increases up to a limit and then decreases with addition of bitumen. Addition of fiber leads to decrease in Bulk Density.

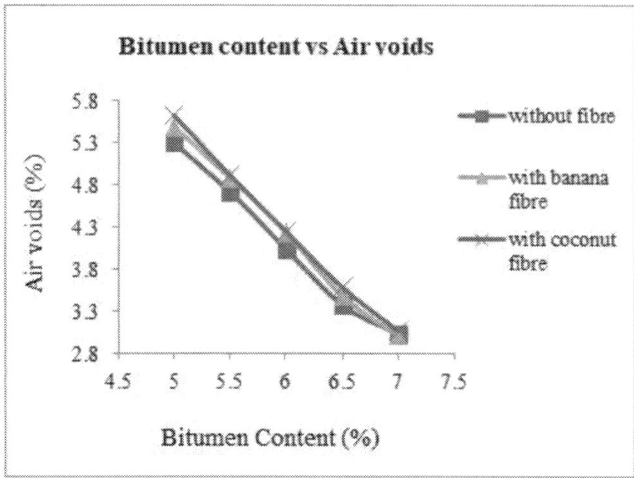

Figure 4. Percentage bitumen vs Air voids.

Figure 4 shows that due to networking effect of the fibre there is an increase in air voids in case of samples containing fibers. For all the cases the bitumen content and the Air voids are inversely proportional.

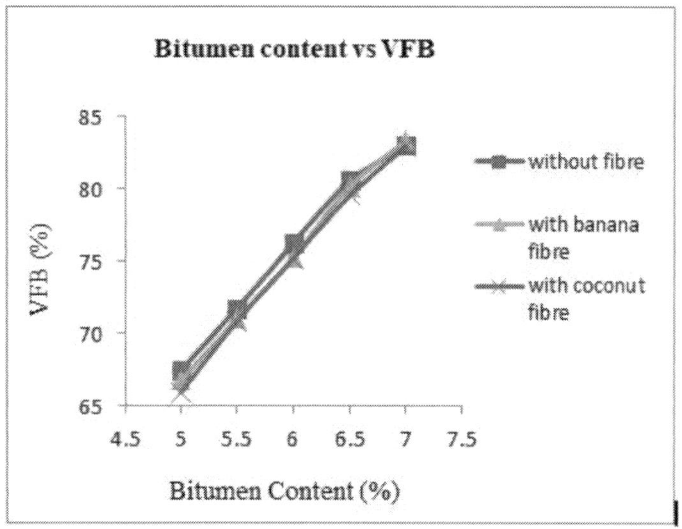

Figure 5. Percentage bitumen vs. VFB.

Figure 5 shows that with rise in binder content, the voids filled with bitumen goes on increasing and VFB is less for stabilized sample.

4.2. Optimum Bitumen Content

Observation shows that optimum binder content for SMA without fibre is 6.36% by weight of aggregate which is 5.95% by weight of the total mix, but it becomes 6.06% of total weight of mix when 0.3% of coconut fibre is added or the OBC becomes 6.46% by weight of aggregate. But for addition of 0.3% of banana fibre the OBC becomes 6.43%. All these values have been obtained by calculating the average value of bitumen content corresponding to 4% air voids, maximum density and maximum stability.

5. Conclusion

After proper analysis, it is concluded that bitumen pavement receives extra strength for use of well graded aggregates, shortage of such aggregate affects the paving industries. But use of coconut fibre can be taken as a solution to this shortage, at the same time its easy availability makes the process easy. Banana fibre can also be used with same purpose. From Marshall test it is noticed that 0.3% of fibre can bring adequate improvement in Marshall characteristics.
Conducting analytical study on the use of gap graded materials like SMA in Bituminous paving, it is concluded that adding coconut fiber in the mixture as a stabilizer to check the possibility of drain down of Bituminous in one hand and supplementing strength to the mixture in the other, is more efficacious in all respect. Both banana fiber and coconut fiber are easily acquired from the waste in profusion but on comparison after use, the experiment result shows that coconut fiber stabilized SMA has higher specific gravity and higher tensile strength. Though it is deciphered that whatever is the fiber the maximum Marshall Quotient, the maximum value of stability, and bulk specific gravity of SMA mixture is found at 0.3% fiber content, even the flow values of the mixture are located within the specified range of 2 to 4 mm, still the coconut fiber stabilized mixture is better as it provides a higher Marshall stability, hence its use is feasible and preferable in Indian context and condition.
Further studies are required to test the validity of the additives in the bituminous mixtures along with its effect on fatigue behavior and rutting characteristics.

References

[1] Asasutjarit C, Hirunlabh J, Khedari J, Charoenvai S, Zeghmati B and Shin U C 2007 Development of coconut coir-based lightweight cement board *Construction Building Materials* **21** 277-88
[2] Aziz M, Paramasivam P and Lee S 1981 Prospects for natural fibre reinforced concretes in construction *International Journal of Cement Composites Lightweight Concrete* **3** 123-32
[3] Bose S, Kamraj C and Nanda P 2006 Stone Matrix Asphalt (SMA)-A Long Life Pavement Surface
[4] Cook D, Pama R and Weerasingle H 1978 Coir fibre reinforced cement as a low cost roofing material *Building Environmental Technology* **13** 193-8
[5] Li Z, Wang L and Wang X 2006 Flexural characteristics of coir fiber reinforced cementitious composites *Fibers Polymers* **7** 286-94
[6] Ali M, Liu A, Sou H and Chouw N 2012 Mechanical and dynamic properties of coconut fibre reinforced concrete *Construction Building Materials* **30** 814-25
[7] Ali M 2011 Coconut fibre: A versatile material and its applications in engineering *Journal of Civil Engineering Construction Technology* **2** 189-97
[8] Ali M, Li X and Chouw N 2013 Experimental investigations on bond strength between coconut fibre and concrete *Materials Design* **44** 596-605
[9] https://skmobi.files.wordpress.com/2017/04/morth-specifications-for-road-bridge-works-5th-revision-by-sk.pdf
[10] Paramasivam P, Nathan G and Gupta N D 1984 Coconut fibre reinforced corrugated slabs *International Journal of Cement Composites Lightweight Concrete* **6** 19-27
[11] Reis J 2006 Fracture and flexural characterization of natural fiber-reinforced polymer concrete *Construction building materials* **20** 673-8
[12] Subramani T 2012 Experimental Investigations on Coir fibre reinforced bituminous mixes *Strain* **2** 4-5

IOP Conf. Series: Materials Science and Engineering **970** (2020) 012022 doi:10.1088/1757-899X/970/1/012022

Modification of Bitumen Using Sustainable Materials for Pavement Design

Hemanta Kumar Behera, Debabrata Giri, Sudhanshu Sekhar Das

Department of Civil Engineering, Veer Surendra Sai University of Technology, Burla, Odisha, India

E-mail: beherahemanta5@gmail.com

Abstract. Construction materials will be in the extinction stage in the future. Use of the waste materials to ensure environmental protection is the need of hour. For the construction of essential infrastructure like flexible pavement requires bitumen. Bitumen obtained as a petroleum refinery by-product used for construction of flexible pavement is on the verge of extinction, since natural oil petroleum is a non-renewable resource. Again low life of bitumen due to oxidation is another problem. Hence, it is important to use potential modifiers to partially substitute and enhance the life of bitumen and flexible pavements. In this paper use of numerous modifiers and their possible effects on bitumen and bituminous mix strength and toughness are described by researchers. Different modifiers such as polypropylene, Nano montmorillonite, Nano-silica, Nano-clay, Low-Density Polyethylene (LDPE), Styrene Butadiene Styrene (SBS), High-Density Poly-Ethylene (HDPE), Carbon Nano-Tubes (CNT), graphene, Graphene Oxide (GO), fly ash, cloisite, evotherm, Cecabase Rt, Nano fill, rubber waste, crumb rubber etc. were used for modification of bitumen. Binder proportions are approximately similar on the other hand percentages of modifiers used differ mostly by researchers. It is evident from the literature that the optimal proportion of modifier would only yield improved performance. However, the quality of the modifiers shows a key part, to fulfil the requirements of the ideal pavement and increased life carefully chosen based on the soil conditions and environmental circumstances.

1. Introduction

Sustainable construction materials are the necessity of the era. As a replacement for using all available natural resources of construction material like bitumen leads to focus of researchers on modification of the bitumen and evaluation of the strength and durability of modified bitumen an essential area of research. The strength and longevity improvement without environmental effects makes bitumen modification an essential element of material research. The bitumen modification would also bring a significant reduction in costs from construction point of view if partly replacement of bitumen with other enhancing the properties with addition of additives possible. Hence, the purpose of this paper is to elaborate various bitumen-modifiers evaluated by the researchers.

1.1 Need for modification of Bitumen

Yousefi [1] emphasized on the inadequacy of that pure bitumen for modern roads and growing up traffic, forcing the engineers to bitumen modification for enhancing performance throughout service life. The altered binder is steadier and demonstrates less susceptibility to permanent deformation in hot climate under heavy traffic, braking and accelerating forces. It counterattacks the fatigue loads and has an excellent bond between binders and aggregates. Therefore, the material which can change the chemical configuration of bitumen can be a modifier. Modified bitumen has more adhesion at execution temperature than pure bitumen and lower thermal sensitivity in the range of operation temperatures and adequate viscosity. Furthermore, its sensitivity to the loading time is poor, and its resistance to plastic deformation, fatigue, and cryogenic cracks are high. Ultimately, the properties are perfect for efficiency and service after aging. Section 2 of the paper describes the materials used in modification; section 3 demonstrates the literature study for different modifiers used in bitumen followed by conclusion in section 4.

2. Materials used for Modification

2.1. Bitumen
Bitumen is described as a virtually non-volatile at ambient temperatures and softens gradually when heated, with adhesive and water resistant properties originated by means of fractional distillation from crude oil or available in natural asphalt lake, totally or else almost fully soluble in toluene, and very viscous or almost hard at ambient temperatures. This is well known that the original characteristics of bitumen, as well as the characteristics of bitumen crude oil, are highly dependent on its production and processing method. Higher quality crude oils and proper processes of distillation will enhance the bitumen properties. The heavier crude oil typically produces greater bitumen yields. Therefore, it is of utmost importance to have adequate information on the bitumen characteristics from various aspects. This knowledge proves to be more significant as other difficulties such as phase discontinuity, poor dispersion, and instability with polymers/additives make it challenging in the manufacture and application of bituminous materials for certain bitumen applications.

2.2. Polymers
After its use, the polymers are thrown into the motherland as waste which obstructs rainwater infiltration owing to its non-biodegradability. Nearly 70 % of the plastic manufactured is most frequently used as waste, including single use, for instance carrying bags, etc.
From the reports of Central Pollution Control Board, waste plastics are generated 15342 tons/day in India and are 5.6 million tons per annum. Consequently, numerous polymer forms like Low-Density Polyethylene (LDPE), Polypropylene, and High-Density Polyethylene (HDPE) are used as modifiers for bitumen. The application of these modifiers safeguards the bituminous mixture along with the underneath layers by inhibiting water percolation.

2.3. Used/waste Rubbers
Thanks to the additive and fire retardant material present in tires, tires are thrown out in the atmosphere polluting the soil for the use of tires. Those emit methane gases when burned. Overall, fifteen million tons of scrap tires are manufactured annually, of which India contributes one million tons. Thus, we have to use those tires technically for our requirements without making any impact on the environment.

2.4. Clay and Nano clay
Clays are mineral deposits that occur naturally and exhibit element inconsistencies. Naturally obtainable clay is significantly decreased to Nano size to accomplish great surface area in order to adhere to the asphalt and bind together with the aggregates. The Nano clay alteration when applied even in small amounts, yields multi-folds of improvement.

2.5. CNT and Graphene
Carbon nanotubes (CNTs) consist of cylinder-shaped molecules, together with single-layer rolled-up carbon atom (graphene) sheets. This can be single-walled (SWCNT) or multi-walled (MWCNT), with both the SWCNT having a diameter of a lesser amount of 1 nanometer (nm), and MWCNT comprising of several dense nanotubes approaching more than 100 nm in diameter. Its dimension can exceed a few

micrometers or even millimeters. Due to the immense physical properties graphene seems to have become a new super-material. Due to its lower size, super strength with enormous mechanical, thermal, electronic, electrical, and optical properties graphene is actually the most striking nanomaterial. It has high specific surface area, chemical stability, optical transmittance, elasticity, porosity, biocompatibility, adaptable band gap, and simple chemical functionality that basically help to modify its properties. Both of these materials were dispersed with bitumen for improvement of its rheological characteristics, viscosity, softening point, penetration and resistance towards long term as well as short term aging. These two Nano modifiers help the binder to resist moisture susceptibility, which makes the pavement stronger and durable [1-36].

3. The Study of Literature

3.1. Polymers

The plastomeric and elastomeric polymer of both forms can be used in asphalt. Asphalt has many applications for paving, maintaining seals such as filling, patchwork, recycling, and slurry seal. For all of these uses, polymer evolves to blend with bitumen. Prasad K et al. [15] recommended that the binder must first be heated to 260° C-280° C after it should be added to waste plastics, or else the blend will not be accurately mixed. And if this occurs, the binder may likely catch fire before it hits its point of fire and a loss of bitumen weight can also happen. Yazdani et al. [5] added polypropylene, Styrene-Butadiene Styrene (SBS) elastomer, Nanoclay and plastomer. All the compressive strength of changed bitumen and the softening point are improved by 55 %. Among the various percentages tested, it was calculated that 3 % (SBS), 5%polypropylene (PP), and 1.5 % nano-clay (NC) gave a better quality outcome in improving the properties. Vamshi [12] established that a 10 % plastic-modified bitumen stripping value was zero compared to 0.4 % virgin bitumen. There is also a 7.99 % reduction in material costs by using plastic waste on the highway. As compared with regular bitumen, modified bitumen reduces penetration.

Rajasekaran et al. [20] established thermal analysis characteristics using the thermogravimetric examination demonstrates that polymers such as polyethylene, polypropylene and polystyrene are becoming softer from 130 °C – 140 °C lacking gas growth. It is harmless when used below 150 °C with molten waste plastic. The increasing order of strength also for the modified bituminous block is polystyrene, polyethylene, and polypropylene. In their paper, Dixit et al. [6] indicated that a limit of 0.6% of plastic fiber modification was only sufficient. The stability of the Marshal increases by 5.4 % with a binder material. Rokdey et al. [19] reported a 160° C mixture of the waste plastics with hot bitumen. Plastic bitumen alteration improved binder melting point. Thermoplastics, thermosets and elastomers were used as modifiers. They have also learned the bitumen does not leach the polymer. Kazmi et al. [29] has shown that 9 % of LDPE by bitumen weight is an efficient binder ratio. They received a 32.5 % increase in the stability performance relative to sleek bitumen. Void in mineral aggregates was 5 %. Up to 66.7 % of the estimated air voids. In their analysis, Menaria et al. [22] found that 8 % of waste plastic is the optimal quantity to be used to alter bitumen. They suggested plastic roads with high temperature zones around 50°C can also be laid. The binder's optimum value was 4.5 %. The dry method is observed for this kind of work.

Sahu et al. [18] observed the change in bitumen and indicated that polymer has an excellent adhesion characteristic, so there are zero stripping values in conventional bitumen for almost 72 hours against 5 % for 24 hours. For 10 % of plastics, moisture sensitivity of polymer altered bitumen covered aggregates is determined to be 0.12 %; the value for regular bitumen is 4 %. Cheshmehgol et al. [35], 2 to 14 % LDPE specimens prepared. They reported that indirect tensile strength rises by 15.2 % for unconditioned specimens, and is 23 % higher for conditioned specimens. The tensile strength ratio for unconditioned samples is 88 % and 94 % with a 10 % LDPE alteration. Singhal et al. [14] determined that overall material costs are reduced by 7.99 % when bitumen is replaced with waste plastic (6 %), and the Marshall stability value reached is 1161 kg. Once the plastic content has increased, the value of ductility and penetration decreased and the value of softening points increased.

Table 1. Percentage of modifiers used by different authors

Percentage of modifiers	Type of modifier	Author
1,2,3,5,10	Crumb rubber	Amit Kumar Sahu
10	Plastic waste	Avula Vamshi
2,4	Polymer	Ali Jamshidi
4,6	Nano silica	
0.1,0.5,1	Carbon Nano tubes	Ezio Santagata
2,4,6	Nano silica	Farhad zafari Rhi
2,4,7	Cloisite-15A	Ghaffarpour
2,4,7	Nanofil-15	
3,6	Cloisite	Van de ven
6	Nanofil	
3,6	Organic MMT	Jeroen Besamusca
2	Natural rubber	Krishnapriya
0.3, 0.4	Evotherm	Manjunath
0.3, 0.4	Cecabase Rt	
3,7,9	Organic MMT	Mahdi Lamya
1,2,3,5,10	Waste plastic	Minakshi Shingal
2,4,6,8,10,12,14	LDPE	Mehdi Karimi Cheshmehgol
4,6,8	LDPE	Prakash somani
2,4,6	Crumb rubber	
6,8,10	HDPE	Rajeshkumar
5,7,9,11	PE	Rao
2,4,7	Montmorillonite	Saeed Ghaffarpour Jahromi
5,10,20,25	Polyethylene (PE), Polypropylene (PP), Polystyrene (PS)	S Rajasekaran
0.1,0.2,0.3,0.4,0.5,0.6,0.7,0.8,0.9	Plastic bag fibres	Sandhya Dixit
2,4,6,8,10,12	Polyethylene Terapthalate, LDPE, HDPE	Shiva Prasad
10,20,30,40	Waste plastic	Shwetha Rokdey
7	Nanofil-15 Cloisite-15A	S. Ghaffarpour Jahromi
3,6,9,12	Fly ash	Saad Issa sarsam
1,2,3,4	Silica fumes	
1,2,3,4,5,6,7,8,9	Waste polyethylene, waste crumb rubber	Yuqiao Yang
6,8,10,12,14,	PE,PP,PS	Yash Menaria
2, 4	Nano montmorillonite	Zhanping You

3.2. Rubbers

Rubber is collected from unwanted tires which are not usable in cars. As we go for the use of waste material, as our material costs are raising and we are paving the way also for environment cleaning. The adjustment can be applied to both crumb rubber and natural rubber. This can be put in to bitumen in powdered form. Rubber adsorbs the bitumen's aromatic substance, and therefore covers one another.

Kumar et al. [37] adjusted bitumen by crumb rubber of 10 % and found an optimal HDPE content of 8 % for both unmodified and modified bitumen, resulting in an improved Marshall Stability value. The 6 % optimum binder content is decided in their work. Krishnapriya [27] researched the effectiveness of modified mixes made from Natural Rubber. She determined that the optimal binder content was 5%. Fatigue life raised the resilient modulus by 0.9 sec for the rest period. Natural modified bitumen in rubber has outstanding resistance to rutting. Yang et al. [6] found that 8 % of the polyethylene additive was good in penetration associated to the similar percentage of the polyethylene and rubber crumbs compound waste. The softening point was set at 8% for waste polyethylene but the ductility is lower. The crumbs of rubber waste have low temperature resiliency. Somani et al. [21] saw an optimal 6% bitumen content as they reached a 945 kg Marshal Stability value. The stability value falls above 6 percent of the binder. So, the alteration of 8 per cent LDPE and 12 % crumb rubber resulted in a Marshall stability value of 1250 kg and a flow value of 3.86 mm. Manjunath et al. [23] decided to add 0.3 % and 0.4 % of the at a temperature of 130 °C, the great stability is achieved and the stability values improved to 24 % and 22 % for Evotherm, and cecabase grade 2 respectively.

3.3. Nanoclay

Besamusca [10] studied two Nano clays of organic montmorillonite (MMT). He stated that the modifiers are influenced by bitumen's aging characteristics, which may be related to changes in viscous properties due to modifiers being included. Viscosity is reliant on the shear rate, and must be considered highly reliant on temperature. Using Nanofill-15 and Cloisite-15A, Jahromi et al. [32] altered bitumen by 0.2 %, 0.4 %, and 0.7 % more dynamic creep efficiency as compared to unmodified bitumen as showing modified bitumen. Tests related to Nanofill-15A, such as resilient modulus test, dynamic creep test, indirect tensile strength (ITS) test, and fatigue resistance test performed between MMT and cloisite-15A shows good performance. Jahromi et al. [31] analyzed samples at 5°, 25° and 40 °C at a loading frequency of 0.5 Hz. Pulse times were 500 ms and retrieval time was 1500 ms. Modification is accomplished using 2 to 7 % Nano clay. The stiffness of both Cloisite-15A and Nanofill-15 ranged from 8 % to 40 %and 3 % to 18 % respectively, depending on the test temperature and the Nano clay content. Creep tests performed at 40 °C and 60 °C for altered mixes revealed a strong pulse. Van de Ven et al. [11] evaluated the impact of addition of Nano clay modifiers in new and aged bitumen by means of Penetration, softening point and dynamic shear rheometer; cloisite Nano clay enhanced resistance to rutting, ITS, and stiffness for 40/60 bitumen. In both short and long-term aging, Nano fill additive improved resistance to aging for 70/100, bitumen. Jahromi et al. [33] reported that the fatigue-resistant parameter improved by incorporating cloisite in 7%compared with unmodified bitumen 1.2 to 1.4 times in unexploited condition. Modification of Nano fill reduces fatigue life but aging shows the same fatigue life. He also observed that improved bitumen properties depend on the form of clay pre-treatment, Nano clay, and polymer variety.

You et al. [4] used bitumen by weight between 2 %and 4 %. Rotational viscosity grew because of 41 % and 112 % on average. Complex dynamic shear modulus improved by 66 %with the addition of 2 % Nano clay A and 125 % with the addition of 4 % Nano clay A. Similarly, Nano clay B increases 184 % and 196 % of the dynamic shear modulus. Using both the Nano clays (NC) the risk of strain failure can be diminished in the tensile strength test. Yang et al. [7] noticed that both organophilic as well as sodium MMT can enhance resistance towards aging and these alterations have a higher side considering the bitumen quality at low temperatures, where snow susceptibility is present. Mahdi et al. [24] took 3 %, 7 %, and 9 % of organic MMT NC (N3 and N4) adding into bitumen. All N3 and N4 have X-Ray equivalent diffraction properties of 2.28, and 3.8 respectively. Complex shear modulus has evolved in dynamic shear rheometer (DSR) and reduced phase angle for modified binders as opposed to original ones. Thus altered binders have improved shear deformation resistance. Jamshidi et al. [30] recorded a major change in asphalt rheology with different kinds and amount of Nano-materials (2 % and 4 %). Centered on a rutting factor gradient analysis, the critical temperature for the NI44 P2 binder found to be 64 °C. Silica is well recognized for adhesion and the exceptional surface area of nanosilica material and stiffness modulus diminishes temperature sensitivity. Silica fumes improve the softening point and viscosity when fly ash raises the softening point and reduces viscosity. Rhi et al. [2] observed from their inspection that Nano silica penetration declined after 6 % addition, softening point improved and ductility lowered. This may be that more binder is absorbed from a high surface area. Between the several proportions, 2 %, 4 %, and 6 %, it was the 6 % nanosilica that soon led the binder not to age.

3.4. CNT and Graphene Oxide (GO)

Earlier work has shown that Nano modifiers provide a beneficial means of improving asphalt material properties. Most of the studies present have concentrated on the mechanical properties of modified asphalt at moderate temperatures. Xiao et al. [8] proposed that carbon nanoparticles may be useful in improving the rheological characteristics of bituminous mixes in the short and long term after aging treatments. Amirkhanian et al. [36] suggested the use of a moderately large proportion of Nano-particles (> 1 per cent) to enhance permanent deformation resistance at elevated temperatures in their work on unaged binders. For some phase angle value of 1 percent CNT binder, Santagata et al. [17] found an unspecified 0.5 % mix phase indicating a higher Complex Shear Modulus. Earlier in the Rolling Thin Film Oven Experiment (RTFO), there were fewer differences between the aged and the elderly, but then the Pressure aging vessel came back when a 1 % mixture was stiffer than 0.5 %. Khattak et al. [28] tried to modify neat bituminous binders with carbon Nano-fibers to a degree that was affected by the mixing technique used to disperse Nano-fibers in the binder. Few studies have investigated the mechanical properties at low temperatures, which are critical in harsh winter conditions for asphalt paving. Some of the most significant challenges to applying advanced Nano-materials to the asphalt industry are the relatively high cost of materials.

Wu et al. [9] showed that GO application would help to boost the resistance of bitumen to ultraviolet along with thermos-oxidative aging because of preventing the aging cycle. Yu et al. [3] inspected the microscopic study of the relationship among the GO and asphalt binder using various methods. This experiment dictated far enough that the addition of GO does not alter gel-soluble arrangement to gel-type arrangement and keeps the asphalt binder step of original maltene solubility. Results of the investigation conveyed by Liu et al. [25] specified that GO application does not alter the glass transfer temperature of the asphalt binder; though, such addition may upsurge the amount of crosslinking to bitumen. They also explored the usefulness of GO's modified bitumen for pavement applications. In addition to three other additives, Sasobit, Waste Cooking Oil (WCO), and Sasobit + WCO, they followed their tests by applying 0.05 % of GO to warm mix asphalt binders (WMABs). After the study, it was established that the GO substantially increased the viscosity, permanent deformation resistance and high temperature elasticity of unmodified bitumen (PG 64-22). Li et al [26] examined the boundary among the GO and bitumen, and the effect of adding GO to the composite output based on asphalt. During the mixing process they said CO_2 gas was released; there was no chemical reaction among the GO and bitumen. The CO_2 was precipitated by the degradation of the GO. It was found that the penetration and ductility of the binder was subsequently diminished by mixing with GO but the viscosity and softening point of the material was increased. After analysis, it also determined that the GO can improve the anti-rutting efficiency at both high and low temperatures, with thermal stability. Table 1 shows the details of work done by various researchers/authors.

4. Conclusions

The following conclusions were drawn from the works of searchers on bitumen modification using waste polymer and rubber.

- Modifiers help in improving the water-resistant property of the binder.
- The main aim of using modifiers is that the pavement bitumen should be less susceptible to the temperature at the operational condition and that workability need preservation at construction condition and the modifiers do this magically.
- The cost of an altered binder in comparison with the conventional binder is reduced.
- Compared to neat binders, modified binders possess improved stability and longevity.
- Modifiers enhance the thermal stability on bitumen alteration.

- Bitumen is an organic substance that gets older as it comes into contact with UV light and oxygen on the other hand it gains resistance to aging through altering by means of Nano clay.
- Predicting permanent deformation associated with complex shear modulus and phase angle exhibits enhancement for Cloisite characteristics but with Nano fill very less change occurs.
- Mixing the modifiers in specific work rate or rotation for a definite time plays a vital part in quality improvement.
- Most researchers have focused on pre- and post-modification, viscosity test, softening point, penetration, and Marshall Stability test that yield a quick result for the properties.
- Short-term rutting, long-term fatigue, and water resistance overall times are regions to be explored and therefore innovative research is being conducted in these areas only by few researchers, it needs to be focused even more.
- Further, the modification must be examined in the field even subsequently all laboratory tests have been carried out. As live traffic and environmental situations represents the real-time challenges for at least a minimum width and length.
- The proportion of the application of modifiers must not be moreover high because it reduces efficiency.
- Suitable application of modifiers grounded on neat bitumen with adequate amount transforms the road pavement by means of improved strength, longevity, and budget reduction.

References

[1] Yousefi A A 2002 Rubber-modified bitumens *Iranian Polymer Journal* **11** 303-9

[2] Zafari F, Rahi M, Moshtagh N and Nazockdast H 2014 The improvement of bitumen properties by adding nanosilica *Study of Civil Engineering and Architecture* **3** 62-9

[3] Yu R, Zhu X, Hu J, Zhao W and Fang C 2019 Preparation of graphene oxide and its modification effect on base asphalt *Fullerenes, Nanotubes and Carbon Nanostructures* **27** 256-64

[4] You Z, Mills-Beale J, Foley J M, Roy S, Odegard G M, Dai Q and Goh S W 2011 Nanoclay-modified asphalt materials: Preparation and characterization *Construction and Building Materials* **25** 1072-8

[5] Yazdani A, Pourjafar S and Rahimpour A 2012 Optimization of asphalt binder modified with PP/SBS/Nanoclay nanocomposite using Taguchi method. In: *Proceedings of World Academy of Science, Engineering and Technology*: World Academy of Science, Engineering and Technology)

[6] Yang Y and Cheng Y 2016 Preparation and performance of asphalt compound modified with waste crumb rubber and waste polyethylene-Article ID 5803709 *Advances in Materials Science and Engineering* **2016** 1-6

[7] Yang J and Tighe S 2013 A review of advances of nanotechnology in asphalt mixtures *Procedia-Social and Behavioral Sciences* **96** 1269-76

[8] Xiao F, Amirkhanian A N and Amirkhanian S N 2011 Long-term ageing influence on rheological characteristics of asphalt binders containing carbon nanoparticles *International Journal of Pavement Engineering* **12** 533-41

[9] Wu S, Zhao Z, Li Y, Pang L, Amirkhanian S and Riara M 2017 Evaluation of aging resistance of graphene oxide modified asphalt *Applied Sciences* **7** 702

[10] Van De Ven M, Molenaar A, Besamusca J and Noordergraaf J 2008 Nanotechnology for binders of asphalt mixtures. In: *Proceedings of the 4th Eurasphalt and Eurobitume Congress Held May 2008, Copenhagen, Denmark,* (Copenhagen, Denmark

[11] Van de Ven M, Molenaar A and Besamusca J 2009 Nanoclay for binder modification of asphalt mixtures. In: *Proceedings of the 7th International RILEM Symposium on Advanced Testing and. Characterization of Bituminous Materials, Rhodes, Greece,* pp 133-42

[12] Vamshi A 2013 Use of waste plastic in construction of bituminous road *Journal of engineering research studies* **2** 123-8

[13] Sojobi A O, Nwobodo S E and Aladegboye O J 2016 Recycling of polyethylene terephthalate (PET) plastic bottle wastes in bituminous asphaltic concrete *Cogent Engineering* **3** 1133480
[14] Singhal M, Yadav Y and Mandal R 2016 Use of modified bitumen in highway construction *Journal for Innovative Research in Science and Technology* **2** 176-382
[15] Shiva Prasad K, Manjunath K and K VR P 2012 Study on Marshall Stability Properties of BC Mix Used In Road Construction by Adding Waste Plastic Bottles *Journal of Mechanical and Civil Engineering* **2** 12-23
[16] Sarsam S I 2013 Effect of nano materials on asphalt cement properties *International Journal of Scientific Research in Knowledge* **1** 422-6
[17] Santagata E, Baglieri O, Tsantilis L and Dalmazzo D 2012 Rheological characterization of bituminous binders modified with carbon nanotubes *Procedia-Social and Behavioral Sciences* **53** 546-55
[18] Sahu A K and Singh R 2016 Application of waste plastic materials in road construction. In: *2nd International Seminar On—Utilization of Non-Conventional Energy Sources for Sustainable Development of Rural Areas ISNCESR '16, 17th & 18th March, 2016',*
[19] Rokdey S N, Naktode P and Nikhar M 2015 Use of plastic waste in road construction *International Journal of Computer Applications* **7** 27-9
[20] Rajasekaran S, Vasudevan R and Paulraj S 2013 Reuse of Waste Plastics Coated Aggregates-Bitumen Mix Composite for Road Application—Green Method *American Journal of Engineering and Research* **2** 1-13
[21] Parakash S 2016 Strengthen of Flexible Pavement by Using Waste Plastic and Rubber *International Journal of Civil Engineering* **3** 246-50
[22] Menaria Y and Sankhla R 2015 Use of waste plastic in flexible pavements-green roads *Open Journal of Civil Engineering* **5** 299
[23] Manjunath K, Dheeraj Kumar N and Thippeswamy G 2014 Performance and Evaluation on Marshall Stability Properties of Warm Mix Asphalt Using Evotherm and Cecabase Rt®-A Chemical Additive *International Journal of Engineering Trends and Technology* **12** 406-10
[24] Mahdi L, Muniandy R, Yunus R B, Hasham S and Aburkaba E 2013 Effect of short term aging on organic montmorillonite nanoclay modified asphalt *Indian Journal of Science and Technology* **6** 5434-42
[25] Liu K, Zhang K, Wu J, Muhunthan B and Shi X 2018 Evaluation of mechanical performance and modification mechanism of asphalt modified with graphene oxide and warm mix additives *Journal of Cleaner Production* **193** 87-96
[26] Li Y, Wu S and Amirkhanian S 2018 Investigation of the graphene oxide and asphalt interaction and its effect on asphalt pavement performance *Construction and Building Materials* **165** 572-84
[27] Krishnapriya M 2015 Performance evaluation of natural rubber modified bituminous mixes *Int. J. Civ. Struct. Environ. Infrastruct. Eng. Res. Dev, Trans Stellar Journal Publications* **5** 121-34
[28] Khattak M J, Khattab A, Rizvi H R and Zhang P 2012 The impact of carbon nano-fiber modification on asphalt binder rheology *Construction and Building Materials* **30** 257-64
[29] Kazmi S and Govardhana D 2015 Utilization of waste plastic materials as bitumen--blends for road construction in Oman *Scholars Journal of Engineering and Technology* **3** 9-13
[30] Jamshidi A, Hasan M R M, Yao H, You Z and Hamzah M O 2015 Characterization of the rate of change of rheological properties of nano-modified asphalt *Construction and Building Materials* **98** 437-46
[31] Jahromi S G, Andalibizade B and Vossough S 2010 Engineering properties of nanoclay modified asphalt concrete mixtures *Arabian Journal for Science and Engineering* **35** 89-103
[32] Jahromi S G 2011 Rutting and fatigue behavior of nanoclay modified bitumen. In: *EACEF-International Conference of Civil Engineering,* pp 277-81
[33] Ghaffarpour Jahromi S and Khodaii A 2009 Identification effect of nanoclay on engineering properties of asphalt mixtures *Journal of Modeling and Simulation* **41** 49-57

[34] Dixit S and Rastogi D 2013 Studies on the improvement of characteristics of bitumen with use of waste plastic *International Journal of Emerging Technology and Advanced Engineering* **3** 895-900

[35] Cheshmehgol M K and Ahmad M S 2016 Construction of Bituminous Macadam using Plastic waste *12th Transportation Planning and Implementation Methodologies for Developing Countries* **1**

[36] Amirkhanian A N, Xiao F and Amirkhanian S N 2011 Evaluation of high temperature rheological characteristics of asphalt binder with carbon nano particles *Journal of Testing and Evaluation* **39** 583-91

[37] Kumar K R and Mahendran N 2014 Experimental studies on modified bituminous mixes using waste HDPE and crumb rubber *International Journal of Emerging Technology and Advanced Engineering* **4** 587-97

IOP Conf. Series: Materials Science and Engineering **970** (2020) 012023 doi:10.1088/1757-899X/970/1/012023

Provision of Percolation Pits for Embankment Safety: A-State-of-Art-and-Review

Swarnima Subhadarsini, Debabrata Giri, Sudhanshu Sekhar Das

Department of Civil Engineering, Veer Surendra Sai University of Technology, Burla, Odisha, India

E-mail: swarnima1993subha@gmail.com

Abstract. An embankment is a vacillating wall of earth or concrete, which directs the river or flood water to its actual target and prevents farmlands. During heavy flood, water overtops the crest and silts clog sluice gate. This creates wave and extensive silt pressure on the wall. In order to avoid such situation, it needs to diverted floodwater through a water channel to a silt trap. Silts will settle in the silt trap and clean water will lend into the injection well passing through the multilayer of injection media. The floodwater is accommodated in a specially designed storage well accompanied with percolating cloak. The cleared water columns injected into the deeper layers of semi-impervious soil strata through the non-clogging percolating screens attached to the well. This diverts the water to percolates to deeper earth strata by capillary action and helps to recharge ground water. The paper aims on deliberating different eco-friendly design and construction technique of high raise embankment. Apart from this, an attempt has been made to showcase the increasing trends in research publication and activity in the area of eco-friendly construction practices to direct and absorb water without affecting the artesian spring.

1. Introduction

Embankments are the most common type of compacted structure found around the globe that relay on their own mass to withstand sliding and overturning. Sometimes it plays the role of dykes for flood control but in another way use as abutments in roadway, railway and runway. An embankment it seizes and diverts river water along bank and prevent from overflow. Failure of such structure basically grouped in three categories i.e. hydraulic, seepage and structural. Hydraulic failure occurs due to uncontrolled flow of water over adjacent embankment. Unrestrained watercourse causes erosive action on the embankment slope. Hydraulic failures may directly or in directly related to overtopping, wave erosion, top erosion and gullying. Most of embankment exhibits some seepage. Within limits of velocity and quantity of floodwater, seepage may not distress the embankment to that extent. Under unrestrained seepage condition, first downstream slope or foundation get eroded then it moves towards the upstream side slope by forming pipe like cavities. This underground erosion progress and create open pathway for the flow called piping. Internal erosion under the embankment carries soil particles away from it and results catastrophic breaching. Internal erosion discharge muddy water along with clear-water sometimes no discharge at all depending on head and flow. Occasionally internal erosion takes place even if there is no visible discharge of water from downstream side of embankment. On other hand chemicals, slats, dissolved solids and dispersive clay also causes erosion without any visible notice. Lab testing of such soil is the only solution to control invisible erosion.

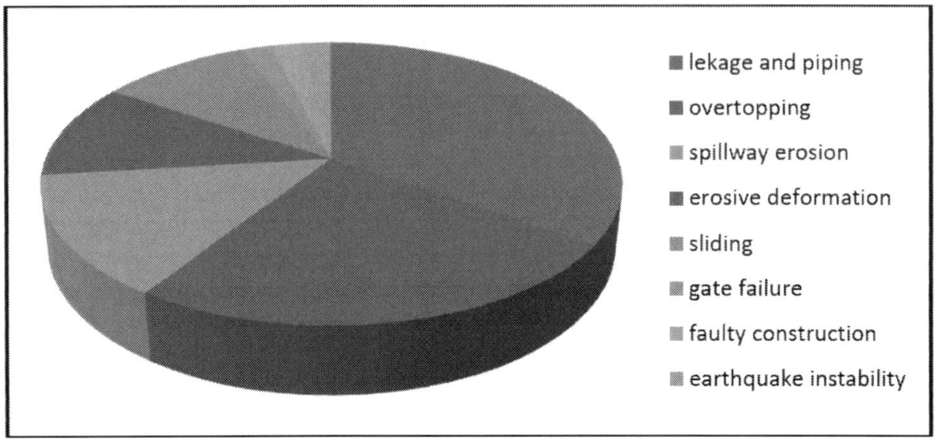

Figure 1 Causes of Embankment Failure [7]

Figure 1 represents that, approximately 40 percent seepage is responsible for the causes of embankment failure. Seepage causes weakening the adhesive properties of soil and leads to formation of slope stability failure. Sometimes tree roots can also provide smooth surface for seepage to travel along. Dead roots of decaying plant form pipe like structure beneath the embankment and provide a smooth surface and pathway for seepage. Structural failure involves rupture of embankment and rift of foundation. Such failure is eminent in large embankment or dams. Hence, in our article we consider seepage, which becomes the primary factor for its failure. Along with seepage, deferential settlement found in majority case of abutment, rupture.

Researchers like [1-9] reported their investigations and represents their observations regarding the causes and remedies of embankment failures. A detailed study of their records about the design and construction of percolating embankment were presented in a concise form.

Figure 2. Hydraulic failure of embankment

Figure 3. Seepage failure

Figure 4. Piping failure

AKAI, Koichi (1958) conduct an experimental study on seepage and its effect on the stability of earthen embankment. Seepage was found to be preponderant near its upper portion results from experimental study of percolating flow through sand models. The percentage of occurrence of seepage was found to be maximum in cohessionless soil. An equation has been set up for the theoretical investigation of the experimental study. The proposed equation gives the critical hydraulic gradient for such kinds of failure. Whereas experimental method is the only way to analyse the critical hydraulic gradient of such failures. Experimental research further extended towards disaster preventive embankment design. However, failure due to rapid draw down can only be analysed through site investigation.

S. A. Aiban *et.el* in (1999) used local materials in the construction of embankment and focused on the recharging ground water aquifer with treated wastewater effluents. Arabian Scientists installed 10 aquifers in an agricultural filed (located in AJ-Aziziyah in eastern Saudi Arabia) using spreading basin. The basin consists of a confined sand dunes filter. Then embankment was compacted from all the sides by sand/marls. The proposed sand filter is constructed form sand dunes having area 25 x 25m and has an effective of nearly 7.0 m. Prime importance given to the permeability and stability of embankment. For error-free and economical construction, they use locally available materials.

Katuyki kimura *et.al* (2000) studied the behaviour of seepage flow occurs due to rainfall. According to him, rainfall aids the expansion of saturated wet zone beneath embankment. Due to the enlargement of saturated wet zone slope eroded and instability occur. For the work, he conducted a centrifuge seepage model testing. Constant rainwater supplied to the model apparatus over the slope surface. Finite element unsteady saturated seepage analyses are made on the prototype of the model fill for the same rainfall conditions as adopted in the tests. The centrifuge seepage test give quit reliable results in uniform rainfall data. These data used effectively for finite element analysis.

CHEN Shan-xiong *et.al* (2001) analysed the unsaturated soil slope under the permeation of rain water. Experiment conducted on expansive soils and residual soils of tropical and subtropical region. Rainwater percolates into the fissures, causes swelling of saturated wet zone. This results in the reduction of shear strength in that zone beneath the embankment. When intensity and duration of rainfall exceeds its limits slope failure occurs. Slope failure is a results of combine phenomena of seepage and quantity and duration of rainfall. With limit equilibrium method, the stability analysis method of unsaturated soil slope due to raining permeation were developed, in which the shear strength parameters were considered as the function of saturation degree and the instantaneous distributions of water content within slope after rainfall. The result shows that the slope failure due to raining permeation usually is characterized by shallow layer failure.

Vivek Tandon et.al (2007) shows a performance monitoring the construction of embankment by using tire chips. Shredded tires have been successfully use as a filler material in the earthen embankment construction. However, it leads to certain problems like spontaneous combustion, settlement and so on. Tire chips can be use as fill materials in error free construction of embankment. Experimental study has been conducted by taking three types of amalgamation (1) mixture of soil and shredded tire, (2) shredded warped in a geotextile, (3) an embankment made with conventional soil. Performance based on vertical settlements, temperature change, air and water constitutions are all the three variations in different ways. Soil with tires chips shows greater stability against vertical settlements but have less percolating factor. P. Henneber et.al (2014) assessed the environmental impact of shredded tires as embankment filling material in its construction. Experiments like percolation, lysimeter and lixiviation have been conducted on the polluted water. Examination of 200 physicochemical parameters of collected liquid was commenced. The toxicity of the collected liquid sample was assessed in a small shellfish and bacterium. The results then compared with the existing data. Along with the experiment fire propagation risk in the core and surface of embankment were also inquired. Experimental study concluded certain limits of use of shredded tires and listed their consequences on environment. The study presumes lastly that toxicity from the residue of the shredded tire considered to be hazardous waste.
Arul Arulrajah et.al (2016) use recycled waste foundry sand as a sustainable sub grade fill and pipe bedding construction material. Waste foundry sand (WFS) is a byproduct of metal and non-metal casting. For the creation of metal parts by using molds, sand along with phenolic resin was use. The waste generated after the casting process is called as Waste foundry sand. Due the presence of metal and negative public reception, WFS commonly discard to landfill as a waste material. However, WFS efficiently use as a civil construction material which includes recycled glasses. The geotechnical and environmental properties of WFS satisfies and hence quantize for fill materials in embankments and bedding pipe. Leachaet from such metal deposit is not suitable for drinking and domestic use purpose. Pollutant in the leachate under go diffusion and dispersion process before it reaches to the ground water aquifer (means concentration of pollutant significantly reduced). Such transport of any pollutants can be precisely calculated by using ground water flow models. Moreover, use of WFS will save embodied energy and reduce carbon footprint.
Daljeet Singh *et.al* (2017) drew the effect of canal water level on stability of embankment and side slope. The effect of capillary and water supply level always keeps the phreatic line moving up and down. The flowing water exerts the pressure on sides of canal banks thus increasing the chances of instability during drawdown. After decrease of water level, the pore water created rapid draw dawn condition. This rapid draw down along with vehicular load in rainy season put whole embankment and pavement structure into a vulnerable situation, as the factor of safety against slope failure comes well below to unity. The work focused on the stability of canal bank and find out effect of these forces on it. Suitable measures are also adopted against percolation. For study, four different conditions were tested.

Table 1. Experimental conditions

Conditions	Explanation
Dry condition	This condition is similar to construction of canal.
Wet condition at full supply level (FSL)	The canal bank supposes to experience a long spell of rain.
Rainy weather + canal is running at Lowest Water Level (LWL).	Canal bank experienced long spell of rain with flow at LWL.
Rainy weather + surcharge of traffic load + canal is running at Lowest Water Level (LWL)	This was considering for worst service life of road pavement.

The embankment material has been tested for its in-situ properties; the Standard Penetration Test (SPT) was conducted to draw the profile of canal embankment layers and location of phreatic line at various stages of water supply level in Sekoru woreda Jima zone of Oromia region in Sekoru Woreda Jima zone of Oromia region dry as well as rainy seasons. For stability analysis a software named as SLOPE/W was developed by GEO-SLOPE International Canada.

2. Methodology

From above studied cases seepage seems to be more hazardous to the stability of embankment. By using shredded tires [6] and locally materials [2] like pebbles, gravels cannot recharge ground water as per the requirements. The article has the focal point on recharging ground water to deeper soil strata. Percolating pit with percolating cloak may help to meet the objective of the topic. Percolating pit is a simple hole dug into the ground to facilitate ground water recharge through infiltration of surface runoff into the soil layers. First locate the best place for construction of percolating pit according to the contribution of watershed, available depression, water sources, soil profile, non-interference of farming operation, required distance from embankment, mode of digging, locally available manpower. For better and increased percolation of flooded water approximately 4' x 3' x 3' sized pit-dogged m around the borehole of depth 15'-30'. Then arrange the pit with layers of pebbles and sand as per given in Figure 5. The figures represent the basic manuals for the construction of percolating pits. In addition, Figure 6 represents the percolating pipes installed in a pit. The sand and pebbles layers will help to filtrate the dart particles from floodwater. Floodwater penetrate bottom through perforated RCC slab cover. The vent helps to regulate air pressure in the percolating pit so that water flows unimpeded through percolating pipe. Sand and pebble layer helps to filter floodwater and clean one send to recharge ground artesian.

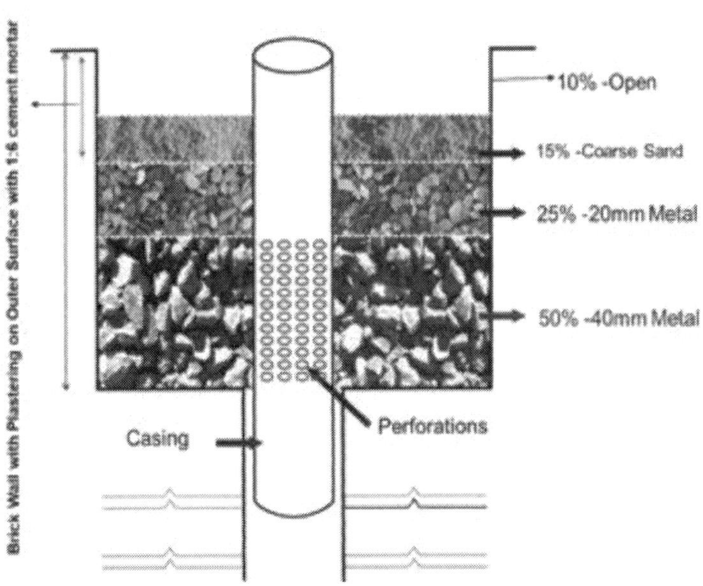

Figure 5. Soil profile arrangement for percolation pit instalment

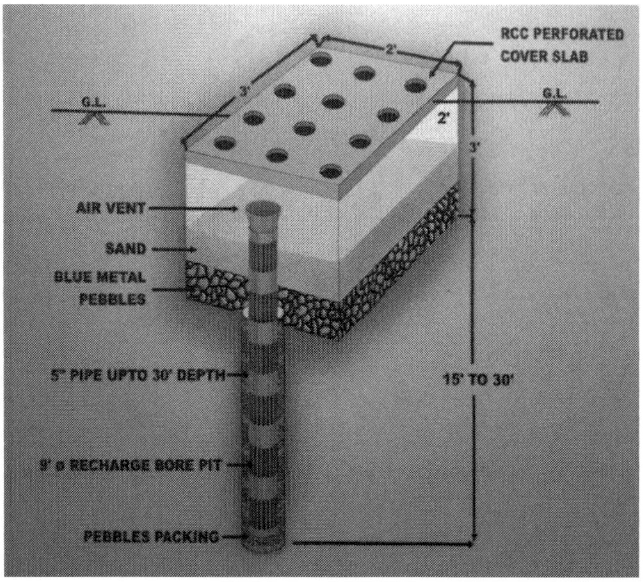

Figure 6. Percolation pit

3. Significance

Use of such percolating pits along flood affected side of embankment or the side which facing failure due to seepage, may give better result for percolating floodwater simultaneously recharging ground water. The percolating pit helps floodwater to flow down and there may be a chance of reduction of flood level during disaster. Installation of percolating pit in regular interval helps in better water storage capacity to a greater extent. At the same time, it helps for better stability of slope, which is going too recommended for near project.

4. Conclusion

According to the views of past records embankment faced two type of fracture i.e. settlement on the upper surface and seepage in the embankment body. Heavy rainfall weathered the upper surface simultaneously heavy flood create seepage problem. Use of pervious and cohessionless materials for the embankment construction may deduced the critical hydraulic gradient. Whereby providing stone or gravel pitching on the opposite side of embankment helps to keep the phreatic line within the body. This can reduce the seepage problem to some extent. Fly ash regularly use to compacting and strengthening the upper surface of embankment but it fails to protect it from seepage. Geosynthetic is only helpful for stabilizing the embankment surface but cannot helps in recharging ground water. Hence to full fill the basic and required objective for recharging ground water without affecting the quality of underground and to provide stability to embankments artesian percolating pit may be proved beneficial.

References

[1] Aiban S, Ishaq A and Al-Suwaiyan M 1999 Utilization of local material in the construction of an embankment for recharging groundwater aquifer with treated wastewater effluents *Water science technology* **40** 25-32

[2] Akai K 1958 On the Character of Seepage Water and Their Effect on the Stability of Earth Embankments

[3] Arulrajah A, Yaghoubi E, Imteaz M and Horpibulsuk S 2017 Recycled waste foundry sand as a sustainable subgrade fill and pipe-bedding construction material: Engineering and environmental evaluation *Sustainable cities society* **28** 343-9

[4] Chen S-x and Chen S-y 2001 Analysis of stability of unsaturated soil slope due to permeation of rainwater *Rock and Soil Mechanics* **22** 447-50

[5] Hennebert P, Lambert S, Fouillen F and Charrasse B 2014 Assessing the environmental impact of shredded tires as embankment fill material *Canadian Geotechnical Journal* **51** 469-78

[6] Kimura K, Okumura T, Narita K and Ohne Y 2000 Behavior Of Seepage Flow In Embankments Developed By Rainfall. In: *ISRM International Symposium*: International Society for Rock Mechanics and Rock Engineering)

[7] Shrivastava P, Patel A and Patil R 2013 Design of percolation pit for rain water harvesting in clayey soils of South Gujarat

[8] Singh D, Jha J and Gill K 2017 Effect of canal water level on stability of its embankment and side slopes *International Journal of Engineering Science and Technology* **9**

[9] Tandon V, Velazco D A, Nazarian S and Picornell M 2007 Performance monitoring of embankments containing tire chips: case study *Journal of Performance of constructed Facilities* **21** 207-14

IOP Conf. Series: Materials Science and Engineering **970** (2020) 012024 doi:10.1088/1757-899X/970/1/012024

Energy Conservation of Speed Breaker with Suitable Design

Abhin Tribikram Tripathy, Santosh Kumar Tripathy

Department of Mechanical Engineering, GIET University, Gunupur, Odisha, India

E-mail: abhintribikramtripathy@gmail.com

Abstract. In this project we have tried to generate electricity through speed breakers present on roads. As we know that vehicles on road are increasing day-by-day which will help us to generate electricity as these vehicles pass through the speed breakers. This electricity generated can be used for different purpose such as lighting of signals and streetlights on road etc. This set up requires very basic mechanical components such as gear shaft bearing. There are also some electrical components such as battery, rectifier etc. The main objective of the work is to design the speed breaker with failure analysis.

1. Introduction

Energy is treated as the most essential part of human beings and it can transform from one form to another. In other words, it describes the expression of the flow of energy is one of its forms. Machines can operate by the input of energy and it can be available in the form of crude fuels and electric power. The consumption of crude fuels and electric power creates a void in the world. Electricity plays a vital role in day-to-day life and we need to take several steps to generate more electricity. The generation of electricity creates some challenging steps to develop a suitable designed project. A proper modernized electric project with the utmost accuracy can give us more electricity.

Power is the major need for living beings, so consumption of power is necessary to fulfil the need of human life. The scientific study says the non-conventional resources will lead to developing power. The kinetic energy developed from the moving vehicles wasted and it can utilize by a set of speed breaker arrangements consist of gear and shaft mechanism. We can utilize the generated electric power in streetlights, traffic signals or it can store in a battery.

2. Methodology

This project is a combination of system, which converts a group of vehicular movement into electricity as well as a road hump. It serves dual work at a time. Here the linear motion of the vehicle converted into electric power with a suitable transform of rotational motion. The whole system covered with a semi-flexible cover strip, which is lying across the path of vehicular movement, the rest of the system enclosed within a casing inserted in the ground. The whole unit serves the dual work of the generation of electricity and speed breaker or road humps. The cover attached to a generator with a suitable gear mechanism containing a system of gear, spring, bearing, shaft, and motor, when a vehicle passed over the cover strip, it moves down which creates a rotational motion on gears to generate electricity. Here rotor drives an electromagnetic generator and the electromagnetic generator converts the rotational motion into electric power.

While a vehicle moves over the speed breaker, it pushed down the cover plate. A set of springs attached to the cover plate, it compressed while pushed down. The gears are used to transfer rotational motion to the electromagnetic generator and this generator converts the rotational motion into electric power. The generated electric power can utilize immediately or it can store in a battery.

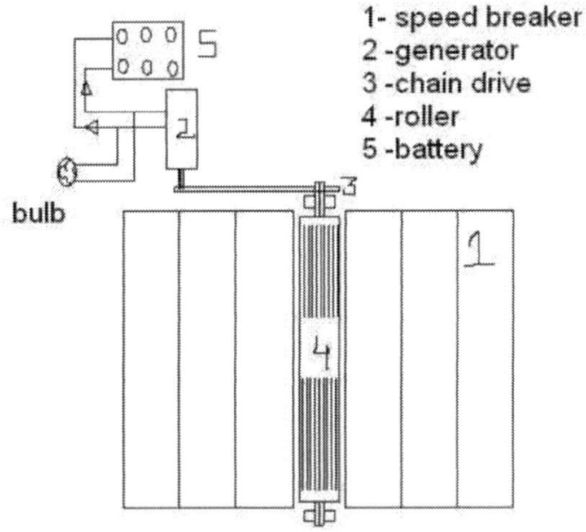

Figure 1. Model of speed breaker

2.1 Components list with their specification

Motor : Voltage: 6V, D.C. Generator, RPM: 1000 rpm
Gear : Mild steel material, No. of teeth: 62(big gear), 44(small gear), Type: Spur gear, No. of gear used: 2
Spring : Load bearing capacity: 6-7kg, Material: Mild steel
Bearing : Type: roller housing bearing, Bearing no: N40
Shaft : Diameter: 8mm, Material: mild steel, Length: 381mm

2.2. Design and calculation

Assuming a load of 1000N

2.2.1 Design of springs

No. of springs = 4
Load on each spring = 1000/4 = 250N
Assuming, modulus of rigidity (G) = 80KN/mm^2
Spring index (C) = D/d
= 30/3
= 10
No. of turns of spring (n) = 14
Allowable shear stress = 350N/mm^2
K_s = 1+1/2C
= 1+1/2*10
= 1.05

But,
Allowable shear stress = Ks*8*W*D/3.14*d^3
= 350 = 1.05*8*250*10*d/3.14*d^3
= d = 4.37mm
D = 4.37mm×10
= 43.7mm

Where,
D: Diameter of spring
d: Diameter of wire
Now,
Deflection (ẟ) = 8*W*D3*n/G*d4
= 8*250*43.7^3*14/80*4.37^4*10^3
= 80.09mm

2.2.2. Design of Shaft
Twisting moment of the shaft (T) = W*ẟ
= 1000*80.09*10^-3
= 80.09N-m
Assuming,
Allowable shear stress of the shaft = 45N/mm^2
So, T = π*d^3*τ/16
⇨ 80.09 = π*d^3*45/16
⇨ d = 20.85mm~22mm

2.2.3. Design of gears
Assuming, the rack moves 80mm in 3secs linearly
Diameter of the pinion = 25mm
Circumference of the pinion = π*d
= π*25 = 78.5mm~80mm
The pinion rotates 1/3 revolution per sec when the rack moves downward.
Therefore,
Speed of the pinion = 1/3 rev/sec
= 1/3*60 = 20 r.p.m
N1 = 20 r.p.m
N2 = Speed of the spur gear.
i.e., N1 = N2 = 20 r.p.m
N3 → Speed of the pinion
N2/N3 = T3/T2
⇨ N3 = N2*T2/T3
⇨ N3 = 20*320/32
⇨ N3 = 200 r.p.m
Total power generated (P) = 2*π*N*T/60
= 2*π*200*80.09/60
=1677.4 Watts
Assuming total losses in the system = 30%
Actual power generated in the system = (1-0.3)*1677.4
= 1174.18 Watts
This output power can directly utilize or can store in a storing device.

3. Conclusion
This suitable designed project gives an idea of the utilization of wasted kinetic energy into electric power. The whole system transformed the unused energy into usefulness. The maintenance cost can be reduced by a proper design of springs, shaft, and gears based upon the applied load. This designed set up helps to increase the life period of speed breaker or it can increase the efficiency level of a speed breaker.
The whole system comprises of external parts, which is a semi-flexible cover, and internal parts, which is a combination of gear and shaft mechanism. The internal part requires a proper design

because it not only wears the applied loads but also developed electric power. The whole setup was developed according to the applied load and the speed of the moving vehicle. However, the generated energy is no harm, so it can be treated as a "Green energy".

References

[1] Srivastava S and Asthana A 2011 Produce electricity by the use of speed breakers *Journal of engineering research studies* **2** 163-5

[2] Fatima N and Mustafa J 2011 Production of electricity by the method of road power generation *International Journal of Advances in Electrical & Electronics Engineering* **1** 9-14

[3] Das C, Hossain S M and Hossan M 2013 Introducing speed breaker as a power generation unit for minor needs. In: *2013 International Conference on Informatics, Electronics and Vision (ICIEV)*: IEEE) pp 1-6

[4] Ahuja D and Tatsutani M 2009 Sustainable energy for developing countries *SAPI EN. S. Surveys Perspectives Integrating Environment Society*

[5] Jalihal S A, Ravinder K and Reddy T 2005 Traffic characteristics of India. In: *Proceedings of the Eastern Asia society for transportation studies*: Citeseer) pp 1009-24

[6] Sharma A K, Trivedi O, Amberiya U and Sharma V 2012 Development of speed breaker device for generation of compressed air on highways in remote areas *International Journal of Recent Research Reviews in Environmental Science Bio/Technology* **1** 11-5

[7] Mukherjee.D and Chakrabarti.S 2005 *Non-conventional Power Plants*: Public printing service, New Delhi)

IOP Conf. Series: Materials Science and Engineering **970** (2020) 012025 doi:10.1088/1757-899X/970/1/012025

Treatment of Industrial Waste Water using Single-Stage Rotating Biological Contactor

Lusi Subhadarsini, Rakesh Roshan Dash

Department of Civil Engineering, Veer Surendra Sai University of Technology, Burla, Odisha, India

E-mail: lusi.iter9@gmail.com

Abstract. This paper presents treatment of industrial waste water using a single-stage Rotating Biological contactor (RBC). A RBC unit typically consists of a series of closely spaced large flat or corrugated discs that are mounted on a common horizontal shaft and are partially or completely submerged in wastewater. The shaft continually rotates by a mechanical motor or a compressed air drive and a bio film is established onto the entire surface area of the media, which metabolizes the organic materials contained in the wastewater. Microorganisms get attached to and grow on these discs. The discs mounted on the surface of a rotating shaft are usually made up of different forms of plastic such as polypropylene or polyurethane. For this study, shaft of RBC is wrapped with polypropylene foam as bio-film medium. From the analysis of results obtained for the present study it has been found that, polypropylene foam as a bio-media can remove organic matter up to 91.2%, nitrate up to 86.2% and phosphate up to 96.3% for 48 hr of hydraulic retention time.

1. Introduction

Industrialization played an important role for scio-economy of the country. Generally, a lot of wastewater are generated from industries due operational processes. Biological processes are best suited for the treatment of effluents that are rich in nutrients, and using these processes, 80–95% reduction in the effluent organic matter can be achieved. Amongst the biological processes, attached growth wastewater treatment processes have shown to be more energy efficient than suspended growth processes. Rotating Biological Contactor (RBC) is one of the attached growth processes. These processes can be carried out aerobically as well as anaerobically. Aerobic rotating biological contactors (RBC) comprise closely spaced circular discs mounted on a rotating shaft. Microorganisms get attached to and grow on these discs. They have been found to be twice more efficient in treating domestic as well as industrial wastewater as compared to the suspended growth. Advantages of using aerobic RBC include short hydraulic retention time (HRT), high biomass concentration, high specific surface area, low energy consumption, operational simplicity, insensitivity to toxic substrate, less accumulation of sloughed bio film and partial stir. Efficiency of the RBC, however, gets affected by various factors such as loading rate, rotational speed, HRT, staging, temperature and disc submergence. A RBC unit typically consists of a series of closely spaced large flat or corrugated discs that are mounted on a common horizontal shaft and are partially or completely submerged in wastewater. A drum filled with some lightweight packed supports can also be used in place of conventional discs. The shaft continually rotates by a mechanical motor or a compressed air drive and a bio film is established onto the entire surface area of the media, which metabolizes the organic materials contained in the wastewater

Researcher M. Wallis et. al (2000) [1] had studied the capacity of the bio-film of a 3-stage rotating biological contactor (RBC) for the treatment of wastewater containing cadmium, copper, and zinc. The system effectively removed the metals, in the order Cu > Zn> Cd with removal capacities of 73, 42 and 33% respectively. Vinage et.al (2003) [2] studied a lab-scale RBC model and the author has found out that, the rotational speed of shaft plays a vital important role in enhancing the overall performance of RBC. The author had developed a mathematical model of RBC and compared the output with the experimental data obtained. It is found out to be very effective for organic matter removal.SaikalyP.et.al (2003) [3] developed a lab-scale three stage RBC model for studying its effectiveness for the removal of ammonical nitrogen(NH3-N). The step-feed method was adopted for the above experiment and the results obtained were compared with the results of regular- fed RBC model in terms of ammonical nitrogen removal, dissolved oxygen conditions, and HLR and OLR. The result analysis indicates that the step-feed unit has better removal efficiency than the regular-fed one. The percentage of NH3-N removals in the RBC batch-feed stages 2 and 3 were higher than those of the RBC control although the OLR was higher at the RBC step-feed stages and the DO concentrations were found out to be lower. The existence of greater influential NH3-N concentrations in step-feed RBC stages two and three is the main reason for the higher removal percentages. A Tawfik et. al. (2006) [4] conducted an experimental investigation on the treatment of domestic wastewater in the RBC system at a temperature of 12-24 ° C. The RBC set-up consisted of a series connected 2-stage system. The prototype was operated at different OLR and HRT to get the best performance of RBC under varying conditions. An increase of 11 g / m2/d in OLR was observed. The effluent quality of soluble COD, however, remains unchanged. Most of the COD was removed in the first stage and in the second stage, nitrification took place. Samir A. et. al (2006) [5] proposed and explains the design of a rotating biological system on a lab-scale. To determine design parameters, RBC design criteria were identified and implemented. Locally available stainless steel and polyethylene pipes were used for the fabrication of the model. The designed RBC's performance proves adequate for the industrial purpose. There was an average rate of loading and removal. Increases in ammonia concentrations improved removal efficiency up to 3.75 mg/1 ammonia, after which removal efficiency remained about 40%. Syed E. et., al (2008) [6] conducted a test on a two-stage Rotating Biological Contactor (RBC) laboratory model with provision for varying rotating blade speeds. Gray wastewater was taken for the analysis. The experiment was performed for different COD loadings and varying blade speeds. From the result analysis, it was found out that 3rpm speed gives maximum COD removal that is, 95.07% as compared to the 4.5 and 6rpm speed that is, 95.04% and 94.96% removal efficiencies respectively. A. Ebrahimi and M. Asadi (2008) [7] had analyzed a three-stage RBC system for the removal of organic content from whey. The analysis was done under different detention periods such as 8h, 16h, 24h, and 36h and also under varying disc surface area. The acrylic sheets were used as disc material and submergence of 33% was kept during the experiment. The treated effluent was analyzed for COD, TS, VS, Phosphate, pH, and proteins. From the result analysis, it was found out that, the COD removal efficiency was 80 and 83% at the detention period of 16 and 24h respectively and at HRT 36h the removal efficiency was increased to 92%. The COD removal efficiency was increased to 96% when the disc surface area was increased by 10%. Dolores et, al (2009) [8] conducted the technology of set biomass, and bio disk, for sewage treatment in big and medium-sized reactors. The structure of bio-films and the interaction of microorganisms were vital parameters for effective system function and regulate. Bio-film size and complete dry weight were commonly used for characterizing bio-films but were not enough to define bio-film pattern. The availability of nutrients in the bio-film is one of the controlling parameters for the effectiveness of bio-film. So it is better to develop a small, stable and active bio-film for getting the best treatment of water and sewage. Joanna Rodziewicz et. al (2010) [9] had prepared a lab-scale RBC model and conducted the research under both conventional conditions (without current) and under different current flow conditions such as 0.2A, 0.8A, and 1.5A. The results were analyzed for organic content and ammonical nitrogen on municipal wastewater. The highest nitrogen removal efficiency was observed with the passage of electrical current at a density of 0.2A. The system was found out to be 64% efficient for denitrification and 93.4% efficient in nitrification. Kapoor et.al (2011) [10], they researched the common use of RBC technology for municipal and industrial effluent therapy. In the mining sector, RBCs have also been used effectively to remove pollutants such as ferrous iron (Fe2+). cyanide and its different species, oxalate, and selenium from mining waste effluents. This article represents the result of two continuing study projects being conducted at Natural Resources Canada's Mining and Mineral Sciences Laboratory. Using a three-stage RBC, the research efficiently oxidized Fe2 + to Fe3 + at 80C. Oxidation levels of Fe2 + varying from 63.0 g to 146.0 g Fe2+/m2/day were noted under different flow rates, temperature, and Fe2 + concentration working circumstances. Kader et. al. (2011) [11], had developed a mathematical model to analyze RBC's capacity to treat grey water. The whole setup was consisting of three units such as the RBC unit, settling tank unit, disinfection tank unit. After the optimization of the RBC model, three different gray water batches were prepared for the treatment such as low, medium and high levels of gray water. Using the data obtained from the mathematical model the lab scale set-up was done. The result analysis showed that the removal efficiency of BOD was between 93 to 96% and TSS removal was between 84 to 95%. Pathan et al. (2011) [12], Grey water's features differ from nation to nation and depend on the respective country's cultural and social behaviour. The grey water had to be characterized and recycled considerably. Grey water has been isolated from black water in this respect and analyzed for different physiochemical parameters. RBC (Rotating Biological Contactor) is a more efficient therapy method for COD (Chemical Oxygen Demand) removal and organic matter removal, among numerous grey water recycling treatment techniques. But in Pakistan, this technology has not been implemented and tested. Before implementing on a mass scale, there was a considerable need to measure the effectiveness of the RBC technology for small scale greywater recycling. A single-stage RBC reactor was fabricated for the treatment of grey water in the laboratory. The rotational speed of simulator was 1.70 rpm. The RBC's disk region was submerged in the grey water by about 40%. At each HRT (Hydraulic Retention Time) water specimens were gathered and analyzed using conventional techniques for parameters such as "pH, conductivity, TDS (Total Dissolved Solids), salinity, BOD5, COD, and suspended solids". The findings are encouraging with 53 % and 60 % removal of BOD5 and COD respectively. Kadu et. al (2013) [13], the author had analyzed the effectiveness of the RBC unit for the removal of soluble biodegradable substances from wastewater. The result analysis showed that the RBC successfully removes BOD up to 89.8% and a COD removal up to 80% had been achieved. The TSS removal rate was found out as nearly 80. Jayamala and Nitin A. Deshpande (2016) [14], had developed a partially submerged three stage rotating biological contactor for treatment of sewage. The RBC model was operated at a speed of 6 rpm and the effluent was tested for BOD removal efficiency. The results obtained, showed a BOD removal of 98.32% and indicated that rotating biological Contactors can treat organic and inorganic waste efficiently, cost-effectively and effectively. Rana et. al [15] had researched the removal of phenol from the industrial wastewater by using the RBC system. Parameters such as the concentration of mixed liquor suspended solids (MLSS), original phenol concentration, hydraulic retention time (HRT), disk submergence, rotation per minute (RPM) were optimized to improve the efficiency of phenol removal. The test result showed that, when synthetic wastewater containing 1g/l C6H12O6 and 250g/l phenol was treated through the reactor with 40% disc submergence, 30rpm rotational speed and 14h retention time, more than 56% phenol removal was achieved.

From the detailed study on previous work done it has been clear that the RBC system is very effective for the treatment of wastewater containing of organic waste products. The shaft rotational speed plays a very important role in achieving the best removal efficiency. Other important parameters that influence the overall performance of RBC are the detention period, bio-film material, percentage of submergence, and availability of oxygen for the microorganisms. From the above study, it also has been clear that the shaft rotational speed of 2-6rpm and submergence of 35-45% gives the optimum removal efficiency. As few research has been carried out on bio-film materials that are naturally available, so for this project polypropylene has been considered as a bio-film medium. The wastewater has been treated separately in batches by using both polypropylene as bio-film medium.

2. Details of Experimentation

2.1. Methodology

A prototype of single-stage RBC reactor was fabricated and the whole assembly consists of different parts such as reactor, rotating shaft, motor for rotation of the shaft, cylinder for placing the media profile, media profiles, pulley for achieving the required rotational speed and stand for supporting the whole set-up. The properties of the lab-scale model are tabulated below

Table 1. Properties of reactor

Specification	Values
Number of stages	1
The diameter of the central cylinder	0.0762 m
Length of the central cylinder	0.3048 m
The total surface area of the central cylinder in use	0.14 m^2
Working volume of the reactor	22.5 litre
Number of pulleys	2 (used for controlling the rotational speed)
Media profile and its thickness	Polyurethane foam with 4mm thickness
Rotation per minute	4 rpm
Submergence (%)	43%
Motor used	One single phase induction motor

2.2. Properties of polypropylene foam

This is otherwise called as biochemical sponge. This is a porous material and very flexible in nature and provides the breeding ground for the cultivation of fine nitro bacteria. This foam can be reused after every set of treatment up to approximately six months and is easily available in the market. It is very economical in nature.

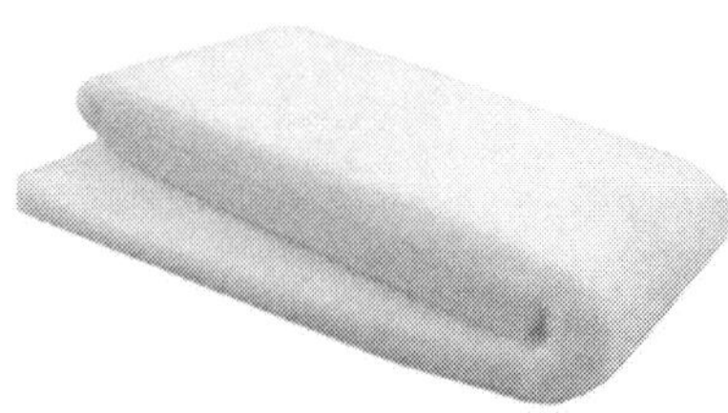

Figure 1. Polypropylene foam

Figure 2. RBC set-up

2.3 Working principle of the reactor
The reactor works on the principle of the attached growth system. The central cylinder has been wrapped up with the media profile. The central cylinder was kept 43% submerged throughout the experiment and the rotation of the cylinder with media profile was kept at 4rpm. The rotation was achieved by the setup of two pulleys and a single phase motor.

2.4 Mechanism
When the central cylinder with active bio-film in 43% submerged state rotates at 4rpm speed through the wastewater, the active bacteria present in the bio-film consume the organic matter as their food and grows in presence of nutrients thereby leaving the wastewater-free from organic loading and excessive nutrient. The rotation helps in oxygen transfer to the microbes which help for their growth. When the organic matter is subjected to degradation in the presence of oxygen, then carbon dioxide, water, and nutrients are released and new cells are formed. If the bio-film layer is brown in colour then the microbes present are healthy and if the colour is grey, then they are unhealthy in nature.

3. Reactor operation

3.1 Acclimatization
Before processing the synthetic wastewater through the reactor, the reactor was subjected to acclimatization. It is the process by which the microorganism grows on the media surface to form an active layer of bio-film. The reactor was supplied with tap water, glucose, cow dung (20%) and drain water (10%) for this purpose.
The acclimatization was done for 5days. After the formation of active bio-layer, the water present in the reactor was released and the prepared synthetic wastewater samples were supplied to the reactor for treatment in batches one after another.

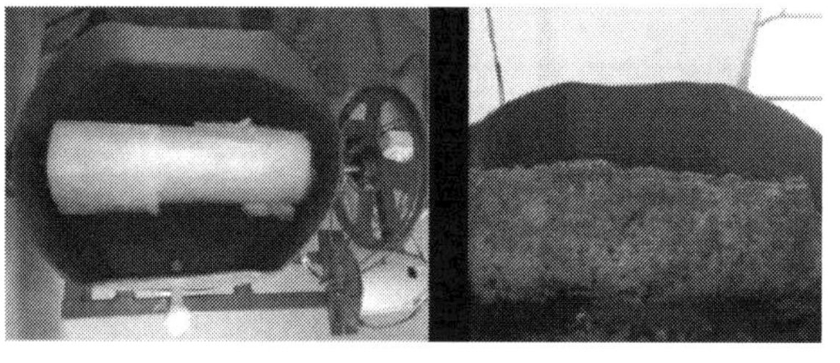

Figure 3. Processing of waste water through the reactor

All total of five batches of synthetic wastewater was processed through the reactor. The sample waters were collected at an interval of 1h, 3h, 8h, 24h and 48h for each batch separately. The reactor was constantly in running state up to 8h and after that, the wastewater was left for settling and the sample water was collected in 48h.
The water samples were then tested for various parameters such as COD, nitrate and organic content, phosphate, ammonia, pH, and TDS. Then that batch wastewater was released from the reactor and the reactor was provided with the next batch of synthetic wastewater and so on. The procedure has been followed for the media profiles such as polypropylene. When all the five batches were processed by using polypropylene as bio-media, the polypropylene media was taken out.

3.2. Characterization of synthetic waste water
The synthetic wastewater was prepared by using chemicals such as glucose (C6H12O6), potassium dihydrogen phosphate (KH2PO4), and potassium nitrate (KNO3), starch etc. Stock solutions were prepared according to the Indian standard code and diluted as per requirement. The properties of synthetic wastewater are tabulated on the below in table composition of synthetic wastewater

IOP Conf. Series: Materials Science and Engineering **970** (2020) 012025 doi:10.1088/1757-899X/970/1/012025

Table 2. Properties of synthetic wastewater

Batch name	Organic loading (mg/l)	Phosphate(mg/l)	Nitrate (mg/l)
1	100	5	15
2	300	10	20
3	600	20	25
4	900	30	30
5	1200	40	45

4. Nitrate Removal

Nitrate is the nutrient that is required for the growth of microbes. The removal efficiencies for polypropylene bio-medias for different batches are represented below.

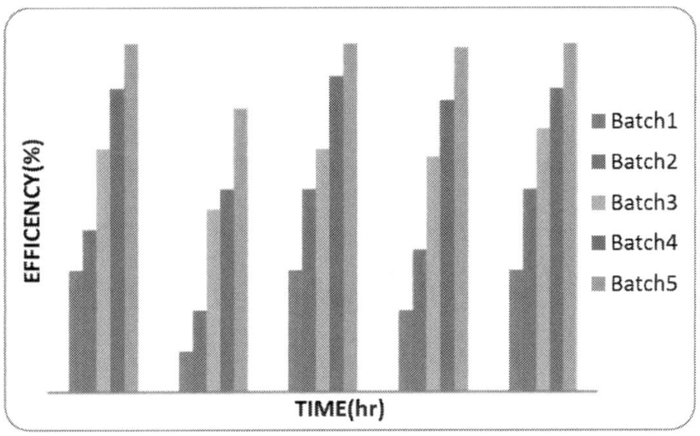

Figure 4. Nitrate removal efficiency for polypropylene bio-media

The results of the graph in Figure 4 interprets that maximum removal efficiency has been achieved at 48 hours and day-5. A maximum nitrate removal of 86.2% has been achieved.

5. Phosphate Removal

The phosphate removal efficiencies of for different batches with polypropylene media profile have been represented below graphically. The results obtained in Figure 5 shows maximum phosphate removal efficiency of 96.3% and have been achieved at 48 hours for most of the batches.

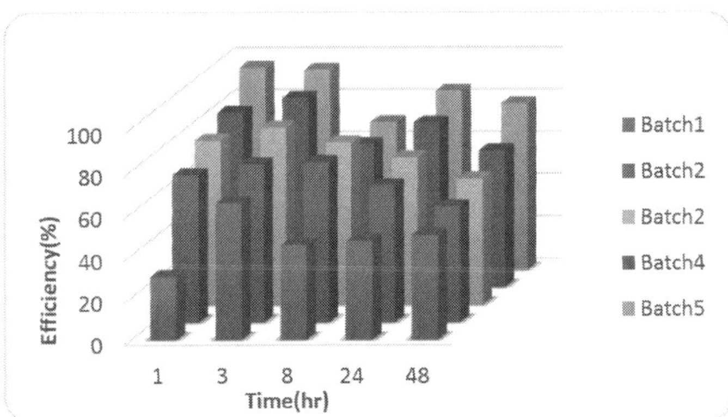

Figure 5. Phosphate removal efficiency for polypropylene bio-media

6. COD Removal

The COD removal efficiencies for different concentrations media profile have been graphically represented. A maximum COD removal efficiency of 91.2 % has been achieved and 48-hour duration is proved to be appropriate for all the batches

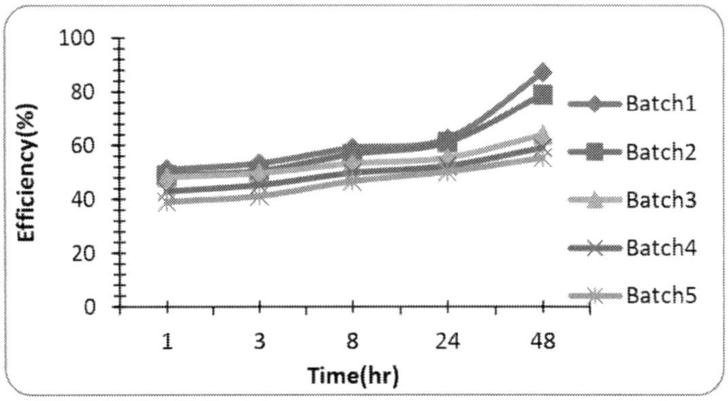

Figure 6. COD removal efficiency for polypropylene as bio-media

7. Result and Discussion

This section contains information about all the outcomes that has been derived from the experimental studies. The outcomes and reason behind fluctuation are discussed in this section. The results are for media profiles i.e. polypropylene foam on the basis of various parameters such nitrate and organic content, phosphate.

8. Conclusions

From the analysis of results obtained for the present study it has been concluded that, polypropylene foam as a bio-media has greater efficiency for organic matter removal that is 91.2%. The nitrate removal efficiency for polypropylene 86.2%. The phosphate removal efficiency 96.3% is obtained for polypropylene. The 48 hr detention period gives maximum removal efficiency.

References

[1] Manoj R, Sonali B and Jyoti R 2015 Review Paper on Study of Rotating Biological Contactor for Wastewa-ter Treatment Process *International Journal of Current Engineering Technology* **5** 1539-41

[2] Vinage I and von Rohr P R 2003 Biological waste gas treatment with a modified rotating biological contactor. I. Control of biofilm growth and long-term performance *Bioprocess Biosystems Engineering* **26** 69-74

[3] Ayoub G and Saikaly P 2004 The combined effect of step-feed and recycling on RBC performance *Water Research* **38** 3009-16

[4] Tawfik A, Temmink H, Zeeman G and Klapwijk B 2006 Sewage treatment in a rotating biological contactor (RBC) system *Water, Air, Soil Pollution* **175** 275-89

[5] Sahinkaya E and Dilek F B 2006 Biodegradation of 4-CP and 2, 4-DCP mixture in a rotating biological contactor (RBC) *Biochemical Engineering Journal* **31** 141-7

[6] Cortez S, Teixeira P, Oliveira R and Mota M 2008 Rotating biological contactors: a review on main factors affecting performance *Reviews in Environmental Science Bio/Technology* **7** 155-72

[7] EBRAHIMI A, Asadi M and NAJAFPOUR G D 2009 Dairy wastewater treatment using three-stage rotating biological contactor (NRBC) *IJE Transactions B.* **22**

[8] Di Palma L and Verdone N 2009 The effect of disk rotational speed on oxygen transfer in rotating biological contactors *Bioresource Technology* **100** 1467-70

[9] Rodziewicz J, Filipkowska U and Dziadkiewicz E 2011 Electrolytically aided denitrification on a rotating biological contactor *Environmental Technology* **32** 93-102
[10] Kapoor A, Kuiper A, Bedard P and Gould W 2003 Use of a rotating biological contactor for removal of ammonium from mining effluents *European journal of mineral processing and environmental protection* **3** 88-100
[11] Abdel-Kader A M 2013 Studying the efficiency of grey water treatment by using rotating biological contactors system *Journal of King Saud University-Engineering Sciences* **25** 89-95
[12] Patwardhan A 2003 Rotating biological contactors: a review *Industrial Engineering Chemistry Research* **42** 2035-51
[13] Kadu P A, Landge R B and Rao Y 2013 Treatment of dairy wastewater using rotating biological contactors *European Journal of Experimental Biology* **3** 257-60
[14] Deshpande N A and Jayamala 2016 Treatment of Sewage by Partially Submerged Three Stage Rotating Biological Contactor (RBC) *IJSTE - International Journal of Science Technology & Engineering* **3** 401-5
[15] https://nptel.ac.in/courses/105105048/M11_L14.pdf

Effects of Sulphates on the Characteristics of Ordinary Portland Cement

G Aditya[1], V Manohar Reddy[2], Naresh Kumar Gupta[2]

[1] Government of Andhra Pradesh, Narasaraopet, Andhra Pradesh, India
[2] Department of Civil Engineering, Government College of Engineering, Kalahandi, Bhawanipatna, Odisha, India

E-mail: adityagangadhari1@gmail.com

Abstract. This project presents the results of an experimental investigation in which the effects of sulphates on the initial and final setting times of ordinary Portland cement and compressive strength of cement is investigated. It was observed that the type of Sulphate and the ion concentration affect the setting times significantly; the increase in initial setting time is more prominent. The parameters reduce the compressive strength of cement at different ages. In the present study water cement ratio is taken depending on the normal consistency arrived based on different concentrations of sulphates mixed with OPC (Ordinary Portland Cement) an increase in setting times is observed as the concentration of sulphates in the cement paste increased. The soundness test results also showed an increased pattern which means that there is an increase in volume of cement paste because of sulphate attack. Compressive strength characteristics also showed an increased pattern when compared to cubes cast with distilled water

1. Introduction

Cement concrete construction is getting progressively perplexing and the significance of creating structures that are both financially savvy and strong has never been higher. Sulphate attack in cement concrete has been known to happen when sulphates solutions gotten either from a constituent in the concrete, or from outer sources, for example, groundwater, react with the concrete hydrates present in the solidified cement paste to frame the items which can involve more prominent volume than the reactants. The significant wellsprings of these salts are soils (especially Arid Region), groundwater, seawater, modern synthetic substances and squanders, and manures. The three items acquired because of the reaction for example gypsum, ettrignite and Thaumasite has more noteworthy volumes than the reactants and in this manner brings about the arrangement of cracks and inevitable disturbance in the solidified concrete. The fundamental point of this venture is to examine the pernicious impacts of sulphatese exposure to Portland cement and impact of water cement ratios that guarantees strength of concrete against exposure of sulphates salts. Accentuation is put on a basic examination of models that depict the connection among quality and impact of sulphates with curing age. Information on the increase of compressive strength with age and pertinent properties of concrete, for example, soundness and setting times are examined within the sight of sulphates [1-7].

Ettringite development is viewed as the reason for the greater part of the expansion and disruption of concrete structures associated with sulphates attack, anyway not really any sulphates attack is shaped

by ettringite arrangement more over ettringite development can beniicially utilized with no sulphate on concrete this undertaking presents basic examination of connection between ettringite formation and sulphates attack including some special cases of this attack those identified with thaumasite development are "postponed ettringite arrangement". The term sulphates attack utilized here methods decay of concrete including any sort of sulphate communications with cement paste freely on cooling temperature and sulphates source.
The concrete cubes are helpless against both internal and external sulphates attack. It is likewise perceived and examined in the paper about the quality impacting attributes of sulphates on cement with increment in age [8-15]. The adverse impacts depend on the life span of the attack. Subsequently vital safety measures are to be taken in this angle.

2. Materials

2.1. Materials
Test examination materials include:
- 53-Grade Ordinary Portland Cement (OPC)
- Sand
- Distilled Water
- Sulphates (MgSo4.7H2o+Na2So4+(NH4)2So4)

The properties of these materials are given in the accompanying sub-segments

2.1.1 Cement
Experiments like initial setting time, final setting time, soundness of mortar cubes and compressive strength were tested on 53-grade cement. Oxides and their percentages of 53-grade OPC are presented in Table 1.

Table 1. Chemical Configuration of OPC

Oxides	Percentage of contents
CaO	65.4
Na_2O	0.12
SiO_2	21.6
MgO	0.7
Al_2O_3	5.9
SO_3	1.6
Fe_2O_3	3.8
K_2O	0.4

The percentage composition of the major compounds (known as the Bogue compounds) present in the cement is presented in Table 2 [2].

2.1.2. Sand
Sand utilized all through the test is acquired from the Ennore, near to Chennai (Ennore sand IS 650:1991). The sand was acquired from Tamilnadu minerals, Chennai. It is the main association in India, affirmed by Indian Standard Institution to fabricate and flexibly of sand adjusting to IS 650:1991 [8]. The sand utilized in the test work has the accompanying particle size distribution
- Passing through 2mm X retained in 1mm; 1mm X retained in 0.5 mm and 0.5mm X retained in 0.09mm
- The Distilled water is used as mixing water.
- Sulphates ($MgSo_4.7H_2o+Na_2So_4+(NH_4)_2So_4$)

IOP Conf. Series: Materials Science and Engineering **970** (2020) 012026 doi:10.1088/1757-899X/970/1/012026

- Magnesium sulphate heptahydrate ($Mgso_4.7H_2o$) having the molecular weight of 246.48 gm/mol and its P^H is 5-8.
- Sodium sulphate (Na_2So_4) having the molecular weight of 142.04 gm/mol and its P^H is 5-7.
- Ammonium sulphates ($(NH_4)_2So_4$) having the molecular weight of 133.13 gm/mol and its P^H is 5-6.
- The sulphates used are the Analytical Reagents (AR) and is supplied by SDFCL, Mumbai

Table 2. % Composition of the Chief Mixtures Existing in the Test Cement [14, 15]

Sl. No.	Compound	Conversion formulas	% in cement
1	Tetracalcium alumino ferrite ($4CaO.Al_2O_3.Fe_2O_3$)	3.04 (Fe_2O_3)	11.70
2	Tricalcium silicate (3CaO. SiO2)	4.07 (CaO) -7.60 (SiO_2)-6.72 (Al_2O_3)-1.43 (Fe_2O_3)-2.85 (SO_3)	51.49
3	Tricalcium aluminate ($3CaO.Al_2O_3$)	2.65 (Al_2O_3) -1.69 (Fe_2O_3)	9.31
4	Dicalcium silicate ($2CaO.SiO_2$)	2.87 (SiO_2) - 0.754($3CaO.SiO_2$)	23.37

Table 3. Ennore Sand characteristics

Sl. No.	Properties	Unit	Results
1	Specific Gravity	-	2.64
2	Bulk density	KN/m^3	15.54
3	Fineness modulus	-	2.72
4	Particle size variations	mm	0.09 to 2.0
5	Colour	Hazan	Grayish white
6	Absorption in 24 hours	-	0.8%
7	Shape of grains	-	Sub angular

Table 4. Details of Test Program

Sl. No.	Sulphates Concentrations $MgSo_4.7H_2o$+$NaSo_4$+$(NH_4)_2So_4$ (mg/l)	specimens for setting times test	specimens for soundness test	specimens for compression test	aggregate
1	0				
2	333+333+333				
3	500+500+500				
4	667+667+667	3 (for all conc.)	3 (for all conc.)	3x4 (for all conc.)	18 (for all conc.)
5	834+834+834				
6	1000+1000+1000				
7	1167+1167+1167				
8	1334+1334+1334				

3. Methods

The trial strategies embraced were as per the standard systems in BIS. They are quickly introduced in the accompanying sub segments.

The subtleties of the test work are introduced in Table 4. A sum of 24 examples utilized in Vicat's device for initial and final setting time tests. Similar number of tests of standard form was utilized in Le-Chatelier's apparatus to test for adequacy.

A sum of 96 mortar cubes of 50 cm^2 cross sectional area was tested at various ages (3-days, 7-days, 28 days, and 90 days) for compressive strength

4. Results and Discussions

The results of initial and final setting times, soundness and percentage change in compressive strengths of the test samples are shown in the Figure 1 to Figure 4.

The Effect of sulphates $MgSo_4.7H_2o+Na_2So_4+(NH_4)_2So_4$ on IST, FST of cement are shown in Figure 1 from the Figure, it is observed that both IST, FST process got accelerated with the increase in percentage of sulphates in Portland cement. Initial setting times are significantly increased at all the percentages of sulphates when compared with the results obtained with control samples. The increase in initial setting times are 48, 53, 123, 138, 143, 153 & 163 minutes and final setting times are 15, 15, 20, 30, 45, 70 & 105 minutes for 1000, 1500, 2000, 2500, 3000, 3500 & 4000 mg/l respectively when compared with control samples.

4.1. Compressive strength

The effect of opted percentages of sulphates on the compressive strength of cement mortar is presented in Figure 3 and 4. The results indicated that there is increase in compressive strength of the cement mortar cubes when compared to cubes caste with distilled water. Average compressive strengths at 3, 7 and 28 for concentrations of sulphates $MgSo_4.7H_2o+Na_2So_4+(NH_4)_2So_4$ are shown in the Figure 3. The percentage change in the compressive strength of cement shown in the Figure 4. However, from all the results we noted that the compressive strength of cement increased with increase in concentrations from 1000-4000 mg/l.

4.2. Soundness

The Le-Chatelier test results for expansion measurement in cement should not be more than 10mm for Portland cement for passing soundness test. The effect of different mixing proportions of sulphates in cement on Soundness is given in Figure 2.

The expansion measured is 1.95, 2, 2.05, 2, 2, 2.05, 2.05 & 2.05 for 0, 1000, 1500, 2000, 2500, 3000, 3500 & 4000 mg/l respectively. As the measured values are less than 10 mm. All the samples are considered sound.

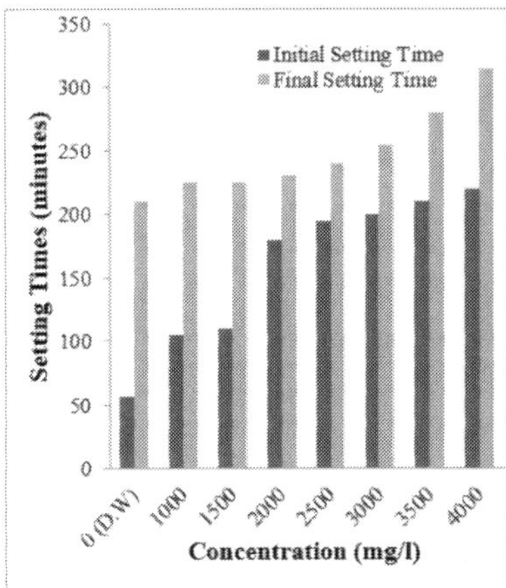

Figure 1. Effect of sulphatesMgSo4.7H2o+Na2So4+(NH4)2So4 on IST, FST of cement

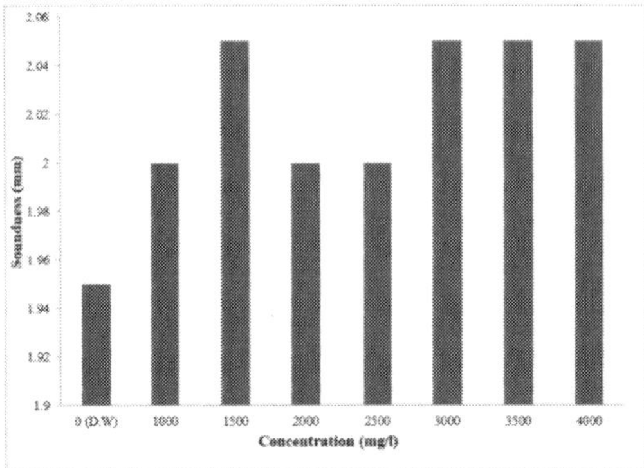

Figure 2. Effect of sulphates $MgSo_4.7H_2o$+Na_2So_4+$(NH_4)_2So_4$ on soundness of cement

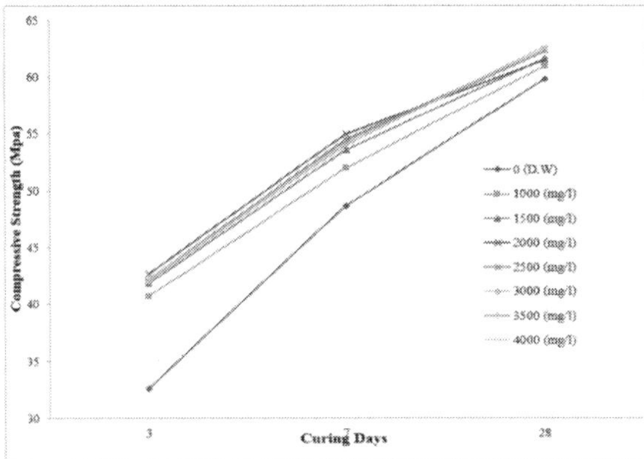

Figure 3. Effect of sulphates $MgSo_4.7H_2o$+Na_2So_4+$(NH_4)_2So_4$ with different concentration on compressive strength of cement

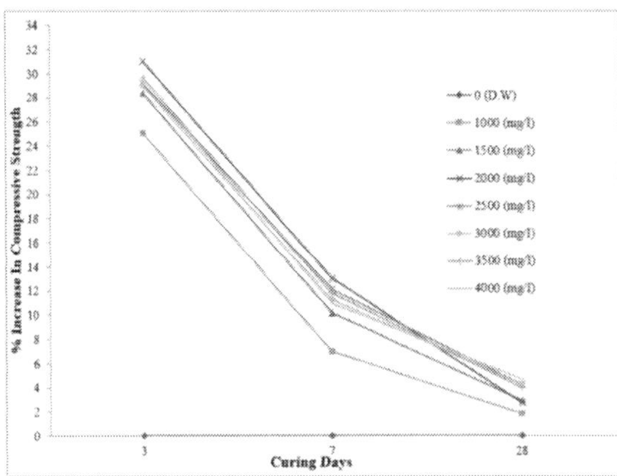

Figure 4. Effect of sulphates $MgSo_4.7H_2o$+Na_2So_4+ $(NH_4)_2So_4$ percentage transistion of compressive strength of cement at different concentration sulphates

5. Conclusions
In light of the trial study the accompanying ends are drawn.

- The setting time of concrete, monotonically increase with sulphate concentration.
- For the same ion concentration % increase in final setting time is less than initial setting time
- The rate increment in initial setting time is more contrasted with that of the final setting time for a similar particle concentration.
- Cations related with sulphate solution impact the level of attack. Among the three sulphate solutions utilized in this examination, the impact of sodium sulphate on setting times of concrete is generally articulated.
- Concrete exposed to sulphate attack, loses compressive strength and this loss increments as a component of sulphate concentration and time of exposion
- Impact of various sulphate solutions on compressive quality of concrete was not similarly serious. The impact of magnesium sulphate was discovered serious.

The cause for change in setting time of concrete within the sight of natural cations (Na+, Mg++, NH4+) and anions (So4- -) credited to the adjustment in solvency of cement paste cations (Ca++) and anions (aluminates and silicates).

References

[1] Biczok I and Blasovszky N 1964 Concrete corrosion and concrete protection
[2] Rogue R and Lerch W 1934 Hydration of Portland cement compounds *Industrial Engineering Chemistry* **26** 837-47
[3] Butt Y M, Kolbasov V and Timashev V 1968 High temperature curing of concrete under atmospheric pressure. In: *Proc. Symp. Chem. Cement Tokyo,* pp 437-76
[4] Cohen M and B. M 1991 *Jour. oT-ACI* **88**
[5] Odler I and Wonnemann R 1983 Effect of alkalies on portland cement hydration II. Alkalies present in form of sulphates *Cement concrete research* **13** 771-7
[6] 1968 IS: 4031-1968, Methods of Physical Tests for Hydraulic Cement, B. I. S., and New Delhi.
[7] Lawrence C 1990 Sulphate attack on concrete *Magazine of concrete Research* **42** 249-64
[8] Institute A C 1991 Proposed revision of: Guide to durable concrete *ACI Mater. J.* **88** 544-82
[9] Lea F M 1970 *The Chemistry of Cement and Concrete, 3rd Ed.*: Edward Arnold Ltd., London)
[10] Mehta P K 1986 Concrete. Structure, properties and materials
[11] Mironov S and Malinina L 1964 Acceleration of Concrete Hardening *Stroiizdat, Moscow*
[12] Moskvin V M 1983 *Concrete and Reinforced Concrete, Deterioration and Protection*: Mir Publishers)
[13] Neville A M 1995 *Properties of concrete* vol 4: Longman London)
[14] Rasheeduzzafar D and Al-Gahatani 1984 *Jour. of A.C.I.* **81**
[15] Scharf H 1990 PhD thesis. Technical University, Clausthal

Bio Cementation Process in the Concrete using Bacillus Subtilis: A Statistical Modeling Study

Dipika Patnaik[1], Akshaya K. Verma[1], Simanchala Panda[2], Naresh K. Sahoo[1]

[1] Department of Civil Engineering, Institute of Technical Education & Research, Siksha O Anusandhan (Deemed to be University), Bhubaneswar, Odisha, India
[2] Department of Civil Engineering, College of Engineering Bhubaneswar, Bhubaneswar Odisha, India

E-mail: akshayakumarverma@soa.ac.in

Abstract. The present study shows the bio-cementation efficacy of Bacillus subtilis incorporating three major parameters namely bacterial concentration, curing time, and nutritional proportion. Batch studies incorporating one variable at a time (OVAT) and central composite design (CCD) along with the response surface methodology (RSM) were undertaken to establish the possible relationship of selected parameters with the mechanical properties (compressive strength, flexural strength and split tensile strength) of cement mortar and bacterial concrete. The investigations performed on the basis of OVAT approach indicated that the extreme level of bacterial concentration, curing time, and nutritional proportion produced the highest improvement in mechanical properties of bacterial concrete producing compressive strength, flexural strength and split tensile strength of 79.5 MPa, 6.22 MPa, and 5.01 MPa respectively. Inline to OVAT approach, analysis of variance (ANOVA) also depicted that the individual as well as some of the possible two factor interactive model terms appeared as significant model terms with p-value was less than 0.05. The response predictive efficacy of the model was ascertained with a high R^2 value (0.99) and high adequate precision.

1. Introduction

Concrete is one of the most widely used materials for construction across the globe due to its durability, longevity and compressive strength. However, the structures made up of such materials are prone to the deterioration due to aging and the action of aggressive agents. There are several means which induce the deterioration including alkali-aggregate reactions, abrasion, fire, corrosion of steel bars etc. However, the corrosion of the steel bars can be arrested for an extended period by employing an intact concrete cover [1]. Moreover, the tensile stresses induced in the structures under certain loads may results in the formation of cracks which create a weak zone, acts as an active site of action for aggressive agents. Such cracks grow in number and magnitude during the service life of structures and thereby affect the durability of infrastructure. Consequently, the infrastructure requires repair within a very short period of time and involves a significant repairing and maintenance cost [2]. Maintenance cost increases exponentially due to concrete deterioration because of aging including the increases in the urban population which reside in such structure [3]. Additionally, in 2008, 8% of world CO_2 emissions (greenhouse gas emissions) were produced by cement and construction industry, which was regarded as the most pollutant generation anthropogenic activity [4]. Thus, the aim of improving the service life of any infrastructure becomes priority considering both economic and environmental prospective [5]. An extended service life generally minimise the requirement of new infrastructure and also diminishes the generation of greenhouse gas emission. On the view of this, the concept of bio-cementation and/or self-healing concrete came into the picture for its possible applicability to improve the sustainability of infrastructure. Here, the structure will regain its strength by auto activation of the crack healing process. In the process of self-healing, the small cracks undergo the repair process automatically under specified environment [6], while bio-cementation generally reduces the resilience of the construction materials by improving the strength.

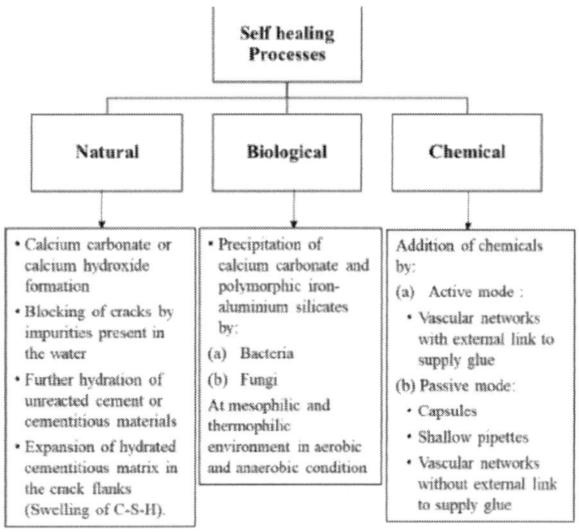

Figure 1. Comprehensive research taxonomy related to self-healing concrete [7, 8]

Several processes (such as natural, chemical and biological) have been practiced for the design of self-healing concrete as shown in the Figure 1. However, developing self-healing concrete using biological processes has received significant attention among the researchers. In this process, the microorganisms, mainly calcifying bacteria are employed in different forms during the preparation of concrete. As a result, the calcifying bacteria induce the production of $CaCO_3$ (calcite) which occupies the openings present in concrete specimen or fills in the cracks, thereby enhance the mechanical properties by reducing its porosity [3]. Several bacterial strains such as Bacillus, Halomonas, Acinobacter, Pseudomonas, Shewanella, Leuconostoc etc. have been utilized to design the self-healing concrete [9-11]. Out of which Bacillus are the widely used bacterial species. Such studies also reported that ability of different type Bacillus bacteria towards the production of calcite after being stimulated by water.

Several studies suggested that nutritional concentration also affect the calcination process in addition to curing time and bacterial composition. However, there are very limited studies related to the preparation of bacterial concrete and its subsequent statistical assessment with respect to the strength properties. Moreover, none of the studies taken nutritional concentration while statistical modelling. Therefore, in the present study, the statistical modelling was performed for optimization of strength properties (compressive strength, flexural strength and split tensile strength) of the bacterial concrete using Bacillus subtilis as bacterial strain incorporating the effect of three major process parameters namely, bacterial concentration (Cells/mL), curing time (days) and nutritional proportion (%).

2. Material and methods

2.1 Bacterial strain and growth conditions

A rod shaped, gram-positive, ureolytic type bacterium of Bacillus subtilis (CT-5) was chosen since it satisfy the essential criteria of survival in harsh physic-chemical conditions by forming the spores under adverse environment. Autoclaved nutrient broth (Himedia, India) with the standard composition was utilized to cultivate the Bacillus subtilis. Additionally, for in vitro precipitation of calcite, the culture was supplemented with sterilized 2% urea (w/v) and 25 mM$CaCl_2$ solution at 37°C in the orbital shaker at 120 rpm maintaining a neutral pH. Routine monitoring of growth was done by measuring the absorbance values using UV-Spectrophotometer at 600 nm. Further, a standard calibration curve was prepared using optical density test to assess the volume of culture solution required for different proportions.

2.2 Ordinary Portland cement

An ordinary Portland cement (43 Grade) conforming to BIS 8112-2013 was utilized for the production of concrete. The properties of OPC 43 were determined as per Indian standard specifications (IS 8112:1989) as given in Table 1.

Table 1. Properties of OPC 43

Physical Properties	Values
Fineness of Cement	2300 cm^2/g
Standard Consistency	31%
Initial Setting Time (min)	35
Final Setting Time (min)	510
Specific gravity	3.25

2.3 Fine aggregate and coarse aggregate

Fine aggregate acts as packing material in concrete to fill in the matrix and provide a condensed structure bound by cement. For this purpose, clean, dry and well-graded natural river sand (passing through 4.75 mm sieves) conforming to Zone II was utilized. Physical properties such as specific gravity and water absorption of fine aggregates were measured as 2.70 and 1.8%, respectively. Coarse aggregates provide the strength and resistance to abrasion, which occupies the maximum volume in the concrete. Crushed gravel with downsize of 20 mm was used as the coarse aggregates. The specific gravity and water absorption of aggregates were 2.65 and 1.39%, respectively.

2.4 Statistical Experimental Design

Design Expert software, version 8.0.7.1 was utilized for the experimental design based on the Central composite design (CCD). The interactive effects of three significant factors A (Bacterial concentration), B (Curing time) and C (Nutritional composition) on the response (Mechanical strength properties), were determined statistically using Response Surface Methodology (RSM). These variables A, B and C were considered as zero and investigated at three different levels (Table 2). The value for α (=1) was chosen keeping in mind that extreme factor ranges are less. The factorial points, axial points and the centre points are coded as (±1, ±1,±1), {(±α, 0, 0), (0, ±α, 0), (0, 0, ±α,)} and (0, 0, 0) respectively. A matrix design containing of 12 experiments produced by the software was performed for optimizing the mechanical strength properties (compressive, flexural and split tensile strength).

Table 2. Original and coded Variable

Variables	Original Variables	Coded Factors		
		-1	0	+1
Bacterial concentration ($\times 10^3$ Cells/ml)	A	5	10	15
Curing time (days)	B	3	15.5	28
Nutritional composition (%)	C	10	20	30

2.5 Experimental Design and Data Analysis for Concrete

A second order RSM experiments using three level full factorial CCD was employed to optimize the major parameters influencing the mechanical strength properties of concrete specimen. The parameters, bacterial concentration, curing time and nutritional composition were selected as the numeric factors affecting the mechanical strength properties of the process. 12 runs including 4 centre points were conducted for preparation of specific specimens randomly (Table 3).

Table 3. Three factor central composite design matrix for Bacterial Concrete

Run	A	B	C
1	15	28	10
2	15	28	30
3	10	15.5	20
4	15	3	30
5	5	28	30
6	10	15.5	20
7	15	3	10
8	5	3	30
9	10	15.5	20
10	5	28	10
11	10	15.5	20
12	5	3	10

The specimens of bacterial concrete were casted with standard procedures, demoulded after a required curing period and subjected to mechanical strength test. The experimental results were subjected to the second order polynomial model employing multiple regression technique followed by analysis of variance (ANOVA). Coefficient of determination (R^2) was assessed to observe the quality of best fit and to explain the variability of the model.

3. Results and Discussion

Batch Study for the Assessment of Mechanical Properties of Bacterial Concrete

3.1. Effect of Curing Time

The effect of curing time on mechanical properties of bacterial concrete was investigated by varying the curing time as 3, 14, and 28 days at fixed nutritional proportion (10%) and bacterial concentration (5×10^3 cells/mL). Figure 2a represents the effect of curing time on compressive strength of bacterial concrete, which can be compared with conventional concrete. A continuous improvement in the compressive strength was observed with increase in curing time for both the conventional concrete as well as bacterial concrete. However, incorporation of bacteria significantly enhanced the compressive strength of concrete specimen producing a maximum strength of 75.9 MPa. The highest value of compressive strength (69.8 MPa) in case of conventional concrete was achieved is 28 days of curing period, while the similar value of in case of bacterial concrete was observed within 14 days of curing time. From these findings, it is clear that the incorporation of bacteria significantly reduced the curing time to produce the similar values of compressive strength. The enhanced strength within the lesser curing period is related to the production of calcite by the utilized bacterial strain leading to the improvement of compressive strength. Similar trends of increase in other strength properties with increase in the curing time were also observed, yielding a maximum flexural strength and split tensile strength of 6 MPa and 5.1 MPa, respectively at 28 days (Figure 2b and Figure 2c). The observed results may be attributed to the fact that introduction of bacteria in the concrete generally improves the strength of concrete specimens by filling the produced calcite into the concrete pores by the process of hydration. However, excess production of calcite may affect the mechanical properties of the concrete [2].

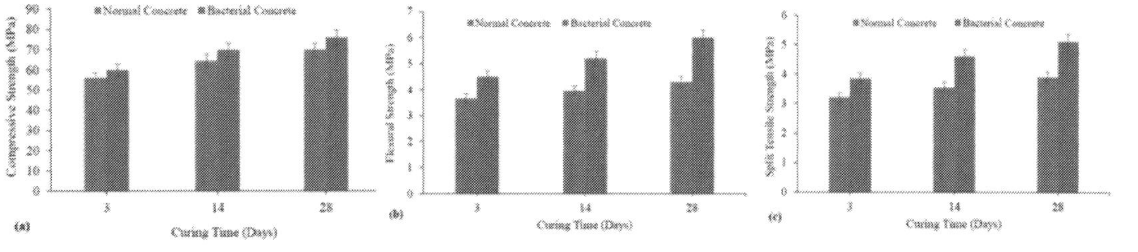

Figure 2. Effect of curing time on; (a) compressive strength, (b) flexural strength, and (c) split tensile strength [10% Nutritional proportion, 5×10^3 cells/mL bacterial concentration]

3.2. Effect of bacterial concentration

Mechanical strength properties of different bacterial concrete specimens were measured by varying the bacterial concentration in three steps (5×10^3 cells/mL, 10×10^3 cells/mL, 15×10^3 cells/mL) at constant curing time of 14 days and nutritional proportion of 10%. A continuous increase in the compressive strength, flexural strength and split tensile strength was observed with increase in the bacterial concentration from 5×10^3 cells/mL to 15×10^3 cells/ml. However, increase in the compressive strength was more prominent as compared to the rest of two properties where only 4-7% of improvement was observed. At extreme bacterial concentration of 15×10^3 cells/mL, keeping other two parameters constant, the value of compressive strength, flexural strength and split tensile strength was obtained as 75.8 MPa, 5.7 MPa, and 4.4 MPa respectively (Figure 3). However, the magnitude of these strength properties was way superior to that of conventional concrete at similar conditions of curing period. One likely reason behind this observation is that the initial amount of bacteria incorporated in concrete mixture (5×10^3 cells/mL) could also produce the sufficient calcite as higher concentration of bacteria in mixing solution at limited curing period.

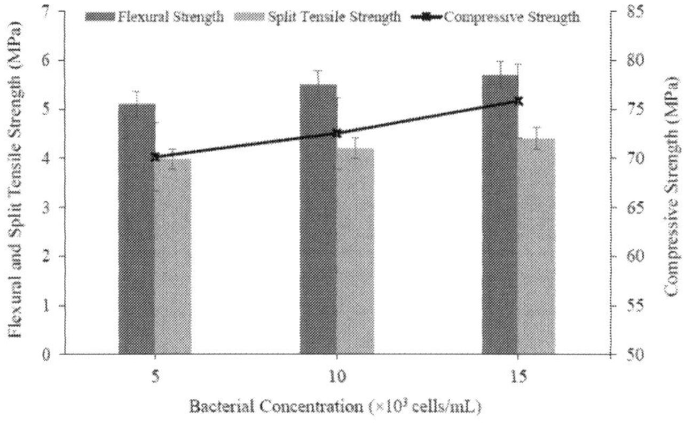

Figure 3. Effect of bacterial concentration on mechanical properties of bacterial concrete (14 days Curing time, 10% Nutritional proportion)

3.3. Effect of Nutritional Proportion

The effect of nutritional proportion on the mechanical strength properties of different bacterial concrete specimens were investigated by employing three step variation in the nutritional concentration (10%, 20%, and 30%) keeping rest of the two parameters constant (14 days curing time, 5×10^3 cells/mL bacterial concentration). As can be seen from Figure. 4, increase in nutritional proportion did not yield any significant improvement in either of the compressive strength, flexural strength or split tensile strength. At extreme nutritional proportion of 30%, keeping other two parameters constant, the value of compressive strength, flexural strength and split tensile strength was obtained as 71.8 MPa, 5.4 MPa, and 4.2 MPa, respectively (Figure. 4). Moreover, the improvement in the mechanical properties with increase in the nutritional proportion was in the range of 2-5%. Addition of nutritional medium during casting of bacterial concrete specimens ensures the enhancement of calcite formation leading to induce the strength of the specimens. However, in the present experiment, it seems that the initial proportion (10%) itself was sufficient to induce the bio-calcification for production of calcite at specified curing period.

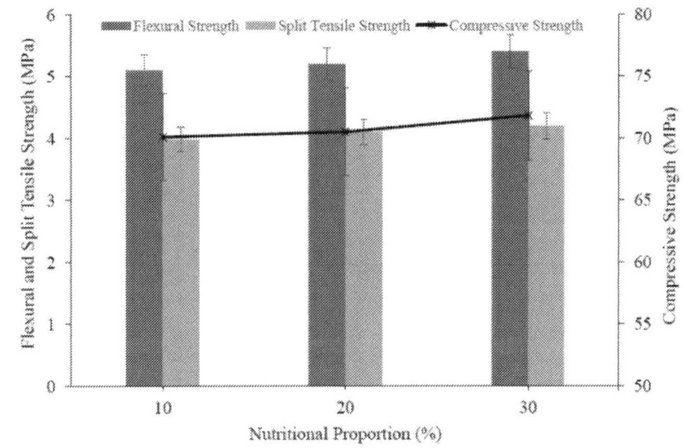

Figure 4. Effect of nutritional proportion on mechanical properties of bacterial concrete (14 days Curing time, 5×10^3 cells/mL bacterial concentration)

3.4. Statistical model development and effect of parameters

RSM model fitting and statistical analysis was also performed on bacterial concrete with the help of CCD experiments. A total twelve number of experimental runs incorporating various combinations were conducted and corresponding compressive strength values were recorded. The result thus obtained shows a wide range of compressive strength (59.9 MPa to 79.5 MPa), flexural strength (4.3 MPa to 6.22 MPa), and split tensile strength (3.7 MPa to 5.01 MPa), therefore warranted an optimization of the incorporated parameters. The empirical equations for the different mechanical strength properties in terms of coded factors are shown as equation 1, 2 and 3 respectively:

$$Compressive\ Strength = 68.57 + 3.17*A + 4.95*B + 1.04*C + 1.70*A*B - 0.51*A*C - 0.052*B*C + 0.64*A*B*C \quad (1)$$

$$Flexural\ Strength = 5.37 + 0.24*A + 0.63*B + 0.086*C - 0.094*A*B + 0.011*A*C - 0.026*B*C - 1.250E-003*A*B*C \quad (2)$$

$$Split\ Tensile\ Strength = 4.37 + 0.084*A + 0.55*B + 0.031*C - 0.011*A*B - 6.250E-003*A*C + 1.250E-003*B*C - 0.011*A*B*C \quad (3)$$

Where A, B, and C are the coded terms for the independent variables. The statistical confirmation of the selected model was accomplished employing analysis of variance (ANOVA) and the results are presented in Table 4, 5, 6, respectively for compressive strength, flexural strength, and split tensile strength. Normally, large F-values, small p-values (<0.05) and R^2 values approaching towards 1 are considered for the validation of a regression model. As can be seen from the ANOVA Tables, a p-value of <0.0001 (significant) and R^2 values close to 1 reflect the appropriateness to authenticate the model.

Table 4. ANOVA Results for Compressive Strength

Source	Sum of Squares	DF	Mean Square	F-Value	p-value Prob>F	
Model	313.13	7	44.73	331.44	0.0003	Significant
A-Bacterial Conc.	80.26	1	80.26	594.70	0.0002	
B-Curing Time	195.82	1	195.82	1450.89	<0.0001	
C-Nutritional Composition	8.65	1	8.65	64.11	0.0041	
AB	22.98	1	22.98	170.30	0.0010	
AC	2.06	1	2.06	15.27	0.0298	
BC	0.022	1	0.022	0.16	0.7132	
ABC	3.33	1	3.33	24.66	0.0157	
Curvature	94.41	1	94.41	699.48	0.0001	Significant
Pure Error	0.40	3	0.13			
Cor Total	407.95	11				
$R^2 = 0.9987$		Adj $R^2 = 0.9957$		Adeq Precision = 61.604		

Table 5. ANOVA Results for Flexural Strength

Source	Sum of Squares	DF	Mean Square	F-Value	p-value Prob>F	
Model	3.80	7	0.54	81.11	0.0021	Significant
A-Bacterial Conc.	0.48	1	0.48	71.03	0.0035	
B-Curing Time	3.19	1	3.19	476.39	0.0002	
C-Nutritional Composition	0.60	1	0.60	8.89	0.485	
AB	0.070	1	0.070	10.51	0.478	
AC	1.012E-003	1	1.012E-003	0.15	0.7233	
BC	5.512E-003	1	5.512E-003	0.82	0.4310	
ABC	1.250E-005	1	1.250E-005	1.868E-003	0.9682	
Curvature	0.36	1	0.36	53.46	0.0053	Significant
Pure Error	0.020	3	0.0067			
Cor Total	4.18	11				
$R^2 = 0.9947$		Adj $R^2 = 0.9825$		Adeq Precision = 27.102		

Table 6. ANOVA Results for Split Tensile Strength

Source	Sum of Squares	DF	Mean Square	F-Value	p-value Prob>F	
Model	2.50	7	0.36	2140.53	<0.0001	Significant
A-Bacterial Conc.	0.056	1	0.056	336.67	0.0004	
B-Curing Time	2.43	1	2.43	14586.07	<0.0001	
C-Nutritional Composition	7.812E-003	1		46.87	0.0064	
AB	1.012E-003	1	7.812E-003	6.07	0.0905	
AC	3.125E-004	1	1.012E-003	1.87	0.2644	
BC	1.250E-005	1	5.512E-003	0.075	0.8020	
ABC	1.012E-003	1	1.250E-005	6.07	0.0905	
Curvature	0.28	1	0.28	1677.02	<0.0001	Significant
Pure Error	5.0E-004	3	1.667E-004			
Cor Total	2.78	11				
$R^2 = 0.9998$		Adj $R^2 = 0.9993$		Adeq Precision = 117.17		

The diagnostics of experimental data was employed to strengthen the model validation, which produced the numerous graphical results such as normal plot of residuals, residual vs run number and predicted vs actual plots for compressive strength, flexural strength and split tensile strength (Figure 5).

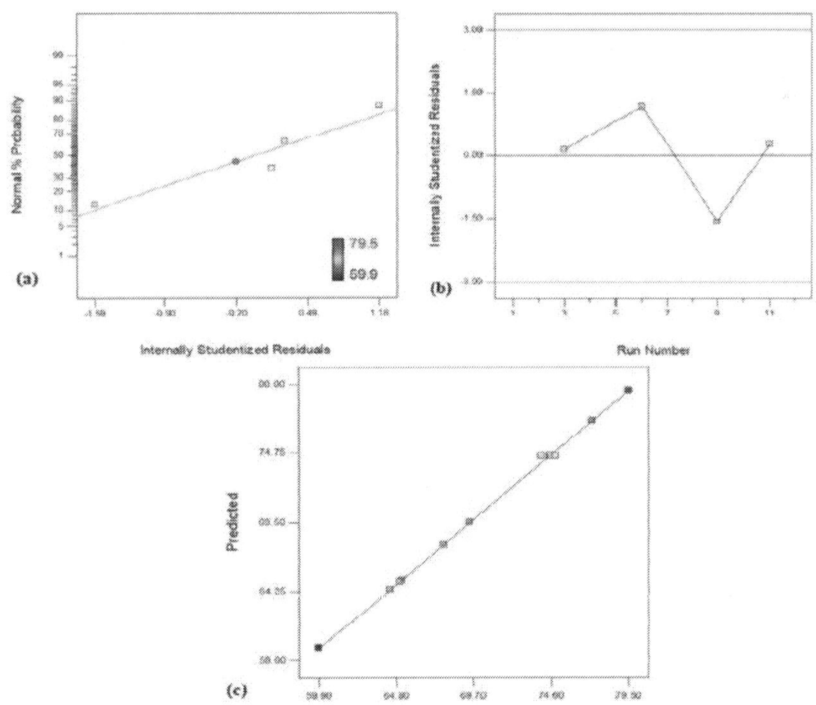

Figure 5. Model Validation (a. normal plot of residuals, residuals vs run, c. predicted vs actual plots)

All the residuals lie on or close to the normal probability line indicating very small errors in experimentation which affirm the reliability of the model. The plot of residual vs run number indicates that the residual lies well within the acceptable range of ±3, which further strengthen the model validation process. Moreover, the experimental values overlap with the predicted values, indicating the validation of the model validation.

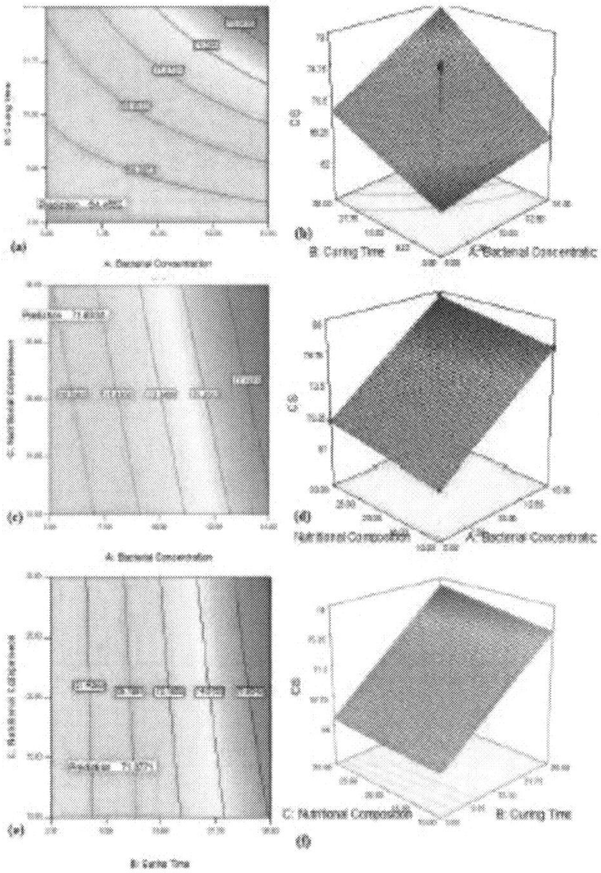

Figure 6. Contour and 3D plots of different combinations for Compressive Strength

3.5 Interactive effect of the factors (variables)

The effects of the selected factors on the mechanical properties of bacterial concrete have been ascertained by the ANOVA. Individual factors A (Bacterial concentration), B (Curing time), and C (nutritional composition) were observed as a significant variable on the basis of p-value (<0.05) for all the studied mechanical properties of bacterial concrete. Moreover, the interactive effect of the model terms AB, AC, and ABC for compressive strength, AB for flexural strength were also appeared to produce the significant effect as compared to other chosen model terms. The interactive effect of the factors was insignificant in case of split tensile strength, since p-value of the two factor model terms as well as three factor model terms are more than 0.05. Therefore, in this case the interactive model terms may be ruled out while deducing the model. The interactive effects of the significant parameters on the mechanical properties was represented by contour and 3D plots (Figure 6), which can be used to recognize the empirical system behaviour in an effective manner. The orange colour region in Figure 6 (on the yellow plane) represents the highest compressive strength. Thus, to achieve it, the range of selected variables should be in this region. The contour and 3D surface plots represent the two factor interaction while the third factor was fixed at centre point. In generally, a high bacterial concentration, high curing time, and high nutritional composition was appeared to produce high values of mechanical strength properties.

4. Conclusions

The following conclusions are drawn from the comprehensive experimental and statistical investigations on the behavior of bacterial concrete. Mechanical properties of bacterial concrete improved at increasing values of curing time, microbial concentration and nutritional proportional where effect of curing time on improvement of mechanical properties of bacterial concrete appeared as more prominent as compared to the rest parameters. Mechanical properties of concrete have been successfully modelled using empirical equation by optimizing the selected parameters. ANOVA output verified the established model by means of a high correlation coefficient ($R^2 = 0.99$) for all the mechanical properties and non-significant of lack of fit. Further, the interactive effect of selected parameters was also explored and presented with the help of 3D surface as well as contour plots. The interactive effect of the model terms AB, AC, and ABC for compressive strength, AB for flexural strength were appeared to produce the significant effect as compared to other chosen model terms. Overall, the application of culture broth of Bacillus subtilis directly into fresh concrete at different nutritional proportion improved the strength of concrete considerably.

References

[1] Han B, Zhang L and Ou J 2017 *Smart and multifunctional concrete toward sustainable infrastructures*: Springer
[2] Vijay K, Murmu M and Deo S V 2017 Bacteria based self healing concrete–A review *Constr. Build. Mater.* **152** 1008-14
[3] Joshi S, Goyal S and Reddy M S 2018 Influence of nutrient components of media on structural properties of concrete during biocementation *Constr. Build. Mater.* **158** 601-13
[4] Benhelal E, Zahedi G, Shamsaei E and Bahadori A 2013 Global strategies and potentials to curb CO_2 emissions in cement industry *J. Clean. Prod.* **51** 142-61
[5] Pan X, Shi Z, Shi C, Ling T-C and Li N 2017 A review on surface treatment for concrete–Part 2: Performance *Constr. Build. Mater.* **133** 81-90
[6] Khaliq W and Ehsan M B 2016 Crack healing in concrete using various bio influenced self-healing techniques *Constr. Build. Mater.* **102** 349-57
[7] Al-Thawadi S M 2011 Ureolytic bacteria and calcium carbonate formation as a mechanism of strength enhancement of sand *J. Adv. Sci. Eng. Res* **1** 98-114
[8] Jonkers H M 2007 *Self healing materials*: Springer pp 195-204
[9] Ghosh P, Mandal S, Pal S, Bandyopadhyaya G and Chattopadhyay B 2006 Development of bioconcrete material using an enrichment culture of novel thermophilic anaerobic bacteria *Indian J. Exp. Biol.* **44(4)** 336-9.
[10] Ramikrishnan V, Panchalan R and Bang S 2005 Improvement of concrete durability by bacterial mineral precipitation" Proceedings ICF 11 *Torino, Italy*
[11] Rivadeneyra M A, Ramos-Cormenzana A, Delgado G and Delgado R 1996 Process of carbonate precipitation by Deleya halophila *Curr. Microbiol.* **32** 308-13

IOP Conf. Series: Materials Science and Engineering **970** (2020) 012028 doi:10.1088/1757-899X/970/1/012028

Effect of Stone dust and Lime in the Geotechnical Properties of Clayey Soil

Rasmiranjan Samal, Ayan Mishra

Department of Civil Engineering, Silicon Institute of Technology, Sambalpur, Odisha, India

E-mail: rasmiranjan.samal@silicon.ac.in

Abstract. Clayey soils are commonly stiff in dry state but lose their stiffness when saturated with water. Soft clays are characterized by low bearing capacity and high compressibility. In this research, stone dust obtained from stone processing area used for Stabilization of clayey soil with lime. Thus, the effectiveness of using stone powder and lime in stabilizing fine-grained clayey soil (CL) was investigated in the laboratory. The soil samples in natural state and when mixed with varying percentages of lime and waste stone powder were used for the laboratory tests that included Atterberg limits tests, standard Proctor compaction tests, unconfined compression tests and California bearing ratio tests. The value of maximum dry density, Unconfined Compressive Strength and California Bearing Ratio decreased with increase in % of stone dust and lime. The Optimum Moisture Content value remain same with increasing stone dust content and lime.

1. Introduction

1.1. Need for Stabilization of soil

It may be defined as the modification of soil done to improve their certain characteristics. It develops the shear capacity and shrinking, swelling of soil. It boosts the load carrying power to assist the construction works. In our project work we have undertaken clayey soil, substituted with stone dust by adding lime

1.2. Materials for soil stabilization

1.2.1 Clayey Soil

Clay may be defined as a soil variant which may be found directly from the Earth's surface which contains many types minerals, also it includes chemical compounds like metal oxides certain types of organic matters. These types of materials show plasticity behaviour because of moisture holding capacity thereby it becomes firm, fragile and non-plastic upon dehydrating. This classification of materials contains tiniest particles among all the other groups of soil variants. Due to this small size the solitary particles cannot be viewed by naked eyes and it requires a microscope to get into the details. This particular attribute permits a bulk of clay particles to accommodate in a small space. As a result of which clay has a very smooth texture as the small particles cannot create unevenness on the surface.

The composition of clay is very murky due to miniature particle proportion. Plant roots cannot perforate due to the strong bonding between the inter particles which is firm. Clayey soils are much thicker and substantial due to high density as a result of which the time taken to warm up is longer after spells of cold atmosphere. Clay substances due to less void spaces are impervious to erosion than other soil varieties.
For the growth of flora in clay type of soil certain alterations are done for the successful growth and development. Clay soil usually has scarce nutrients and micro-nutrients which are essential components for plant development, hence organic matter is added. After the addition of organic matter, the clay may become alkalescent as a result of which certain modifications may be carried out to balance the clay's pH to convert it to a neutral medium. Therefore, the clay soil is examined before planting to determine soil's pH and amount of essential nutrients present. Due to a very high water holding capacity clay soil shows very little permeability. Clay materials are closely packed so water takes much more time to penetrate through the particles as a result of which clay sucks up the water and expands. This issue prevents water from perforating deep and also it hampers plant roots due to unnecessary expansion. [8-11]

Table 1. Properties of clay

Properties	Code referred	Value
Specific Gravity	IS 2720 (Part 3/Sec 1) - 1980	2.62%
Maximum Dry Density (MDD)	IS 2720 (Part 7) - 1980	17.55 KN/m^3
Optimum Moisture Content (OMC)	IS 2720 (Part 7)- 1980	16 %
Liquid Limit	IS 2720 (Part 5)- 1985	72%
Plastic limit	IS 2720(Part 5) - 1985	39%
Unconfined compressive strength (UCS)	IS : 2720 (Part 10)	1048 KPa

1.2.2. Stone Dust
Stone dust is like a gloomy, coarser version of sand. It is an end product of running stones gone through a crushing machine to make crushed stones. Its exact structure will definitely depend on what variety of stone was run through the crushing device. For instance, sometimes granite is pushed through such a machine; in other situations, it could be limestone, for example. The machine has a screen that entraps the bigger material (that is, the crushed stone). The smaller material or "screenings" falls through the screen. Relying on the dimension of the holes in the screen used, the material can be so fine in texture that it basically looks like a powder.

Table 2. Chemical properties of stone dust Source

Component	Weight (%)
CaO	3.5-40
Al_2O_3	0.5-40
MgO	2.5-25
SiO_2	1-12
SO_3	0.23-3
Alkalis	0-4

Stone dust or powder is a type of crushed stone; it consists of similar variants of stones but it converted into powder form. Stone dust is always resistant to water also it acts a binding agent. Stone dust is impervious hence they are used in various civil applications. Due to its impervious quality, it never allows the mould to grow. This is an end product of stone crushing factory. Maximum amounts of stone dusts that are produced goes unused. It can be exploited for concrete works as a result of which it will mitigate the solid waste along with proper recovery. Stone dusts concrete gains about 15% better strength than normal sand.

Table 3. Index properties of stone dust Source

Property	Value
Natural Moisture Content(%)	9.11
Optimum moisture content (%)	11.5
Particle Size distribution	
Silt (%)	2.9
Sand (%)	97.1
CBR Un soaked (%)	26.28
Specific Gravity	2.76
CBR Soaked (%)	11.5
Cohesion (kN/m^2)	0.07
Maximum dry density (g/cm^3)	1.97
Angle of internal friction (degree)	35

1.2.3. Lime

Quicklime is calcium oxide (CaO), which has variable particle size depending on the end use and high available calcium content. After mixing initially, the calcium ions (Ca^{++}) from hydrated lime drift to the surface of the clay materials and displace water and other types of ions. The soil becomes friable and granular, makes it easier to work and compact. The overall procedure, which is called "flocculation and agglomeration," normally occurs in a few hours. Using certain amounts of CaO and water, acidity decreases rapidly and pH becomes more than 10.5 eventually breaking down the clay matters. Products released that are Silica and alumina combine with Ca to produce calcium-silicate-hydrates (CSH) and calcium-aluminates-hydrates (CAH). These are the cementitious by-products. The matrix that is formed by the products improves the strength of the layers of the lime-modified soil. After formation of the matrix, transformation of the soil happens from a void less material to a relatively firm impervious layer with a massive load bearing power. Within the first few hours, initial stage of the procedure begins for a lengthy period of time which may last for years in a properly designed setup. At the end, the whole mix becomes impervious, firm and long lasting structurally.

1.3 Scope and objectives

To study the effect of stone dust and lime on properties of clay.

2. Materials and methodology

2.1 Materials

As a part of this overall investigation, the clayey soil was taken up from the staff quarter's area at Silicon Institute of Technology, Sason, Sambalpur. The clayey soil materials which was procured from the locality was carried to the laboratory for further processes. Some amount of soil was taken and sieved through 4.75 mm sieve and air-dried. Then some other tests such as California bearing ratio test(CBR), unconfined compressive strength test(UCS) and standard proctor test (SPT) were carried out to find the value of California bearing ratio strength (CBR value), unconfined compressive strength(UCS value), maximum dry density(MDD) and optimum moisture content (OMC). The stone dust that is used in the investigation was acquired from Debadihi, Jharsuguda.

2.2 *Methodology*

To examine the impact of stone dust and lime as a additive agent in clayey soils, some of tests were carried out, where the amount of stone dust in the clay was varied in values of 10%,20%,30% and 40%

IOP Conf. Series: Materials Science and Engineering **970** (2020) 012028 doi:10.1088/1757-899X/970/1/012028

(multiples of 10) by weight of the total quantity of soil. Then by adding 1% and 2% lime in each mix, the experiments for various properties were carried out. The experiments were conducted as per following IS codal provision:

- IS 2720 (Part 5) –1985 - Liquid limit & Plastic limit test
- IS: 2720 (Part 7) – 1980 - Standard proctor test (SPT)
- IS: 2720 (Part 10) – 1991 - Unconfined compressive strength (UCS) test
- IS: 2720 (Part 16) – 1987 - California bearing ratio (CBR) test

3. Results and discussions

After conducting the experiments, the various results were obtained and accordingly the graphs were plotted. Where S represents soil, S.D. represents stone dust and L represents lime.

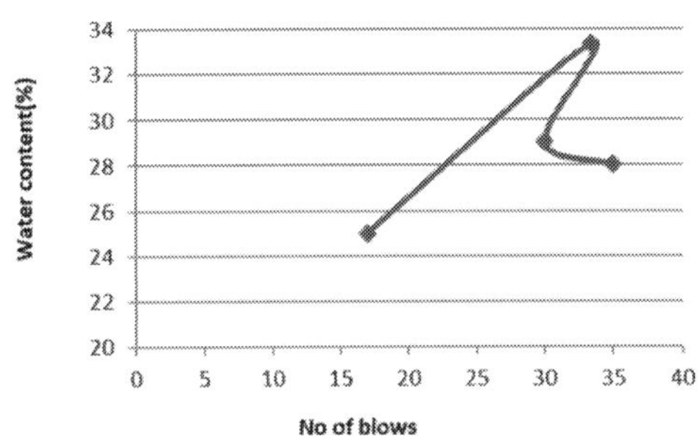

Figure 1. Liquid limit of the soil

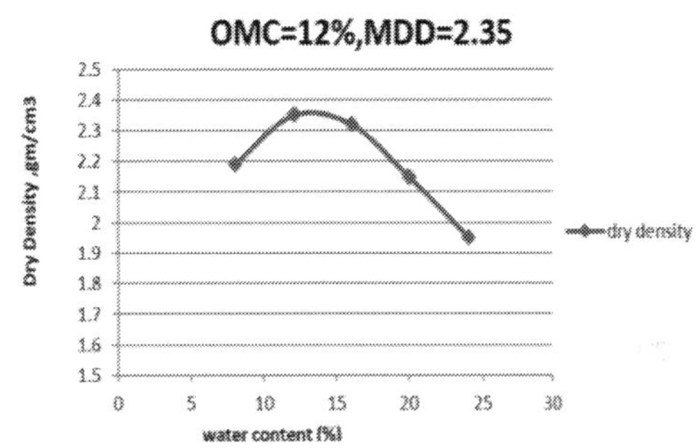

Figure 2. Density-moisture content variation of the soil

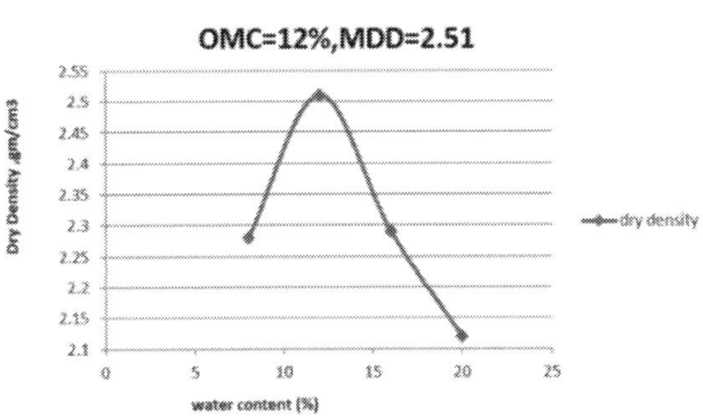

Figure 3. Density-moisture content variation of the soil 90% + stonedust10%

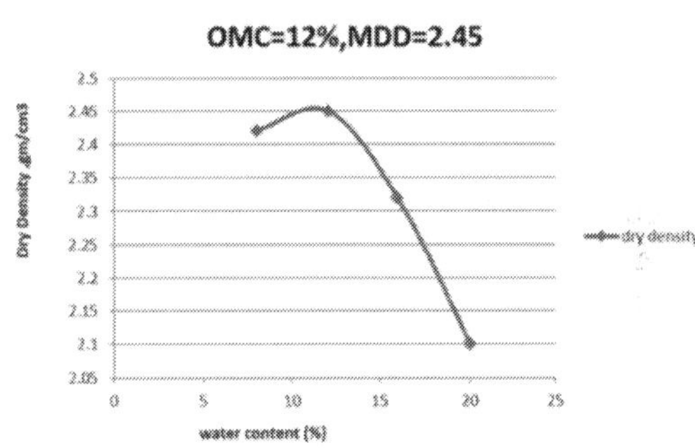

Figure 4. Density-moisture content variation of the soil 80% + stone dust 20%

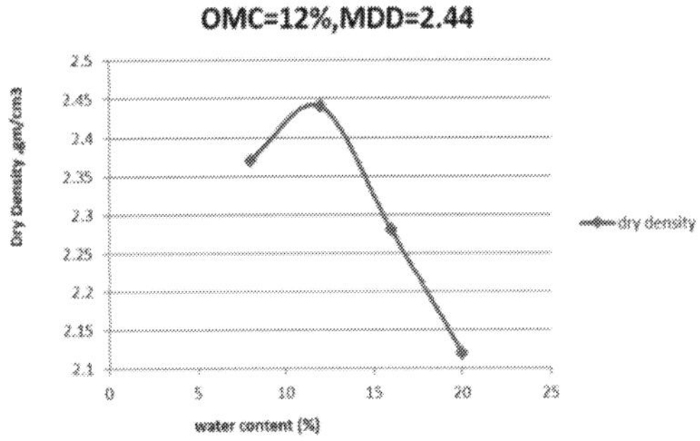

Figure 5. Density-moisture content variation of the soil 70% + stone dust 30%

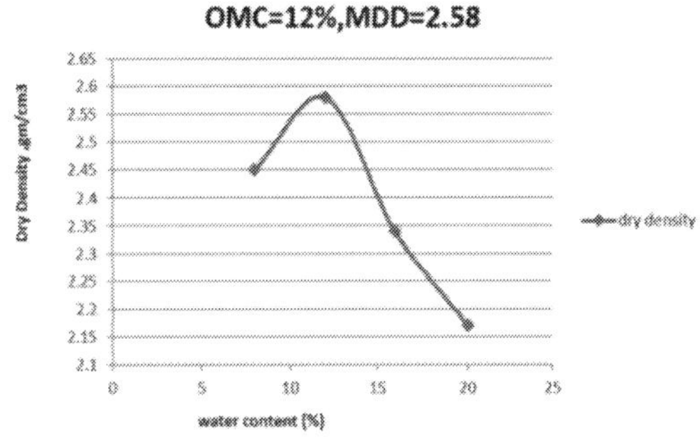

Figure 6. Variation of Density-moisture content of the soil 60% + stone dust 40%

IOP Conf. Series: Materials Science and Engineering **970** (2020) 012028 doi:10.1088/1757-899X/970/1/012028

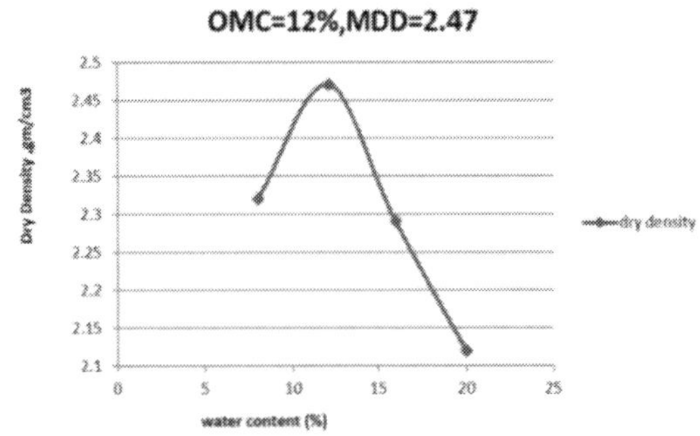

Figure 7. Variation of Density-moisture content of the soil 90% + stone dust 10% + lime 1%

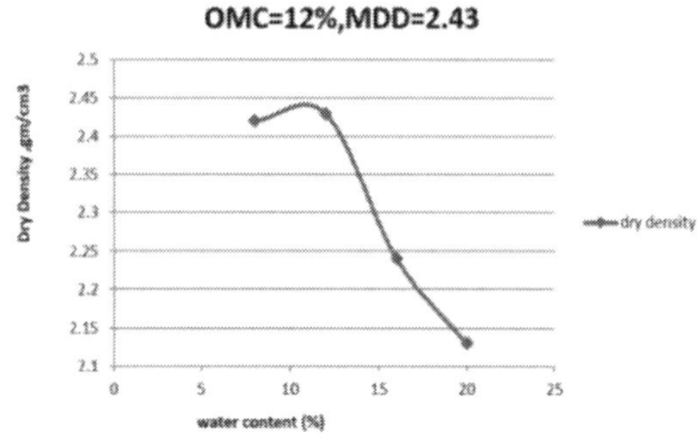

Figure 8. Variation of Density-moisture content of the soil 80% + stone dust 20% + lime 1%

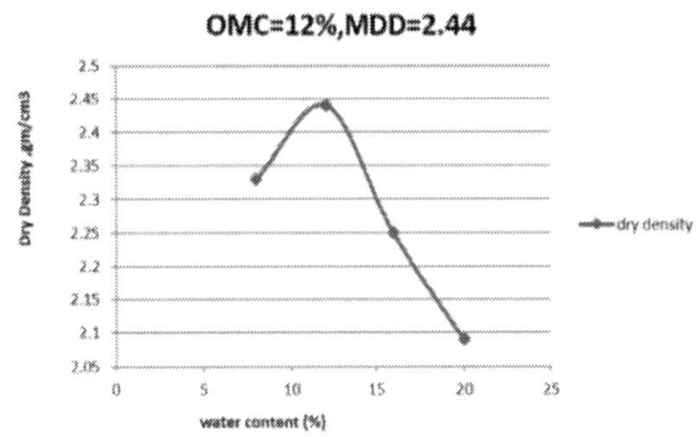

Figure 9. Variation of Density-moisture content of the soil 70% + stone dust 30% + lime1%

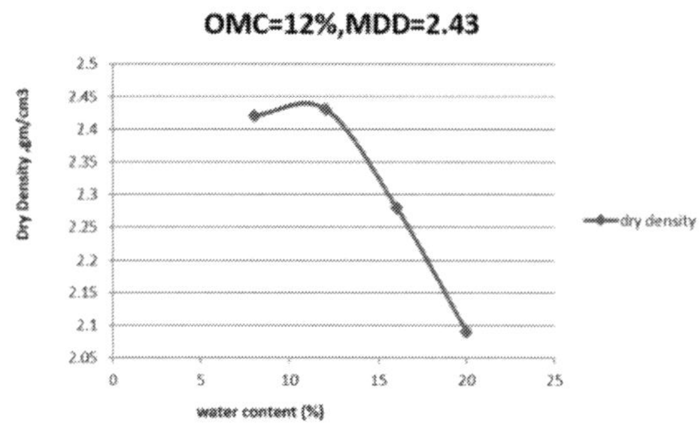

Figure 10. Variation of Density-moisture content of the soil 60% + stone dust 40% + lime1%

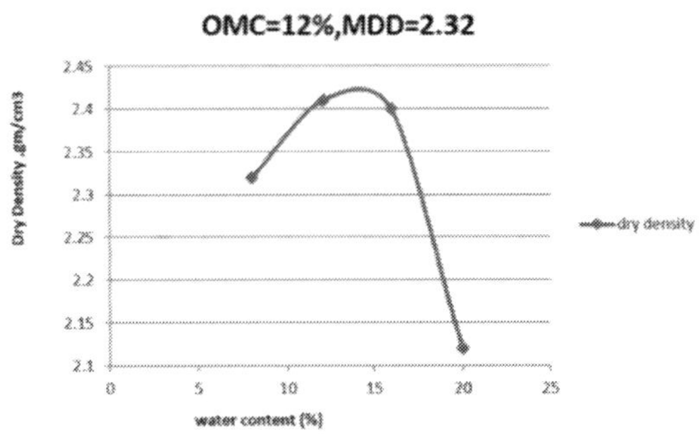

Figure 11. Variation of Density-moisture content of the soil 90% + stone dust 10% + lime 2%

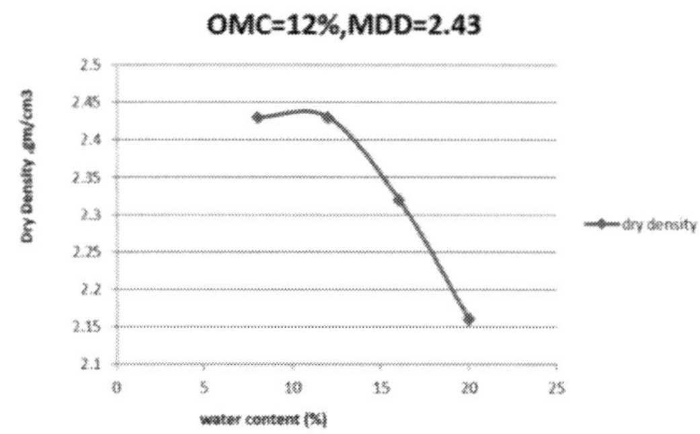

Figure 12. Variation of Density-moisture content of the soil 80% + stone dust 20% + lime 2%

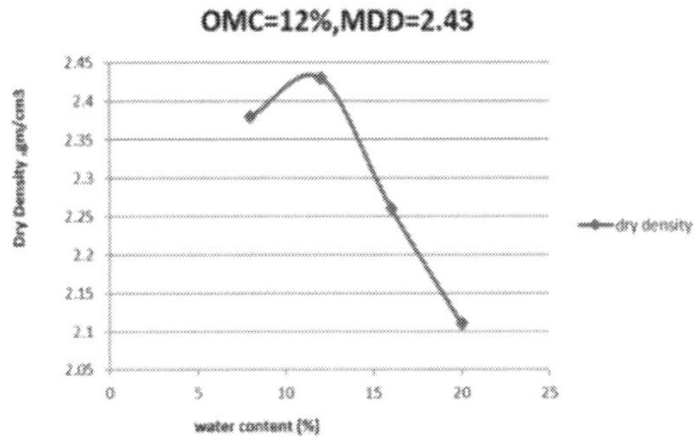

Figure 13. Variation of Density-moisture content of the soil 70% + stone dust 30% + lime 2%

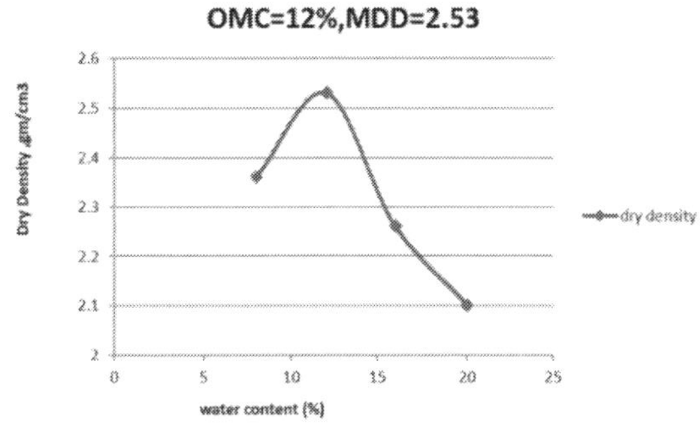

Figure 14. Variation of Density-moisture content of the soil 60% + stone dust 40% + lime 2%

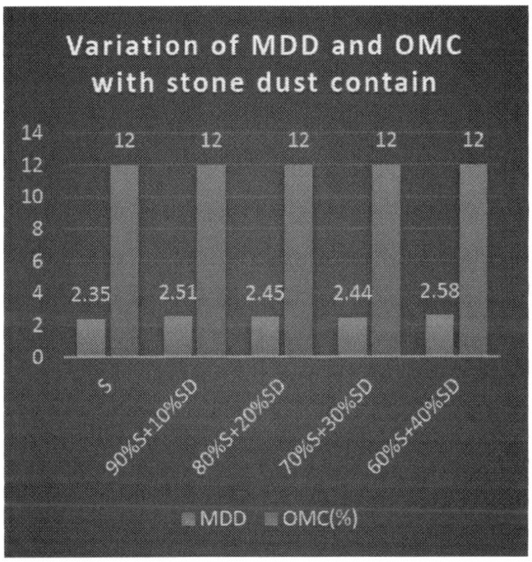

Figure 15. Variation of UCS value of clayey with different stone dust contain

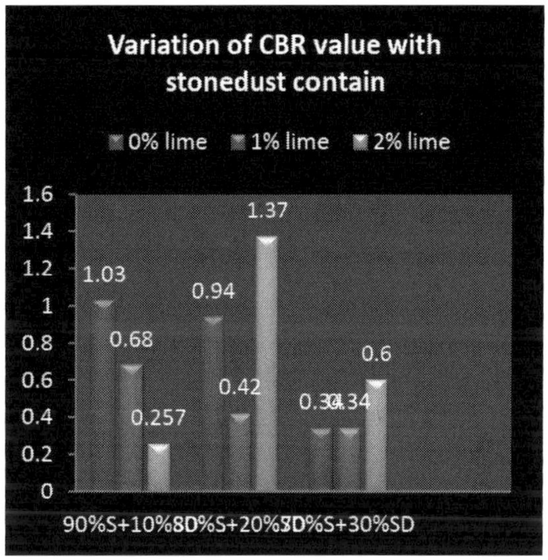

Figure 16. Variation of CBR value of clayey with different stone dust contain

- Clay was substituted with varying different percentage of stone dust (10% - 40% in multiple of 10) and addition of lime of 1% and 2% by weight which affects as an additive on the clay.
- The magnitude for MDD value of clay soil initially increased with the replacement of soil by stone dust. Then it showed decrement with increment of stone dust percent. Maximum Dry Density value was detected for a mixture of 60% soil and 40% stone dust.
- The value of every proportion of Optimum Moisture Content remains constant.
- It was interpreted that the Unconfined Compressive Strength value of soil with replacement of stone dust and addition of lime with different proportion is decreasing, as the percentage of stone dust increasing and the maximum dry density was observed for a mixture of 90% of soil and 10% stone dust and 1% of lime and the minimum value of Unconfined Compressive Strength was observed for mixture of 60% 0f soil 40% of stone dust and 1% of lime.
- It was interpreted that California Bearing Ratio value of soil with replacement of standard matter and addition of lime with different proportion is decreasing except for an addition of 2% lime. The maximum California Bearing Ratio values observed for a mixture of soil 80% and 20% stone dust

and 2% lime. It was observed that, California Bearing Ratio value was minimum for a mixture of 90% soil, 10% stone dust and 2% lime

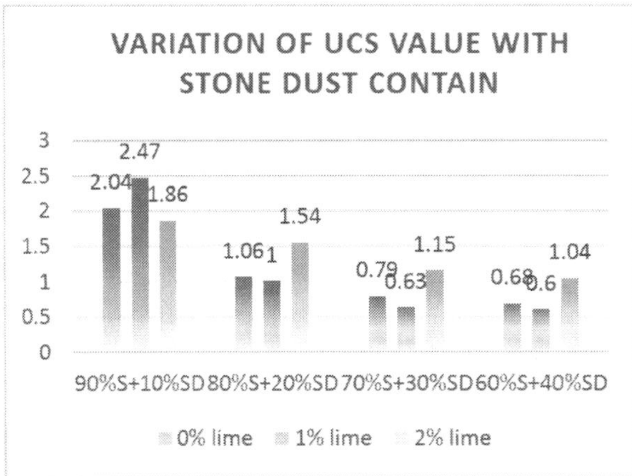

Figure 17. Variation of UCS value of clayey with different stone dust contain

4. Conclusion

On the basis of results procured and after comparing we got following conclusions

- The Optimum Moisture Content value remain same with increasing stone dust content and lime.
- The value of maximum dry density, Unconfined Compressive Strength and California Bearing Ratio decreased with increase in % of stone dust and lime.
- Hence Stone dust is not a proper substitution of clay.

In future scope, Stone dust may be replaced by red mud and using some additive like lime or other substances.

References

[1] Al-Joulani N 2000 Engineering properties of slurry waste from stone cutting industry in the west bank. In: *Proceedings of the First Palestine Environmental Symposium,*

[2] Jahanshahi M 2005 An improvement method for swell problem in sulfate soils that Stabilized by lime *American Journal of Applied Sciences* **2** 1121-8

[3] Kumar A, Walia B S and Bajaj A 2007 Influence of fly ash, lime, and polyester fibers on compaction and strength properties of expansive soil *Journal of Materials in Civil Engineering* **19** 242-8

[4] Malik A and Thapliyal A 2009 Eco-friendly fly ash utilization: potential for land application *Critical Reviews in Environmental Science Technology* **39** 333-66

[5] Nadgouda K and Hegde R 2010 The effect of lime stabilization on properties of black cotton soil. In: *Indian Geotechnical Conference,* pp 511-4

[6] Ali M S and Koranne S S 2011 Performance analysis of expansive soil treated with stone dust and fly ash *EJGE* **16**

[7] Haldar M K and Das S K 2012 Effect of substitution of sand stone dust for quartz and clay in triaxial porcelain composition *Bulletin of Materials Science* **35** 897-904

[8] Phanikumar B and Sharma R S 2007 Volume change behavior of fly ash-stabilized clays *Journal of materials in Civil Engineering* **19** 67-74

[9] Phanikumar B, Naga Reddayya S and Sharma R 2001 Volume change behavior of fly ash-treated expansive soils. In: *Proceedings of the 2nd International Conference on Civil Engineering, Indian Institute of Science, Bangalore, India,* pp 689-965

[10] Phani Kumar B and Sharma R S 2004 Effect of fly ash on engineering properties of expansive soils *Journal of Geotechnical Geoenvironmental Engineering* **130** 764-7

[11] Misra A, Biswas D and Upadhyaya S 2005 Physico-mechanical behavior of self-cementing class C fly ash–clay mixtures *Fuel* **84** 1410-22

MDD & OMC of Black Cotton Soil Reinforced with Randomly Distributed Banana Fibers

L. B. Patil, S. S. Pusadkar

[1]Department of Civil Engineering, Government College of Engineering, Jalgaon, Maharashtra, India

E-mail: lalitalalitapatil@rediffmail.com

Abstract. Expansive soil causes serious problems on civil engineering structures in arid and semi-arid climate regions of the world. Such soils, swell during rainy season and shrink in summer season. In India, expansive soil includes almost the entire Deccan plateau, and causes damage to buildings, roads, pipelines, and other structures every year. The black cotton soil had improved by various methods to enhance the engineering characteristics of soil. However, due to various issues such as leaching, release of mineral, these methods are not sustainable. Hence, an attempt had been made to improve black cotton soil characteristics by using the waste material such as banana fiber. In the present study, the experiments were conducted to understand the density & moisture content of soil by reinforcing the soil with randomly distributed banana fibers. The laboratory tests were conducted on locally available black cotton soil. The banana fiber were added to the soil in varying percentages of 0.0 %, 0.50 %, 1.00 %,1.5 % and2.00 % and in varying lengths of 1 cm,2cm and 3 cm randomly. The standard Proctor tests were conducted to study its effect of banana fiber on dry unit weight and OMC. It was observed that MDD is decreasing and OMC is increasing with increase in % of fibers and length of fiber.

1. Introduction

The civil engineering structures founded on the ground and transfer the load safely to it. The properties of soil below play major role in the stability of the structures. Expansive clays swell due to water absorption and shrink due to water evaporation. Hence, the structures founded on these grounds are severely damaged. The various techniques are used to avoid the problems caused by expansive soils. The improvement of soil is carried out with locally available waste material then it will possible to reduce the cost of structure. The best option is to reinforce the soil with some strong material. To improve the soil. The soil reinforcement is usually carried out with rod, strips, geo-membrane etc. Many researchers also attempted to make use of natural fibres such as jute, coir, and banana as reinforcing materials in soil. The natural fibre materials are available locally and are very cheap. Also they are biodegradable. Hence, use of natural fibres lead to green and sustainable ground improvement [1-9].

2. Literature Review

The reinforcement of soil using natural and artificial fibers is continually studied by various researchers. These studies are discussed below.

IOP Conf. Series: Materials Science and Engineering **970** (2020) 012029 doi:10.1088/1757-899X/970/1/012029

Babu and Vasudevan [1]studied the reinforcement of red soil using coir fibers and reported maximum increase of stress, toughness and stiffness of the soil when the fiber length was between 15 and 25 mm. Chaple and Dhatrak [2] investigated the properties of clayey soil reinforced with addition of coir fibers. The soil and coir fibers were mixed in proportions. The percentage of coir fibers varies from 0.0 %, 0.25 %, 0.50 %, 0.75 % and 1.00 %. It was observed that, OMC increased from 18.1 % to 20.76 % and MDD was decreases from 17.28 KN/m^3 to 16.62 KN/m^3. The increase in OMC may be due to water absorption by the fiber. Subramani and Udayakumar [7] reinforced claye soil with coconut fibers and reported maximum improvement in U.C.S. and C.B.R when 0.5% of coir mixed with the soil. Sunny and Joy [8] reinforced marine clay with banana fibers and reported that banana fiber content with 0.75% shows maximum improvement in the soil properties. The literature review shows improvement in various soil properties with the addition of natural fibres. Since, Jalgaon district is hub for production of banana and lot of banana trunk are available for making the fibres. Also, this area contains black cotton soil which causes many problems to structures. Hence, in order to solve the problem of banana waste and black cotton soil, an attempt has been made to understand the behaviour of banana fibres reinforced black cotton soil.

3. Methodology

The soil was collected from three different villages namely Nashirabad, Takarkheda and Dhamangaon of Jalgaon district from Maharashtra. All the sample were tested for plasticity characteristics and free swell index. The soil having high plasticity and high free swell index was selected for the study i.e. soil from Dhamangaon. The banana fibers were collected from Varangaon village from Jalgaon district of Maharashtra. The fibers were mixed in soil with proportions. The banana fibers vary from 0.0 %, 0.50 %, 1.00 %, 1.5 % and 2.00 %. The lengths of banana fibers were 1 cm, 2cm and 3 cm. The standard Proctor tests were conducted with different % of banana fibers.

4. Materials and Methods

4.1. Black Cotton Soil

The soil was collected from various area within Jalgaon district of Maharashtra for testing. Various properties of soil are shown in Table 1.The soil from Dhamangaon village shows higher PI and DFS, hence selected for the study.

Table 1. Properties of three different village's soil

Village	LL (%)	PL (%)	I_P (%)	FSI (%)
Nashirabad	62.67	51.09	11.58	52.6
Takarkheda	64.67	52.26	9.41	63.40
Dhamangaon	85.33	62.72	22.61	69.53

4.2. Banana Fiber

The banana fiber is a natural fiber obtained from a banana plant. The fiber was obtained mainly from pseudostem, which acts as a strong fiber after properly drying. It is a better fiber with suitable stiffness and good mechanical properties. The fiber was obtained from Varangaon near Busawal in the district of Jalgaon of Maharashtra state, India. The banana fibers used for the present study in the percentage of soil varies as 0.0%, 0.50%, 1.00 %, 1.5% and 2.00%. The length of banana fibers used was 1cm, 2cm and 3cm.

4.3. Test Methods

The initial studies were directed to analyses geotechnical characteristics of black cotton soil i.e. liquid limit, free swell index and plastic limit. For determining maximum dry density and optimum moisture content of original and reinforced soil, standard Proctor test was also done.

5. Results and Discussion

The consequences of tests directed on Dhamangaon soil are presented in Table 2. As per IS soil classification, the soil was categorized as silt of high compressibility (MH). The MDD of sample at standard Proctor test was found to be 14.9 KN/m^3 and OMC as 29.6 %.

Table 2. Properties of black cotton soil

Soil	LL(%)	PL(%)	I_P(%)	FSI(%)	OMC(%)	MDD kN/m^3
MH	85	62	23	70	29.6	14.9

5.1. Standard Proctor compaction test on soil with variation of banana fibers constituent

This was performed on reinforced soil with wavering percentage of banana fibers and length for obtaining maximum dry unit weight (MDD) and optimum moisture content (OMC) as shown in Figure 1. Similar curves were plotted for all the tests. Table 3, 4 and 5 illustrates OMC and MDD obtained from standard Proctor test for soil with wavering percentage of banana fibers for 1 cm, 2 cm, and 3 cm length of fiber.

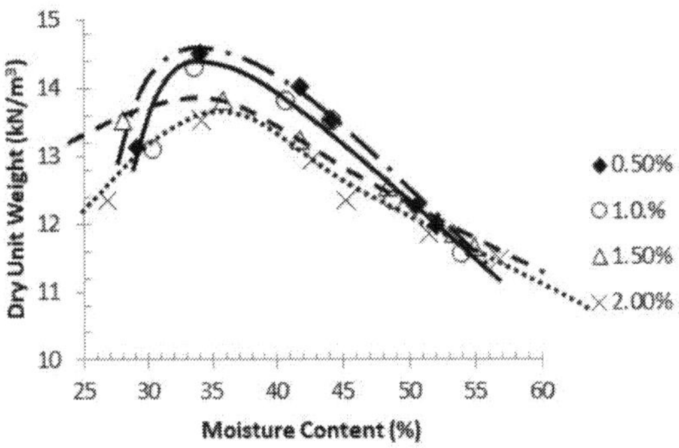

Figure 1. Dry Unit Weight –Moisture Content Curve for 1 cm Length of Banana Fiber

Table 3. OMC and MDD for 1 cm length of Banana Fiber

%Fiber added	0	0.5	1.0	1.5	2.0
OMC(%)	29.6	35.0	35.4	36.0	36.4
MDD(KN/m^3)	14.9	14.7	14.2	13.9	13.5

Table 4. OMC and MDD for 2 cm length of Banana Fiber

%Fiber added	0	0.5	1.0	1.5	2.0
OMC(%)	29.6	32.0	34.0	35.0	36.6
MDD(KN/m^3)	14.9	14.4	14.1	13.7	13.0

Table 5. OMC and MDD for 3 cm length of Banana Fiber

%Fiber added	0	0.5	1.0	1.5	2.0
OMC(%)	29.6	29.4	32.6	33	36.5
MDD(KN/m^3)	14.9	14.1	13.8	13.6	12.74

Figure 2 and 3 illustrates the variation of OMC and MDD attained from customary Proctor test for soil with wavering percentage of banana fibers for 1 cm, 2 cm, and 3 cm length of fiber. It shows that as the % content of fiber and length increases, the maximum dry density goes on decreasing. As the %

banana fiber content increases OMC increases and became same at 2% content while as the length increases, OMC decreases.

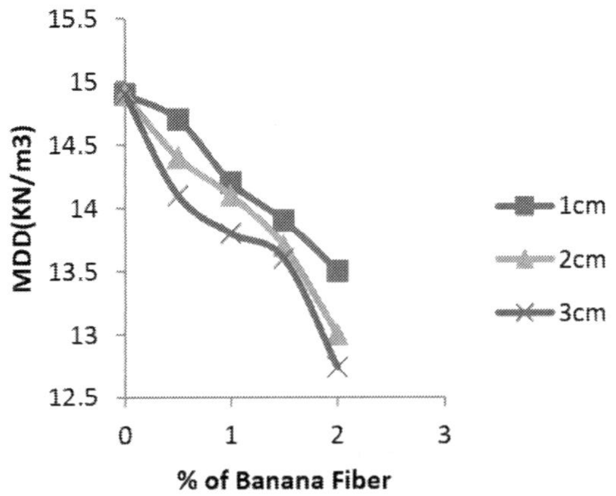

Figure 2. Variation of MDD with % of Banana Fiber

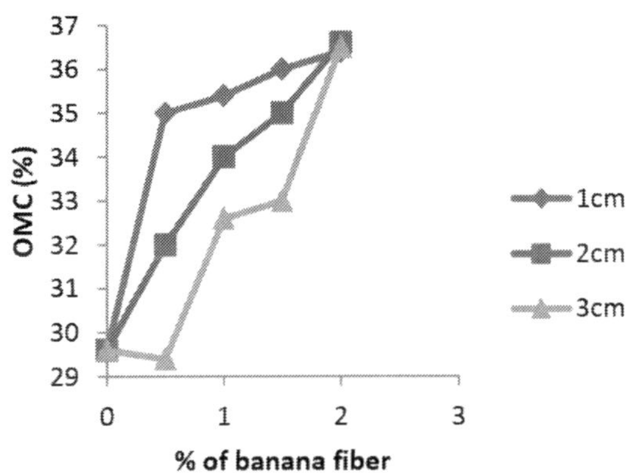

Figure 3. Variation of OMC with % of Banana Fiber

6. Conclusions

The black cotton soil was reinforced with banana fiber of varying length and %. The fibers were mixed in soil in randomly. The inclusion of fibers in soil has been studied for unit weight characteristics of reinforced soil. Addition of banana fibers influences the characteristics of black cotton soil. Some conclusion which can be drawn:

- Optimum moisture content of reinforced soil increased with percentage and length of banana fibers
- Maximum dry unit weight of reinforced soil reduces with percentage and length of banana fibers
- For fiber less than 2 cm length, 0.5% banana fiber shows maximum increase in OMC while for 3 cm length, 1 % banana fiber.
- The decrease in MDD is marginal at 0.5% or 1% banana fiber.

References

[1] Sivakumar Babu G and Vasudevan A 2008 Strength and stiffness response of coir fiber-reinforced tropical soil *Journal of Materials in Civil Engineering***20** 571-7

[2] Chaple P M and Dhatrak A 2013 Performance of coir fiber reinforced clayey soil *The International Journal of Engineering Science* **2** 54-64

[3] Singh H 2013 Effects of coir fiber on CBR value of itnagar soil *International Journal of Current Engineering and Technology* **3** 1283-6
[4] Bairagi H, Yadav R and Jain R 2014 Effect of jute fibres on engineering characteristics of black cotton soil *International Journal of Engineering Research & Technology (IJERT)* **3** 1550-2
[5] Singh R and Mittal E S 2014 Improvement of local subgrade soil for road construction by the use of coconut coir fiber *International Journal of Research in Engineering & Technology***3** 707-11
[6] Kumar D, Nigam S, Nangia A and Tiwari S 2015 California bearing ratio variations in soil reinforced with natural fibres (a case study Bhopal Bypass Road) *International Journal on Emerging Technologies* **6** 95
[7] Subramani T and Udayakumar D 2016 Experimental study on stabilization of clay soil using coir fibre *International Journal of Application or Innovation in Engineering Management Decision* **5** 192-204
[8] Sunny T and Joy A 2016 Study on the effects of marine clay stabilized with banana fibre *International Journal of Scientific Engineering Research* **4** 96-8

IOP Conf. Series: Materials Science and Engineering **970** (2020) 012030 doi:10.1088/1757-899X/970/1/012030

Study of Load & Settlement Behaviour of Expansive Soil Reinforced with Granular Pile Using Natural Aggregates

M J Rituparna, Benazeer Sultana

Department of Civil Engineering, College of Engineering & Technology, Bhubaneswar, Odisha, India

E-mail: benazeersultana@gmail.com

Abstract. Expansive soil such as Black Cotton soil is problematic from geotechnical point of view as it is unstable volumetrically with the variation of seasonal moisture making it sensitive to deformation. Its strength decreases, compressibility increases on wetting, it possesses very small shear strength. There are several geotechnical improvement techniques available expansive soil strata, but the use of granular piles is becoming popular in India as it improves the strength parameters, such as increases load bearing capacity, reduces settlement, takes wide spread loads and improves the rate of consolidation. In this study, experimental investigations have been done to judge the load and settlement behaviour of Black Cotton soil when reinforced with geotextile encased granular pile by varying the percentage of natural aggregates, fly ash and lime at varying L/d (length & diameter) ratio of 1.5, 2, 2,5 3 and 3.5. The soil sample was prepared taking a mould of 150mm diameter, 175mm height and granular pile of diameter 50mm. The test results showed significant reduction in settlement on installation of geotextile encased granular pile when compared to untreated soil sample. The ultimate load increased & settlement reduced with increase in percentage of fly ash & lime with each increase of L/d ratio as well.

1. Introduction

India has large geographical diversities with soils, mountains, forest covers, rivers, terrains, deserts, etc. with that a variety of soil types also exist in different parts of our country. For example, the Black Cotton (BC) soil present in central part of India, marine clay, lithomargic clay, and laterite soil of Southern region are a few soils types. BC soil is one of the expansive soils available in our country covering 21% of land area and generally, available in tropical and arid regions. Its appearance varies from black to grayish brown colour. It is majorly originated in our Central India as well as some parts of South India, Deccan plateaus and fields along with Andhra Pradesh, Gujarat, Maharashtra, and also several areas in Odisha.

BC soil shows high plasticity, exhibit major swelling and shrinkage characteristics due to the presence of clay mineral Montmorillonite. Construction of geotechnical designs and their execution on BC soil is associated with great difficulties. As structures founded on BC soil undergoes huge amounts of settlements and deformation, due to alternate drying, wetting and ground water table variation. They possess very low shear strength, and hence, they could tend to damage structures and foundations during the execution or throughout the life of the project. Different ground improvement techniques are available for various types of soils such as, Soil Replacement, Mechanical Stabilization, Chemical Stabilization, and Stabilization using Reinforcement, Sand Drains, and Prefabricated Vertical Drains, Granular Piles or Stone columns. Several problems are not solved effectively with above solutions due to its time and money constraints particularly when the geological strata of expansive soils extended to large depth [1-6].

Granular piles (GP), also known as stone columns can be constructed in both cohesionless as well as soft cohesive clays by method of partial replacement of undesirable local soil with granular material of strong, dense and highly permeability. This increases the load bearing capacity, helps in significant reduction of settlement and increase consolidation rate of expansive soil. GP gain strength from lateral reinforcement and hence encasing granular piles with geotextile can increase the lateral strength. It effectively controls the swelling pressure as well. Other benefits of GP are it is sustainable, feasible, economical, and efficient. Therefore, considering all the benefits, the areas of utilization GP should be explored more.

Previously, load tests were conducted on floating GP constructed in soft BC soil and effect of pile length on load carrying capacity was studied by Siddharth Arora & Rakesh Kumar [2]. A study was done to improve the load carrying capacity of foundation in expansive soil by including fibers and geogrid encased GP material by Kundan Meshram & P.K. Jain [1]. In 2017, Mohd Furkhan & M. Kameswara Rao [3] conducted laboratory tests to understand the load carrying capacity pattern of GP provided in soft soils by varying percentages of Recycled materials in place conventional materials. Recently, Eswara Reddy Orenkanti [6] studied the response of GP in presence of encasement & combination of encasement & lateral reinforcement under compression tests.

2. Materials Used & Properties

The present work focuses on ground improvement of black cotton soil. Keeping the cost efficiency and environmental aspects in mind, various materials are being considered in the present work. The following materials are used:

2.1. Black Cotton Soil

Black Cotton soil is collected from Nachuni Village, under Balugaon, in Khorda District, from 1m depth below GL. The properties of Black Cotton soil are given in Table 1.

2.2. Natural Aggregates

Natural aggregates considered were collected from CET college campus of size between 4.75-10 mm. The properties of aggregates are shown in Table 2.

Table 1. Virgin Expansive Soil Properties

Properties	Values	Properties	Values
Specific gravity	2.42	Plasticity Index	31.83 %
D_{10}	0.002	Shrinkage Limit	21.38 %
D_{30}	0.0061	IS Soil Classification	CH
D_{60}	0.023	Free Swell Index	100 %
Coefficient of Uniformity (Cu)	10.45	Natural Moisture Content	21.94 %
Coefficient of curvature (Cc)	0.735	Optimum moisture content	17.80 %
Gravel content	0%	Maximum dry density	1.79 gm/cm^3
Sand Content	28.38%	Liquid Limit	60.21 %
Silt & Clay Content	65.85%	Plastic Limit	28.38 %

2.3. Fly Ash

The fly ash used is a Class C Fly ash and is originally from NTPC, Talcher and purchased from a local vendor. The chemical properties were collected from NTPC Talcher. Fly ash is used here in place of cement as it has cementing properties when added with water, it will give workability to the natural aggregates. The physical & chemical properties are shown in Table 3.

2.4. Lime
For this project work I have used powdered hydrated lime, due to its binding and adhesive property. It acts as a binding material and imparts strength in the granular pile.

2.5. Geotextile
Geotextile is purchased from Sun textiles private limited from Kolkata. Woven Polypropylene type of geotextile is used. Properties of Woven Polypropylene geotextile are given in Table 4.

Table 2. Properties of Natural Aggregates

Properties	Values
Specific gravity	2.83
Water absorption	0.202%
Aggregate impact value	25.18%
Aggregate Crushing value	26.46%

Table 3. Properties of Fly Ash

Properties	Values
Specific gravity	1.82
Coefficient of Uniformity (Cu)	3.08
Coefficient of curvature (Cc)	1.02
Clay content	0%
Sand content	14%
Silt & clay content	86%
Grain size	Uniformly graded
Colour	Grey
Moisture content	1.98%
Free Swell Index (%)	0%

Table 4. Properties of Geotextile

Properties	% by Mass
Specific gravity	0.91
Unit weight	500 gm/m^2
Thickness	2mm
Colour	White
Breaking strength	16 KN/m
Elongation Break	25 – 10 (%)
Tear strength	0.42 KN
Opening size	0.07-2 mm

3. Methodology

3.1. Experimental Setup
This experimental setup is similar to the California Bearing Ratio (CBR) test setup. Tests were conducted in cylindrical mould to determine the load and settlement behaviour of BC Soil reinforced with geotextile encased GP. A typical test arrangement is shown in Figure 1. All the experiments were carried out on a GP of diameter 50mm surrounded by BC Soil in a cylindrical mould of diameter 150mm and 175mm depth. The soil sample was prepared at a dry density of 1.79gm/cc and 17.8% moisture content. The soil sample was compacted with no visible air voids and GP was installed at centre of sample through replacement method and filling the hole with pile materials in layers.

The Load tests were conducted varying L/d ratio of GP (where, L is Length & d is diameter of GP) from 1.5 to 3.5 with a difference of 0.5. The diameter of GP was kept constant for all the tests, i.e. 50mm.

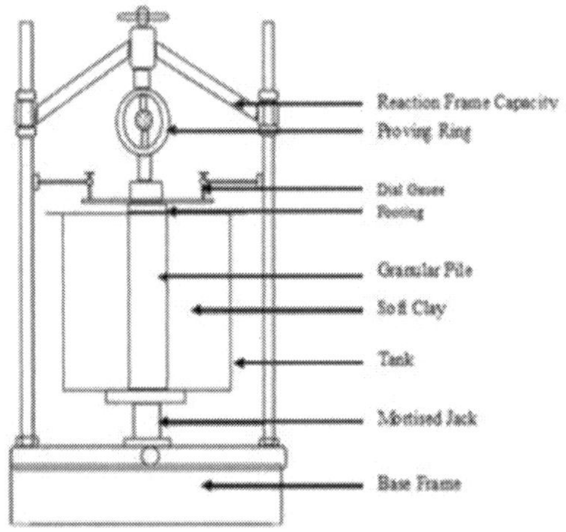

Figure 1. Typical Test Setup for Load-Settlement behavior of GP in BC soil

3.2. Preparation of Soil Sample

A soil sample of 5kgs BC soil was prepared with a moisture content of 17.80% in a C.B.R mould in five layers and each layer was rammed to maintain a desired bulk density of 1.52gm/cc. Soil was added in five layers and 56 number of blows were applied in order to maintain heavy compaction. Sample top was levelled by cutting off the excess. The soil sample hence prepared was kept aside for few hours after installation of GP. After that the load tests are conducted. A 150mm diameter mild steel footing plate is placed on sample top so that the entire setup acts as a composite material. The plunger distributes the axial load uniformly through the proving ring on plate and then to soil sample. The load settlement test is conducted on untreated sample. Then in the similar procedure reinforced soil samples are tested.

3.3. Installation of Granular Pile

After the preparation of soil sample, the GP with natural aggregate of size between 4.75 to 10mm, fly ash and lime paste encased in geotextile were installed by a replacement method. A Stainless steel pipe of 5cm diameter open at both the ends was smeared with grease and inserted into the BC soil sample mould up to desired length of GP. Then the pipe was pulled ensuring that the soil does not fall back. Then a piece of geotextile was rolled and stitched to a diameter of 5cm and placed vertically so as to cover the entire GP. The GP of natural aggregates, fly ash and lime paste was prepared by adding 5% water content (Rakesh Kumar & P.K. Jain 2013). Gradually the water content was increased with the increased percentage of fly ash and lime. The material was charged in three layers and compacted with a tamping rod to maintain uniformity diameter of the GP. It was made sure so that no air voids were left and continued until GP was formed. Two to three soil sample moulds were prepared and was kept for half a day and load tests were conducted in CBR setup.

3.4. Test Procedure

The test setup is similar to CBR test setup. The load was applied with plunger assembly on the mild steel circular footing plate placed on top of soil sample including the granular pile. The footing plate considered here is of the same diameter as that of mould. Through the proving ring the load was applied by a technique of constant displacement rate of 1.25 mm/min. Dial gauge was used for noting the settlement (dial gauge least count= 0.02mm). The load was applied until settlement reached a constant value. The load corresponding to settlement of footing plate was noted.

Figure 2. Installation of Granular Piles in Black Cotton soil

A series of load tests were performed on unreinforced soil sample and soil samples reinforced with geotextile encased GP at different depths. A mild steel circular plate of 15cm dia is taken for all the tests (i.e. keeping diameter equal with mould diameter). It is placed on the soil sample to act as footing plate and all load tests were performed. The test results were compared with the untreated soil sample. The various series of laboratory tests are clearly mentioned in the Table 5.

Table 5. Virgin Expansive Soil Properties

Description	Combination of Tests Conducted On
Test Series 1 at depth of 17.5cm (L/d=3.5)	1. Untreated Black Cotton soil 2. BC soil reinforced with geotextile encased 67% GP with natural aggregate + 30% Fly ash + 3% lime 3. BC soil reinforced with geotextile encased 60% GP with natural aggregate + 35% Fly ash + 5% lime 4. BC soil reinforced with geotextile encased 53% GP with natural aggregate + 40% Fly ash + 7% lime
Test Series 2 at depth of 15cm (L/d=3.0)	1. BC soil reinforced with geotextile encased 67% GP with natural aggregate + 30% Fly ash + 3% lime 2. BC soil reinforced with geotextile encased 60% GP with natural aggregate + 35% Fly ash + 5% lime 3. BC soil reinforced with geotextile encased 53% GP with natural aggregate + 40% Fly ash + 7% lime
Test Series 3 at depth of 12.5cm (L/d=2.5)	1. BC soil reinforced with geotextile encased 67% GP with natural aggregate + 30% Fly ash + 3% lime 2. BC soil reinforced with geotextile encased 60% GP with natural aggregate + 35% Fly ash + 5% lime 3. BC soil reinforced with geotextile encased 53% GP with natural aggregate + 40% Fly ash + 7% lime
Test Series 4 at depth of 10cm (L/d=2.0)	1. BC soil reinforced with geotextile encased 67% GP with natural aggregate + 30% Fly ash + 3% lime 2. BC soil reinforced with geotextile encased 60% GP with natural aggregate + 35% Fly ash + 5% lime 3. BC soil reinforced with geotextile encased 53% GP with natural aggregate + 40% Fly ash + 7% lime
Test Series 5 at depth of 7.5cm (L/d=1.5)	1. BC soil reinforced with geotextile encased 67% GP with natural aggregate + 30% Fly ash + 3% lime 2. BC soil reinforced with geotextile encased 60% GP with natural aggregate + 35% Fly ash + 5% lime 3. BC soil reinforced with geotextile encased 53% GP with natural aggregate + 40% Fly ash + 7% lime

IOP Conf. Series: Materials Science and Engineering **970** (2020) 012030 doi:10.1088/1757-899X/970/1/012030

4. Results and Discussions

From the experimental investigations various Loads-Settlement curves for a series of tests varying L/d ratios are presented below. The tests were performed following the procedures according to IS: 2720(PART-16):1987. With every L/d ratio, the percentage of Fly ash and lime added were varied.

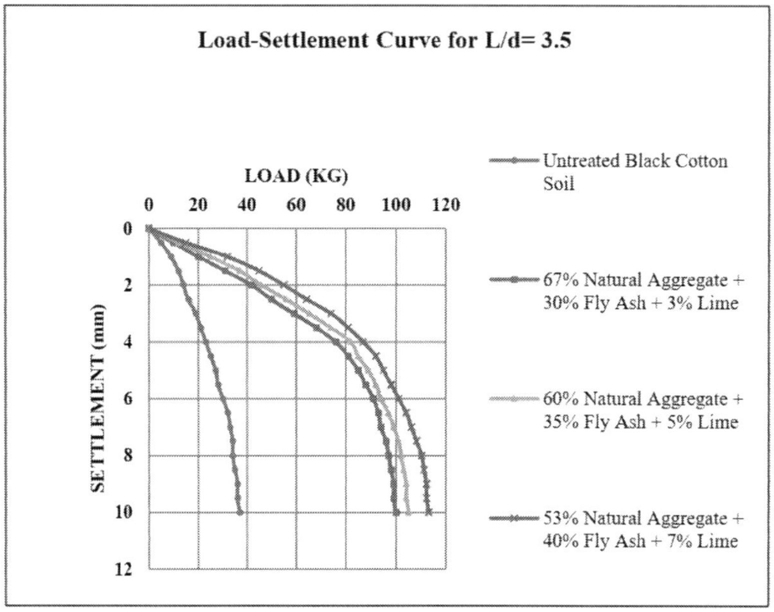

Figure 3. Load-Settlement Curve for GP with L/d= 3.5

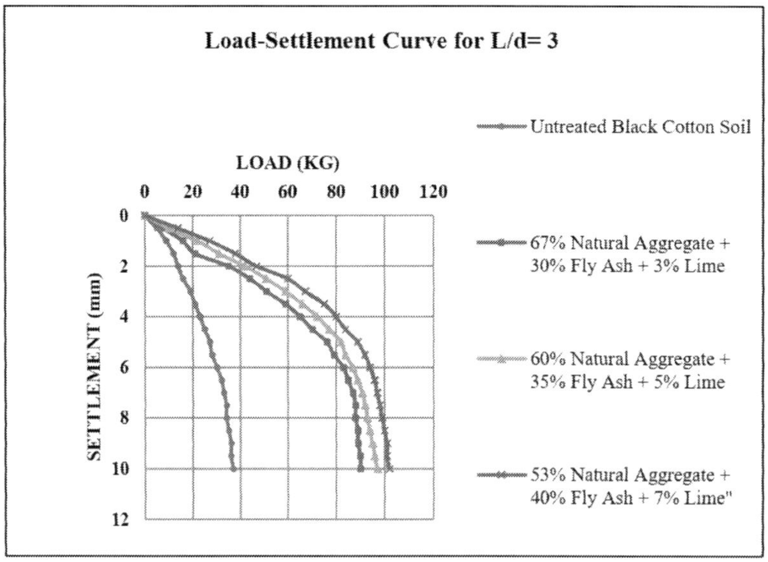

Figure 4. Load-Settlement Curve for GP with L/d= 3.0

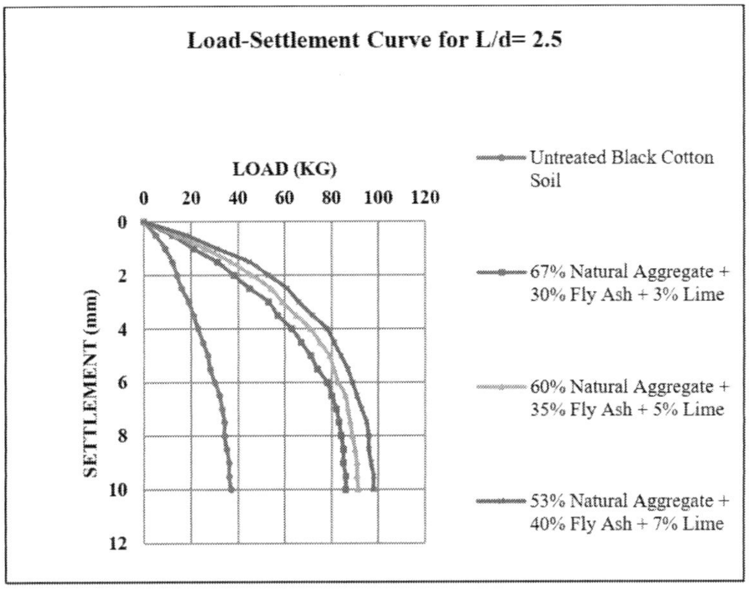

Figure 5. Load-Settlement Curve for GP with L/d= 2.5

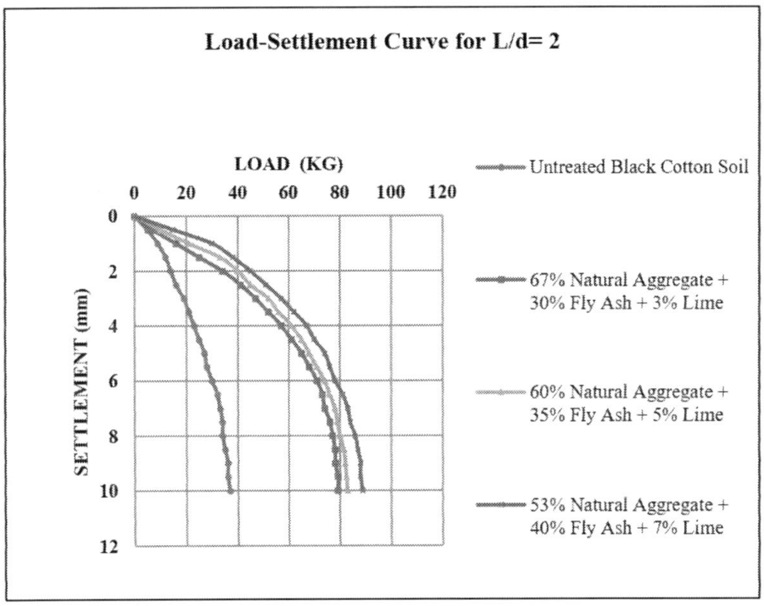

Figure 6. Load-Settlement Curve for GP with L/d= 2.0

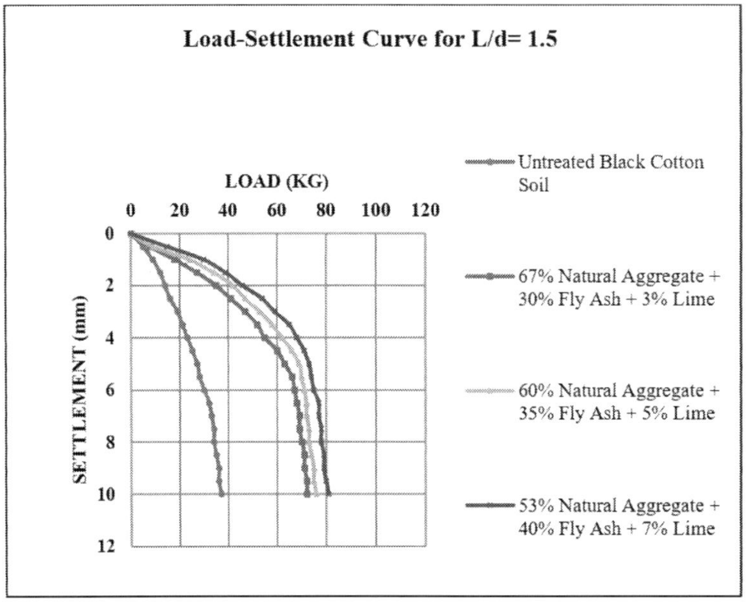

Figure 7. Load-Settlement Curve for GP with L/d= 1.5

The following Load settlement curves are presented by varying the 'L/d' ratio by keeping the percentage of fly ash & lime constant.

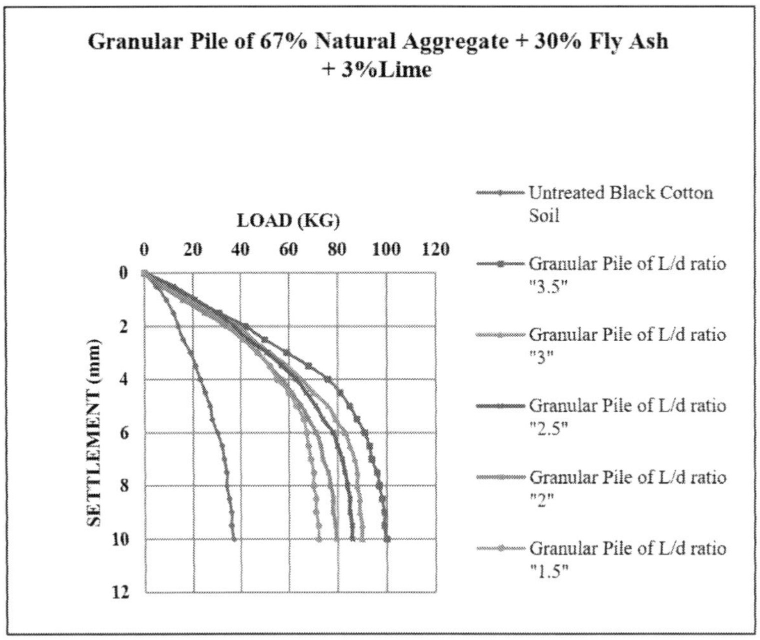

Figure 8. Load-Settlement Curve for GP of NA with 30% Fly Ash + 3% Lime

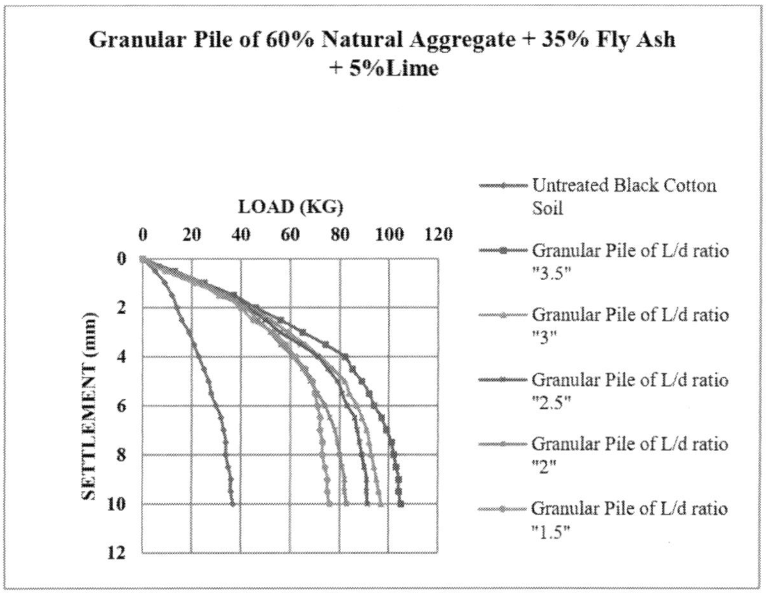

Figure 9. Load-Settlement Curve for GP of NA with 35% Fly Ash + 5% Lime

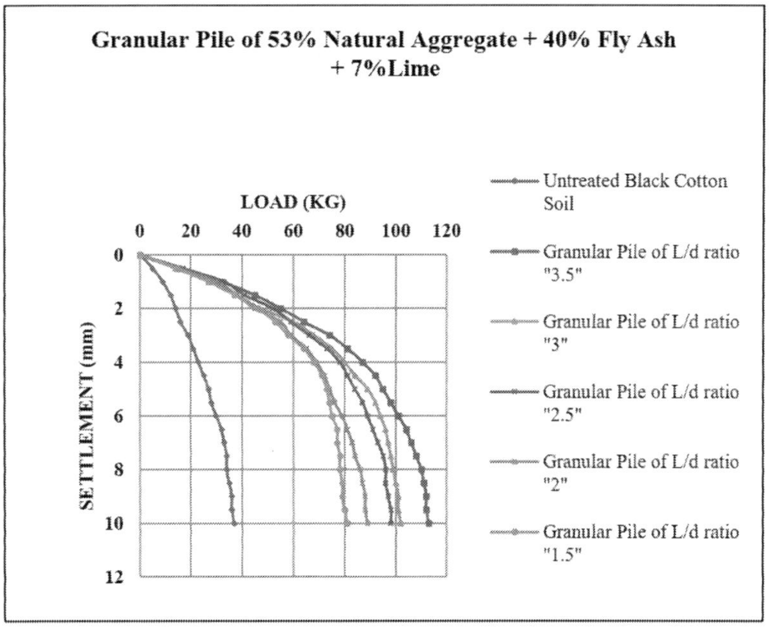

Figure 10. Load-Settlement Curve for GP of NA with 40% Fly Ash + 7% Lime

It is noticed that when the L/d ratio along with percentage of fly ash and lime is increased the settlement is noticed to reduce significantly. The ultimate load of the BC soil for any reinforced soil is noticed to be more than the unreinforced or untreated BC soil sample.

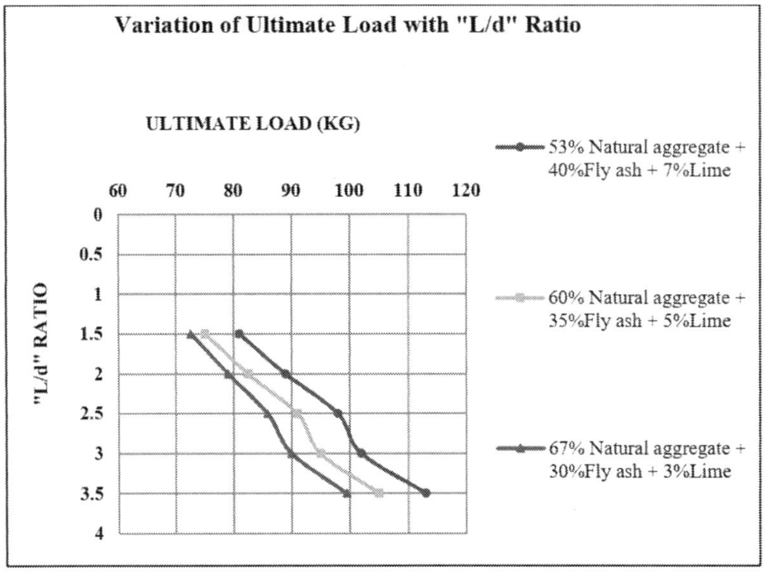

Figure 11. Load-Settlement Curve for Variation of Ultimate Load with L/d ratio of Granular Piles

From the following curves it is also noticed that with increase in L/d ratio of the granular pile as well as the fly ash and lime content the ultimate load is increased. Hence, the Load-settlement behaviour pattern noticed is in accordance with the various literature surveys done by me and falls in place with the model tests conducted by various researchers.

5. Conclusion

The objective of the study is to conduct experiments to study the Load-settlement behaviour of geotextile encased granular pile in BC soil, taking various composition of granular pile with natural aggregates, fly ash and lime at different L/d ratios 1.5, 2, 2.5, 3 and 3.5. Based on the experimental investigations and results obtained it is noticed that:

1. Settlement in reinforced BC soil reduces with increased L/d ratio of GP, i.e. from 1.5 to 3.5. For GP of 53% Natural Aggregate + 40% Fly Ash + 7%Lime, the ultimate load increased by 10% as compared to GP of 67% Natural Aggregate + 30% Fly Ash + 3%Lime.
2. For increased proportion of Fly ash & lime in the granular pile material for each L/d ratio the settlement is noticed to reduce. Although, the rate of reduction of settlement for the increasing proportion of GP material is slow. Here is some observation:
   i) Granular Pile of 67% Natural Aggregate + 30% Fly Ash + 3%Lime at L/d 3.5 takes 37% more load to achieve same settlement as compared to at L/d 1.5.
   ii) Granular Pile of 60% Natural Aggregate + 35% Fly Ash + 5%Lime at L/d 3.5 takes 44% more load to achieve same settlement as compared to at L/d 1.5.
   iii) Granular Pile of 53% Natural Aggregate + 40% Fly Ash + 7%Lime at L/d 3.5 takes 37% more load to achieve same settlement as compared to at L/d 1.5.
3. Ultimate load for geotextile encased Granular Pile of 53% Natural Aggregate + 40% Fly Ash + 7%Lime at L/d 3.5, increased by 192% to achieve the same settlement when compared to unreinforced BC soil.
4. It can also be concluded that granular pile of full length gives more strength and less settlement as compared to untreated BC soil or at any depth.

5. The ultimate load is notice to increase with the increased L/d ratio of GP. Also, it is increasing for all the percentages of Fly ash & lime, even though the rate is slow but the result and pattern is in accordance with all the literature surveys made by me.

According to the laboratory tests conducted, the results comparing the performance of the geotextile encased granular piles in BC soil are in good accordance with all the literature surveys done for this study.

References

[1] Rakesh K and Jain P 2013 Soft ground improvement with fiber reinforced granular pile *International Journal of Advanced Engineering Research Studies* **2** 42-5

[2] Arora S, Kumar R and Jain P 2014 Load-Settlement Behaviour of Granular Pile in Black Cotton Soil *International Journal of Advances in Engineering Technology* **7** 773

[3] Harikrishnan P and Airaj A 2014 A study on load carrying capacity of soft clays provided with granular piles of different materials. In: *Indian Geotechnical Conference IGC,*

[4] Rajput D, Kumar R, Jain P and Chandrawanshi S 2016 Load-Settlement Behaviour of Soft Soil Reinforced with Sand Piles *International Research Journal of Engineering and Technology (IRJET)* **3** 1303-8

[5] Basu P, Samadhiya N K and De Dalal S S 2018 An experimental study on random fiber mixed granular pile *International Journal of Geotechnical Engineering* **12** 1-12

[6] Orekanti E R and Dommaraju G V 2019 Load-Settlement Response of Geotextile Encased Laterally Reinforced Granular Piles in Expansive Soil Under Compression *International Journal of Geosynthetics Ground Engineering* **5** 17

Erosion Studies and Rehabilitation of Beach along West Coast of India

P K Suresh[1], R Sundaravadivelu[1], Manu S Nadesan[2]

[1] Department of Ocean Engineering, Indian Institute of Technology Madras, Chennai, Tamil Nadu, India
[2] Department of Civil Engineering, Adi Shankara Institute of Engineering and Technology, Kalady, Kerela, India

E-mail: sureshpk2000@gmail.com

Abstract. The west cost of India from Kannyakumari in Tamilnadu to Thengapattinam bordering Kerala at Thengapattinam is oriented in the east-west direction compared to other two boundaries which are oriented along north-south direction. In these coastal areas the protection measures executed are frequently getting damaged mainly due to unprecedented waves and south west monsoon waves. Hence it was felt to do a detailed study by taking field profiles at selected locations, design of sections and model studies This calls for a proper understanding of the wave hydrodynamics prevailing as it plays a major role in formation of beaches. The stretch is having high tourist potential and it is important to prevent the erosion of beach and stabilize the coast. Field observations were taken and desk studies were made. Based on the studies suitable model studies were carried out and remedial measures were carried out at one site. Post monitoring of the project was done and based on the post monitoring it was inferred that project was a success. The details are explained in the paper.

1. Introduction

The coastal reach from Kannykumari in Tamilnadu to Neerodi near Kerala is a thickly populated one. Fishing is the sole bread and butter of the coastal community. The coast undergoes tremendous erosion during south west monsoon from June to September resulting heavy erosion of land and infrastructure. Hence studies were made which consists of field visit, desk studies, studies on wave dynamics and remedial measures.

2. Study Area

The total coastline divided into two reaches, based on the field visit (Figure 1). The details of the geometry as inferred from the bathymetric charts indicates reach 1 is from Muttom to Thengapattanam. The coastline is oriented in 125^0 N along east-west direction. The height of the coast is increasing along the reach and it has protrusions. Beaches are observed at selected places along the stretch. The foreshore slope is about 20 in 1000 and it becomes steeper near Thengapattinam. The second reach is from Thengapattinam to Neerody. The value of foreshore slope assessed as 33 in 1000 near Neerodi (Figure 1).

Figure 1. Study Reach

The coastline is oriented in 128^0 N along east-west direction. No beaches are observed at selected places along the stretch. The foreshore slope is about 35 in 1000. The deep foreshore slope results in high wave run up and erosion of armour layer. The coastal road is prone to direct wave action is totally damaged. As a crises management seawalls are constructed to protect the properties on land. The seawall sections are unstable along the places where foreshore slope increases and are in a disturbed state (Figure 2 and Figure 3).

Figure 2. Sliding of existing seawall

Figure 3. Erosion of coastal road

The seawalls are found to give some success in places where the foreshore slope is shallow (Reach1 and Part of reach 2) due to good beach formations. The seawalls, which were constructed earlier, are totally eroded.

3. Desk Studies

The studies include the data on availability of historical shoreline data. Satellite imageries are the best suited sources. The imageries of the coast at Neerodi and Mandaikkadu during 2001 and 2015 were collected from Google earth. The orientation of beaches adjacent to the above coast were identified for both years; on overlapping the imageries of two years it was found that along the coast of Neerodi the shoreline was eroded by about 40 m and adjacent to the coast of Mandaikkadu it was eroded by about 10 m over a period of 15 years (Figure 4).

Figure 4. Erosion along Neerodi

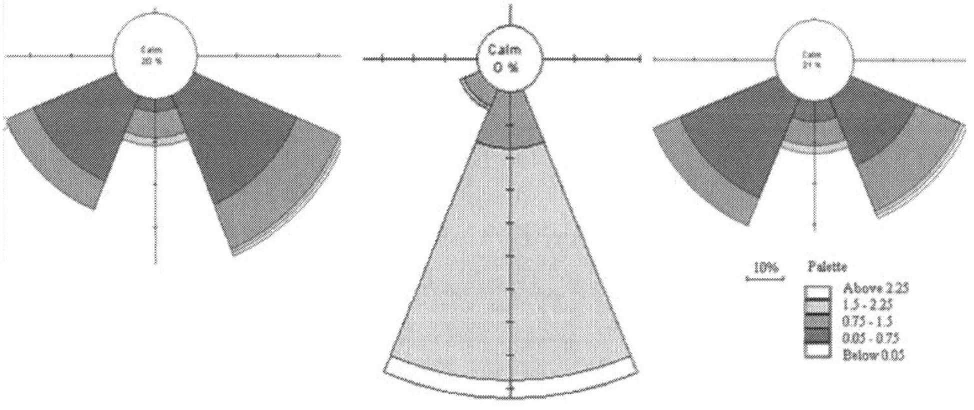

Figure 5. Wave rose diagram

4. Wave Climate

Assessment of wave climate is an important task for any coastal engineering study. Usually the nearest deep water buoy data is collected and the near shore wave climate is derived. Based on the nearshore wave climate, assessment of beach formations carried out. In the present study, the wave data was not available for all locations and except at selected location. In the present work the available data analyses carried out as per Suresh and Sundar [1]. The numerical model MIKE 21 by Danish Hydraulic Institute,

Denmark [2] was adopted. The wave data for the coast is available from Indian National Centre for Ocean Information (INCOIS) website [3]. The wave rider buoy for Colachel for the year 2018 was collected, analyzed and furnished vide table 1 below. The average values were arrived after the analyses of wave data. The wave heights derived based on analyses indicate it is low in non-monsoon and northeast monsoon. The south west monsoon waves are found to be high and maximum wave heights of 1 to 2.m was observed. There were occasional observations of maximum wave heights up to 4 m during the south monsoon (Figure 5).

5. The Protection measures

Usually protection measures adopted in India are seawalls and groins. The seawalls will protect the structures behind whereas the groins are capable of creating beaches. Along the coast of Mandaikkadu a groin field consisting of six short groins installed after detailed studies as described by Sundaravadivelu et al [4]. The protection measures consist of a groin field with six groins. The commencement of the groin field is from the coast adjacent to Church. It consists of six groins G1, G2, G3, G4, G5 & G6 of length 65 m, 81 m, 90 m, 60 m, 63 m & 60 m respectively. The corresponding depth at which the above groins terminate are (-) 2.0 m, (-) 3.0 m, (-) 3.0 m, (-) 2.0 m, (-) 2.0 m and (-) 2.0 *m*. The spacing between the groins G1 to G6 are 100 m, 150 m, 150 m, 150 m and 100 m respectively (Figure 6).

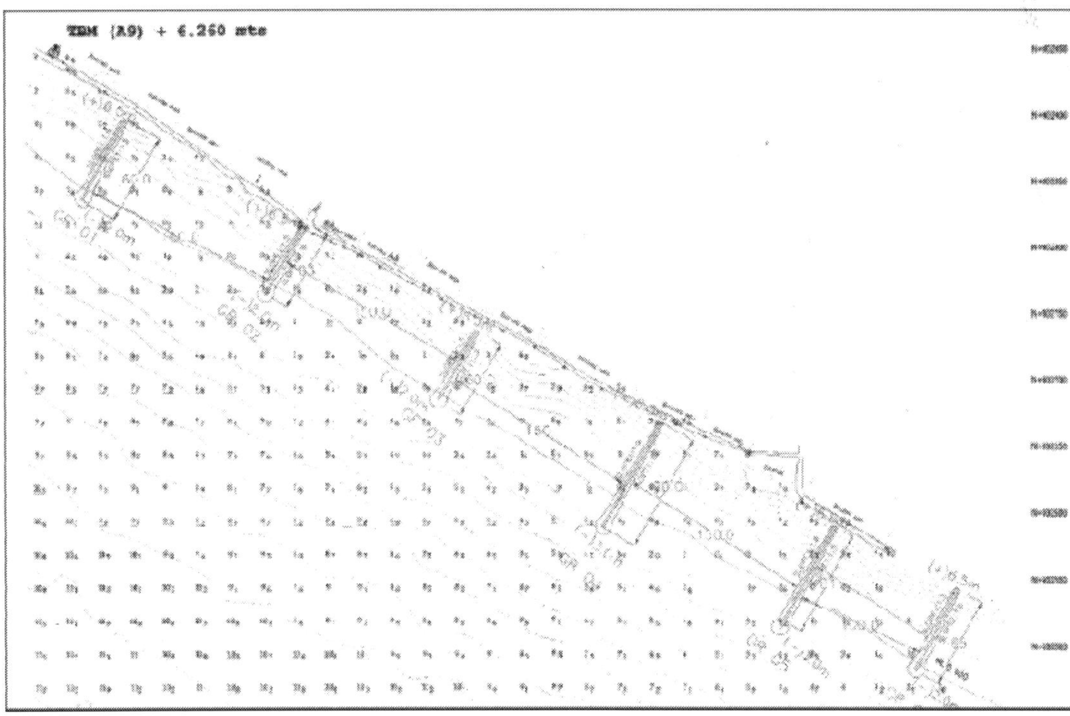

Figure 6. Proposed groin field along Mandaikkadu

6. Numerical modeling of shoreline evolution

The impact of groins on the present shoreline was studied. The predictions indicate that groins will be by- passed in one year after construction. The construction of the groins was completed by the year 2018. Post monitoring of shoreline was carried out. The beach formation up to 80 m was noticed by January 2019. These predictions are as per Kraus and Harikai [5]. The procedures adopted by Suresh & Sundar [6] are adopted for the prediction of shoreline evolution in the presence of shore connected structures. The formulated model basically relates the change in shoreline orientation with the rate of sediment material transported along the beach. The governing equation is given below.

$$(b+Dc)\ \partial y/\partial t = -\ (\partial Q/\partial x + q(x) \quad (1)$$

y = Position of shoreline seaward and a function of 'x' measured along shore and time't'.

b = Height of berm

D_c = Limit of active transport (Hallerimeter [7])

Q = wave induced sediment

Q(x) = quantity of sediment added

In the present case, the term q(x) is not considered; as it represents sediment added from rivers. The long-term shoreline evolution of shoreline because of the introduction of groin field is carried out. This is done with a view to study the effect of groins and whether any adverse changes will be happening. The wave induced sediment component is an important parameter and is carried out based on empirical expressions. Usually the expressions correlates the rate of alongshore transport 'Q' with wave energy flux 'E' along the alongshore direction. The expressions adopted for alongshore transport is explained below. The energy flux Pl along the alongshore direction is given as

$$Pl = Eb\ Cgb\ \sin\alpha\ b_b \cos\alpha\ b \quad (2)$$

The energy density of wave $E_b=(1/8)\ \rho\ g H_b^2$, C_{gb} denotes the wave group velocity; α_b is the angle of wave propagation direction with reference to normal to the shoreline. The suffix 'b' refers to the corresponding variables at breaker point. The breaker angle and breaker wave height H_b is calculated adopting Snell's law.

ρ = Mass density of sea water
G = Gravitational constant

$$Kr = (\cos\alpha_o / \cos\alpha_b)^{0.5} \quad (3)$$

K_r = Refraction coefficient

α_o, α_b = the wave angles at breaker depth and deep water respectively with respect to shore normal as described in Figure 7.

When the coastal modelling point under consideration in the diffraction region, 'Kr' is replaced by 'Kr Kd' and Kd is the diffraction coefficient. Then the alongshore transport is calculated as below.

$$Q = A\,K_r^2\,K_d^2\ \mathrm{Sin}\,\alpha_b\ \mathrm{Cos}\,\alpha_b \quad (4)$$

'A' is called transport parameter calculated as per Van Rijn [8] equation as below

$$Q = 40\ K_{swell}\ K_{grain}\ K_{slope}\ (H_b)^3 \sin(2\alpha_b) \quad (5)$$

Q = Alongshore sediment transport (kg/s),
T_P= Peak period, Swell correction factor, K_{swell} = TP/6, D50 = Particle size (mm),
K_{grain}=Particle size correction factor=0.20/D50, K_{slope}= Slope correction factor = $(\tan\beta/0.01)^{0.5}$
The transport expression is converted in to volumetric units

$$Q = A\,(K_{swell}K_{grain}K_{slope})\,(H_s,b_r)^3\sin(2\alpha_b) \quad (6)$$

When the modelling point is not in the the diffraction coefficient K_d is taken as 1.

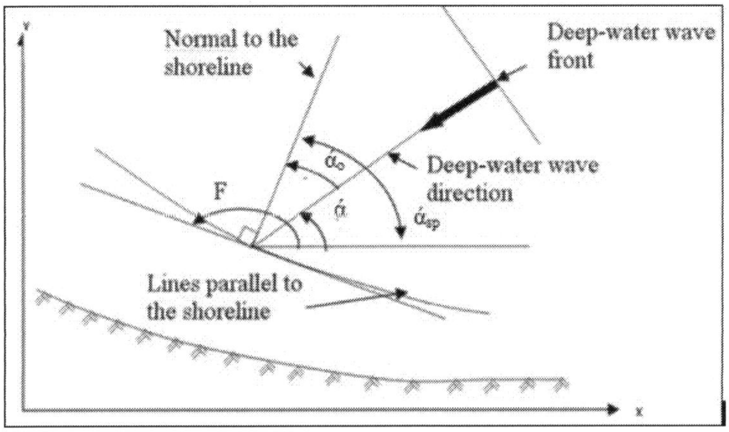

Figure 7. Details of wave angles

If the point is in the geometric shadow region of breakwater / groin, diffraction coefficients adopted as per Dean and Dalrymple [9]. The shoreline variable is made non-dimensional and divided in to 'N' grid point at equal non-dimensional interval 'δx. Then the shoreline changes over non-dimensional time 'δt' are calculated as adopting Crank – Nicholson finite difference scheme. The schematic diagram for finite difference adopted is shown in Figure 8.

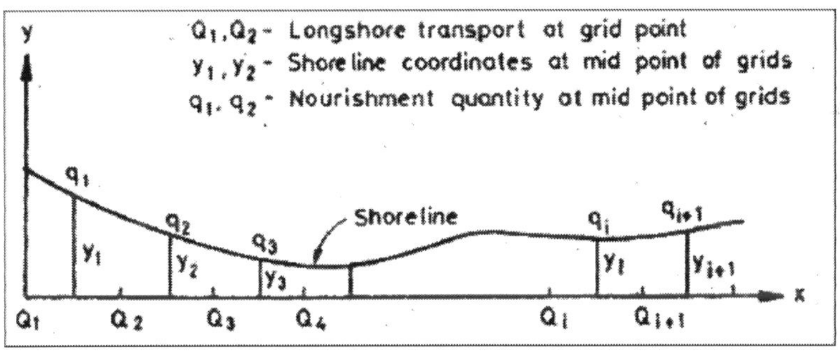

Figure 8. Model scheme diagram

The sample of two-layer groin cross section with a base layer was proposed at (-) 3.0 m water depth and details are shown vide Figure 9.

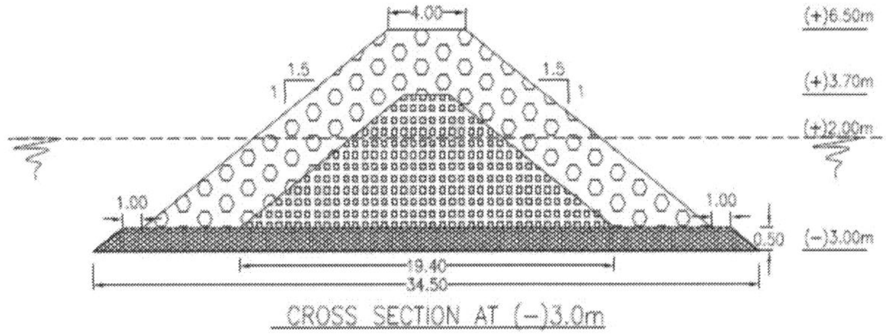

Figure 9. Cross section of the groin

Based on the above model description it was predicated that the beaches will form for the entire stretch of groin field. The structural design of the groin carried out based on the existing bathymetry.

7. Results and discussion

The project of coastal defense work was well planned after detailed studies and field observations. The wave data from INCOIS web site was adopted for initial planning. The project completed in 2018. The post observations indicate good beach formations. The details are furnished vide Figure 10 and Figure 11.

Figure 10. Beach formation along road which was eroded early

Figure 11. Creation of new recreation beach

8. Conclusions

The west coast of stretch from Tamil Nadu to Kerala is subjected to heavy wave energy dissipation during south west monsoon resulting in heavy erosion. The coastal protection measures in the form of short groin field results in good beach formation. The beaches in turn protect the existing roads and land by reducing the energy of waves during monsoon. The predicted and observed observations are nearly matching. Hence such protection measures can be adopted along such eroded stretch after sufficient studies.

References

[1] Suresh P K and Sundar V 2010 Performance of protection measures along the west coast of Tamilnadu. Centre for Earth Science Studies, Thiruvananthapuram February 2010 pp 203-6
[2] Institute D H 2001 User manual and reference guide for MIKE21. DHI Horsholm, Denmark)
[3] Indian National Centre for Ocean Information Services (INCOIS). www.incois.gov.in
[4] Sundaravadivelu R, Sakthivel S and Suresh P K 2019 Remedial measures to combat sea erosion along west coast of India. Mauritius July 2019 pp 203-6
[5] Kraus N C and Harikai S 1983 Numerical model of the shoreline change at Oarai Beach *Coastal Engineering* **7** 1-28
[6] Suresh P K and Sundar V 2011 Comparison between measured and simulated shoreline changes near the tip of Indian peninsula *Journal of Hydro-Environment Research* **5** 157-67
[7] Hallermeier R 1980 A profile zonation for seasonal sand beaches from wave climate *Coastal Engineering* **4** 253-77
[8] Van Rijn L 2001 Longshore sediment transport *Delft Hydraulics*
[9] Dalrymple R A, Kirby J T and Hwang P A 1984 Wave diffraction due to areas of energy dissipation *Journal of waterway, port, coastal, ocean engineering* **110** 67-79

IOP Conf. Series: Materials Science and Engineering **970** (2020) 012032 doi:10.1088/1757-899X/970/1/012032

Stress-Strain Behaviour of Compacted Pond Ash Reinforced with Stone Column

J K Naik

Department of Civil Engineering, Government College of Engineering, Kalahandi, Bhawanipatna, Odisha, India

E-mail: jajatinaik007@gmail.com

Abstract. Ash pond deposits are characterized by its high compressibility and low bearing capacity which is unappropriated for the construction of engineering structures over it. There are several techniques to improve the bearing capacity of ash deposits. Stone column is one of the effective techniques employed to increase the bearing capacity for construction of engineering structures over soft soil deposits. Stone columns are extensively used to improve the bearing capacity of poor ground, time rate of settlements, stiffness, reduce the settlement of structure, and liquefaction potential of soft ground. This paper reports the load settlement behaviour of compacted pond ash beds reinforced with stone columns. Both end bearing and floating stone columns of different length ratios and area ratios were introduced in compacted pond ash beds having different densities. The load-settlement behaviour of footings supported on different configurations of stone columns was studied.

1. Introduction

Fly ash is a derivative of burning minced coal in power plants. In India, around 71 % of electricity expended is generated by thermal power plants, of which nearly 90 % is coal-based generation and remaining 10 % is based on diesel, gas, steam and wind. Usage of bituminous or sub-bituminous coal that produces huge amounts of fly ash in thermal power stations, many captive power plants. The high ash content of Indian coals (30-40%) leads to these large amounts of fly ash. Almost 184.14 million tonnes of coal ash are currently produced in India annually and more than 70,000 acres of land are currently occupied by ash ponds. India is heavily dependent on coal as an electricity source that will continue into the next century, so fly ash management will remain an essential field for national concern and will require vast quantities of land to be disposed of along with water and resources. The high compressibility of pond ash deposit has poor bearing potential for which vast acres of land are lost. Fly ash can be cured using various methods for soil development, one of which is the stone column. By using the condensed stone column, the bearing capacity of the ash pond can be enhanced and structures can be constructed cost-effectively on the ash pond [1-14].

This paper reports the stress-strain reaction of compacted pond ash reinforced with stone column in footing load test. Stone columns were inserted at the center of cylindrical specimens of compacted pond ash. The area ratio of stone columns are varied as of 10 %, 20 %, 40 % and 60 % (Area ratio=$\frac{As}{Ac}$x100, As=Area of stone column, Ac=Area of pond ash specimen). Similarly, the lengths of the stone columns are varied as of 0.25L, 0.50L, 0.75L, and 1.0L, where L is the length of cylindrical specimen.

2. Material Used

Pond ash is collected from Rourkela Steel Plant, Rourkela, Odisha, India which was sieved over 2 mm sieve and dried at temperature of 105 ^{0}C -110 ^{0}C. Stone aggregates collected from local crusher available near to Rourkela which were washed and dried at the temperature of 105^0 C -110 ^{0} C. The two size of aggregate use for preparation of stone column which is having size between 2 mm-4 mm and 1mm-2 mm.

3. Experimental Programme

During this work, two arrangements of research were controlled. The basic arrangement of tests pointed to the determination of the index and engineering properties of pond ash, which joins pond ash 's record properties such as G (specific gravity), particle size distribution and hence indices of consistency. Additionally, with the aid of compaction studies, the similarity of pond ash under different compact energy levels was selected. In addition, direct shear tests and triaxial shear tests have carried out on compacted pond ash specimens at OMC and saturation conditions. The essential aspect of the experiments performed is sketched out below, and hence the methodology. In order to analysis the effects of stone columns embedded on pond ash bed, the second series of tests was controlled. The work was done to calculate the kind of strain reaction of stone column hardened pond ash beds.
The diameter of 2.6 cm, 3.3 cm, 4.8 cm and 5.7 cm stone columns were embedded in pond ash bed which so as to give the area ratio of 10 %, 20 %, 40 % and 60 %. The adequacy of length of stone sections on bearing limit of pond ash beds was concentrated by changing the length proportion as 1.00, 0.75, 0.5, and 0.25. The important part of tests directed and in this manner the technique are laid out underneath.

3.1. Index properties determination

The specific gravity test was conducted for the pond ash according to IS: 2720 (Part-III, Section-1, [1]) and the result is found to be 2.30. To determine the grain size distribution, Sieve analysis and hydrometer analysis was conducted for the coarse and finer particles according to IS: 2720 part (IV) and IS:2720 part (IV) [10] respectively and it is found that the pond ash particles are silt in size and the value of Cu and Cc of the pond ash are as 6.13 and 2.61 respectively.

Table 1. C & ϕ under saturated and unsaturated condition on variation of OMC & MDD

Compaction energy (kJ/m^3)	OMC (%)	Dry density (gm/c^3)	C at OMC (kg/cm^2)	C at saturation (kg/cm^2)	ϕ at OMC (degree)	ϕ at saturation
119	43.23	0.984	0.13	0.018	22	16
357	41	1.031	0.14	0.11	27	25
595	35.5	1.134	0.17	0.13	29.46	27
1604	32.22	1.15	0.21	0.16	36.86	33.45
2674	31.7	1.23	0.25	0.2	38.6	36.76

Where Φ is angle of internal and C is unit cohesion.

Table 2. Footing Load Test by Variation of Area Ratio and Length Ratio

Lr	Ar					
	0		10		20	
	Stress	Strain	Stress	Strain	Stress	Strain
1.0	2.675	11.33	4.816	19.3	5.331	25.33
0.75			4.258	24.	4.876	18
0.50			3.354	22.6	3.954	22.66
0.25			2.844	24.6	3.26	12
0			2.675	11.33	2.675	11.33

Table 3. Footing Load Test by Variation of Area Ratio and Length Ratio

Lr	Ar					
	0		40		60	
	Stress	Strain	Stress	Strain	Stress	Strain
1.0	2.675	11.33	5.963	19.33	6.739	26.66
0.75			5.574	24.66	6.261	33.33
0.50			4.937	15.33	5.658	30.66
0.25			4.133	14	4.767	19.33
0			2.675	11.33	2.675	11.33

Where, Lr= Length ratio and Ar= Area Ratio

3.2 Engineering properties determination

The Protector tests were conducted as per IS: 2720 (Part VII) – [2] and IS: 2720 (Part VIII)-[3]. The blows number of each layer is managed so as to impart energy of 119, 357, 595,593, 1604 and 2674 kJ/m3 of compacted volume. The OMC and MDD is mentioned in the Table.1. The Direct shear test is conducted as per IS: 2720 (Part X)- [4] to evaluate the strength parameter of pond ash. Pond ash specimens were arranged equivalent to their MDD and OMC which results are shown in Table 1.

4. Footing Load Tests

This test was performed for studying load settlement behavior of pond ash which is embedded with stone column conducted on strain regulated loading apparatus. To conduct the test pond ash samples of density 0.90 Kg/cm^3 is prepared within the cylindrical tank size of inside dia. of 39.6 cm and length of 40 cm. At the middle of pond ash, stone column was embedded by cylindrical steel road for offering specified area ratio and for compacting aggregate with the help of iron rod. The test was directed to survey the viability of length ratio 10 %, 20 %, 30 %, 40 % and 60 % the test was conducted to study the adequacy of differing length proportion 0.25, 0.50, 0.75, and 1.0. These specimens were tried during a Strain controlled loading apparatus with strain pace of 1.25 mm/min until the sample fails. Test outcomes were introduced in Table 2 and Table 3 and consequently the example arrangement and instrument were wont to make the samples were given in Figure 1-4 respectively.

Figure 1. Hollow cylindrical pipe to make cavity on pond ash

Figure 2. Compacted pond ash

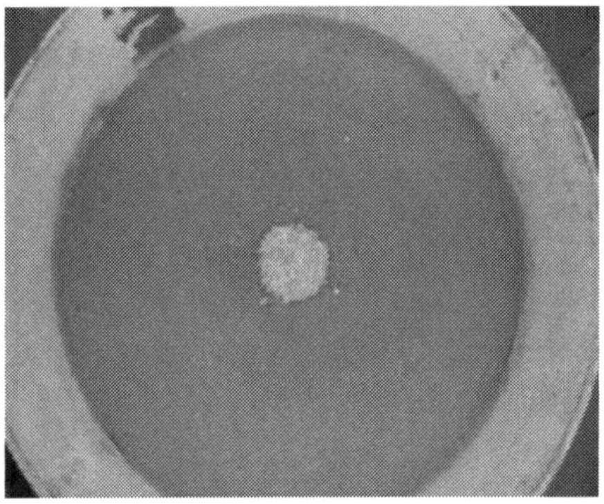

Figure 3. Pond ash reinforced by stone column

Figure 4. Footing load test

5. Results And Discussion

5.1 Stress Strain behaviour of reinforced pond ash bed

Footing load tests were conducted where the untreated pond ash specimens compacted to their corresponding MDD and OMC. This test was conducted to study the load settlement behaviour of pond ash reinforced with stone column in varying length ratio to their corresponding area ratio mentioned earlier and therefore the test result and behaviour has plot in Figure 5 and 6. From these figures, it's shown that,

- For a specific length of stone column the primary stress is maximum at upper area ratio, the initial stress increases linearly with the rise of length ratio. Also, the utmost failure stress depends on the utmost area ratio and length ratio. After reaching the utmost failure stress, the failure zone rises to the upper side of pond ash bed.
- Bearing capability increase proportionally with the increase of length of stone column of a particular area ratio. It shows that the angle of internal friction increases linearly with the rise of stone column diameter and its reinforcing length which leads to increase its compacted density in result bearing capacity get increases linearly. In this case diameter and length of the stone column plays a major part to increase the bearing capacity of stone column.
- At 0.75 length of reinforced pond ash shows closer results with full length ratio stone column so with the consideration of the use of materials, 0.75 length of stone column has more in effect as compared to other.

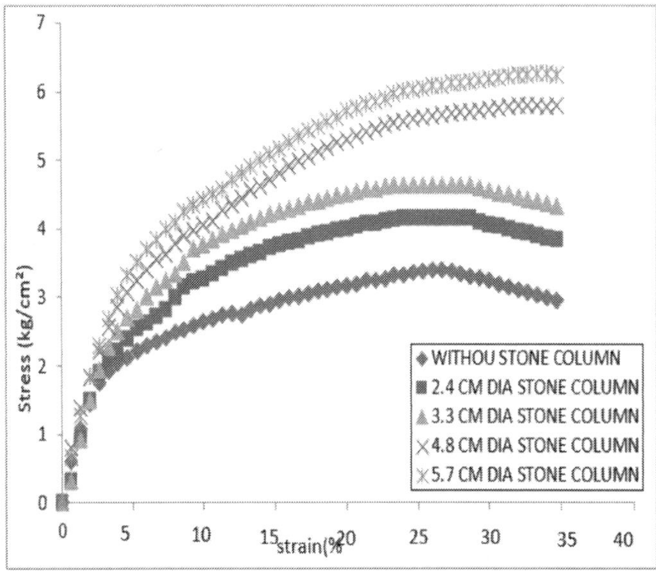

Figure 5. Stress and Strain in 0.5 Length Ratio

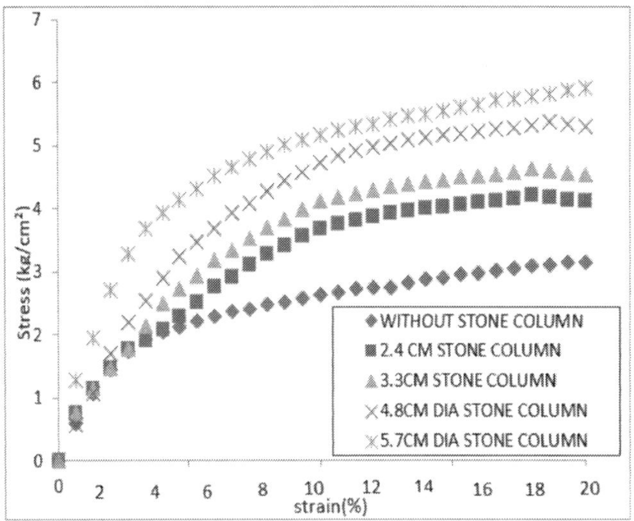

Figure 6. Stress and Strain in 0.25 Length Ratio

6. Conclusions

In this study the behaviour of various compacted pond ash reinforced materials with stone column structure, there are some conclusions are drawn and enlisted below:

i. In the footing load test, failure stress increased linearly with the increase of area ratio and decrease in length ratio.
ii. It is detected that when we consider the low length ratio, stone column particles along with pond ash settles due to load application.
iii. It is witnessed that it gives higher stress to the respective higher area ratios. Correspondingly, it shows maximum stresses for precise area ratio were perceived to its higher length ratios.
iv. It is seen that bearing capacity increases linearly with the expansion of length proportion to their respective area ratio. The expansion of diameter of the stone column and its reinforcing length

result shows that bearing capability increases linearly due to high compacted density of sample specimen which leads to increase its angle of internal friction.

v. It shows closer bearing capacity value when it is considered for the length ratio of full length and 0.75 length ratio reinforced pond ash. Hence the necessity of stone aggregate substantial is more in full length ratio stone column as compare to 0.75 length ratio, so 0.75 length ratio (0.75L) can be considered as more effective and economic as compare to other length ratio.

References

[1] Guetif Z, Bouassida M and Debats J 2007 Improved soft clay characteristics due to stone column installation Computers Geotechnics 34 104-11
[2] Deb K and Dhar A 2011 Optimum design of stone column-improved soft soil using multiobjective optimization technique Computers Geotechnics 38 50-7
[3] Elshazly H, Elkasabgy M and Elleboudy A 2008 Effect of inter-column spacing on soil stresses due to vibro-installed stone columns: interesting findings Geotechnical Geological Engineering 26 225
[4] Deb K 2008 Modeling of granular bed- stone column- improved soft soil International journal for numerical analytical methods in geomechanics 32 1267-88
[5] Deb K, Samadhiya N K and Namdeo J B 2011 Laboratory model studies on unreinforced and geogrid-reinforced sand bed over stone column-improved soft clay Geotextiles Geomembranes 29 190-6
[6] Shivashankar R, Babu M D, Nayak S and Manjunath R 2010 Stone columns with vertical circumferential nails: laboratory model study Geotechnical Geological engineering 28 695-706
[7] Shivashankar R, Babu M D, Nayak S and Rajathkumar V 2011 Experimental studies on behaviour of stone columns in layered soils Geotechnical Geological Engineering 29 749
[8] Vekli M, Aytekin M, Ikizler S B and Çalik Ü 2012 Experimental and numerical investigation of slope stabilization by stone columns Natural Hazards Earth System Sciences 64 797-820
[9] 1980 IS: 2720 (Part-III, Section-1, 1980), Determination of specific gravity, Bureau of Indian Standards, New Delhi, India
[10] 1985 IS 2720, Part 4 (1985a):, Methods of test for soils: Grain size , Bureau of Indian Standards, New Delhi, India
[11] 1980 IS: 2720 (Part 7, 1980), Methods of test for soils: Determination of water content-dry density relation using light compaction, Bureau of Indian Standards, New Delhi, India
[12] 1980 IS: 2720 (Part 8, 1980), Methods of test for soils: Determination of water content-dry density relation using heavy compaction, Bureau of Indian Standards, New Delhi, India
[13] 1973 IS: 2720 (Part 2, 1973), Methods of test for soils: Determination of moisture content, Bureau of Indian Standards, New Delhi, India
[14] 1973 IS: 2720 (Part 10, 1973), Methods of test for soils: Determination of unconfined compressive strength, Bureau of Indian Standards, New Delhi, India

Evaluation of Liquefaction Potential of Coastal Alluvium Using SPT Data - A Comparative Case Study

P K Muduli[1], P Karna[1], M R Samal[2]

[1] Department of Civil Engineering, Government College of Engineering, Kalahandi, Bhawanipatna, Odisha, India
[2] Directorate of Technical Education & Training, Cuttack, Odisha, India

E-mail: pradyut.muduli@gmail.com

Abstract. Earthquake induced soil liquefaction causes large scale damage to the infrastructures, lifeline systems and harbor facilities. Evaluation of the liquefaction potential of a soil subjected to seismic loading is one of the most important steps towards mitigation of liquefaction-induced damages. Nowadays, evaluation of liquefaction potential in case of susceptible soils is considered as a routine practice. The proposed coastal National Highway (NH) connecting Digha of West Bengal to Gopalpur of Odisha under Bharatmala Pariyojana is a major NH project undertaken by the National Highway Authority of India (NHAI). The district Kendrapara lies in 20° 20' N to 20° 37' N Latitude and 86° 14' E to 87° 01' E Longitude and situated in central coastal plain on the Bay of Bengal and comes under the seismic zone III as per the seismic zoning map of Bureau of Indian Standards (IS 1893 (Part 1): 2016). As the coastal alluvium is susceptible to liquefaction, some of the SPT bore log data collected from the two coastal locations of Kendrapara district of Odisha under the above project is used for evaluation of liquefaction potential of soil following available statistical method, artificial neural network (ANN) based method and multi-gene genetic programming (MGGP) based method. It is observed that some layers of the above soil are liquefiable considering a future seismic event of moment magnitude (Mw), 7. All the above three methods are predicting well the liquefaction and non-liquefaction cases.

1. Introduction

Earthquake induced liquefaction of soil causes huge loss of life and damage to infrastructures and lifeline systems. Some microzonation work has been done for Delhi, Kolkata and Bangalore in India but no study has so far been undertaken to establish liquefaction possibilities at local levels in the state of Odisha. In Odisha, the district Kendrapara lies in 20° 20' N to 20° 37' N Latitude and 86° 14' E to 87° 01' E Longitude and situated in central coastal plain on the Bay of Bengal and adjacent to Port City Paradeep. This district has huge potential of growth by way of industrialization in near future. The proposed construction of 2/4 laning with paved shoulder coastal national highway connecting Digha of West Bengal to Gopalpur of Odisha under Bharatmala Pariyojana of Government of India is of 415km. It connects major places of coastal districts of Odisha: Gopalpur – Satpada – Konark – Astarang – Naugaon - Paradip port – Ratanapur – Satabhaya – Dhamra – Basudevpur – Talapada – Chandipur - Chandeswar to the most sought after tourist place, Digha. The main objective of the above project is to reduce the distance and travel time and to give connectivity to remote area. The project lays emphasis on development of these areas and makes them available with the resources [1-20]. The proposed highway would act as the primary arterial route for the socio-economic growth of this region. In this investigation, evaluation of liquefaction potential of soil at two places i.e. Ratnapur and Satabhaya of the district, Kendrapara nearer to Paradip under the above coastal NH project for an assumed future seismic event using available deterministic methods as per IS: 1893 (Part 1): 2016 [12], Juang et al. [4]; Muduli and Das [10] on the basis of collected SPT data.

2. Methodology

In the present study, the Statistical method, ANN based method and MGGP based method as per IS 1893 (Part1):2016 [12], Juang et al. [4]; Muduli and Das [10] respectively based on deterministic approach are followed for prediction of liquefaction potential of costal alluvial soil of the two places, Ratnapur and Satabhaya of district Kendrapara of the state Odisha. Above methods are discussed in brief for completeness and are as follows.

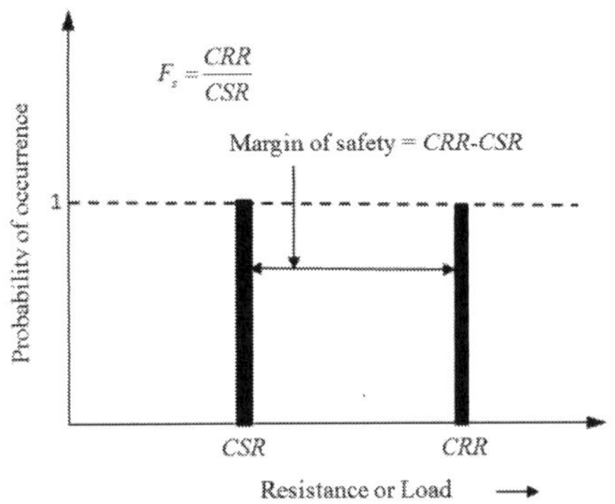

Figure 1. Deterministic approach in liquefaction potential evaluation [13].

2.1. Statistical Method

CSR is employed as an earthquake loading constraint while *CRR* is signified by in-situ test constraints which reveal the density and pore pressure producing assets of top soil [13]. The utmost frequently used deterministic technique to evaluate the liquefaction probable of a site is "simplified procedure" initially established by Seed and Idriss [1]. Factor of safety (F_s) in contradiction of the incidence of liquefaction for any earthquake established on SPT data as prearranged by Youd et al. [2] is adopted by the Bureau of Indian Standards in IS: 1893 (Part 1): 2016.

2.2. ANN based Method

In ANN-based liquefaction method [4], wherein CSR is attuned to the point of reference earthquake, M_w of 7.5 using MSF and is as follows.

$$CSR_{7.5} = 0.65\left(\frac{\sigma_v}{\sigma'_v}\right)\left(\frac{a_{max}}{g}\right)(r_d)/MSF \quad (1)$$

Where, *MSF* is as defined earlier.

$$CRR_{7.5} = -0.485 + 0.289\left[\frac{(N_1)_{60cs}}{35}\right] + \frac{0.739}{\left[35-(N_1)_{60cs}\right]^{0.1}} \quad (2)$$

Above Equation is valid for $(N_1)_{60cs} < 35$ and
for $(N_1)_{60cs} \geq 35$, there will be no liquefaction (NL) in the soil.

$$F_s = \frac{CRR_{7.5}}{CSR_{7.5}} \tag{3}$$

2.3. MGGP based Method

Here in MGGP based method as per Muduli and Das [10], *CSR* is modified from Youd *et al.* [2] and is as given below.

$$CSR_{7.5} = 0.65\left(\frac{\sigma_v}{\sigma'_v}\right)\left(\frac{a_{max}}{g}\right)(r_d)/MSF/K_\sigma \tag{4}$$

where, *MSF* and K_σ are as defined in earlier section

$$CRR = 0.008N_{1,60} - 0.613\cos\left[0.043N_{1,60}\right] + 0.077\cos\left[0.194N_{1,60}\right] + \frac{0.043}{N_{1,60}} + 0.609 \tag{5}$$

$$F_s = \frac{CRR}{CSR_{7.5}} \tag{6}$$

3. Case Study

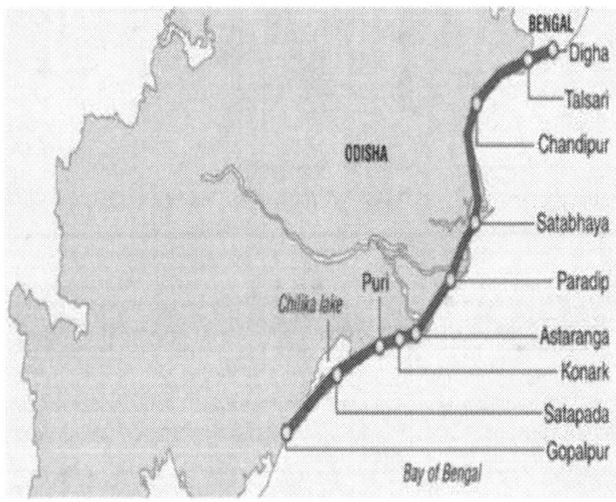

Figure 2. Alignment of proposed coastal NH passing through the state Odisha

3.1 Database Used

In the present study, two coastal areas of Kendrapara district (i) Ratnapur (ii) Satabhaya, through which proposed coastal NH project of NHAI "Gopalpur – Satpada – Konark – Astarang – Naugaon – Paradip port - Ratanapur – Satabhaya – Dhamra – Basudevpur – Talapada – Chandipur – Chandeswar - Digha for up-gradation to two/four lane with paved shoulder and National Highway configuration under Bharatmala Pariyojana" as per the alignment shown in the Figure 2 passes, are selected for assessment of liquefaction probable of soil using the discussed SPT based methods in the earlier sections. These areas are coming under seismic zone-III as per Bureau of Indian Standards code of practice on "Criteria for Earthquake Resistance Design of Structures", IS 1893 (Part 1): 2016. The bore log data pertaining to the geotechnical investigation for construction of some minor and major bridges in the above NH project close to the coast are obtained from the NHAI. Out of 10 collected SPT bore log 06 are selected and the bore log data are processed for analysis range from 3.5m to 23m depth below ground surface with measured *N* values ranging from 01 to 49. Depth of water table is in the range of 0.8m to 2m below ground level for the above two locations. The σ_v and σ'_v values are in the range of 30.02-445.92kPa and

15.31-225.19kPa respectively. Also the a_{max} value is assumed to be 0.16g, which corresponds to the seismic zone-III in accordance to IS 1893 (Part 1): 2016. Further, the future seismic event of M_w=7 (i.e. Equivalent Richter's Scale Magnitude, M_L=6.4) is considered in the present case for the liquefaction potential evaluation.

4. Results and Discussion

Out of the 06 bore log data 03 (BH1, BH2, BH3) are from Ratnapur and other 03 (BH4, BH5, BH6) are from Satabhaya of Kendrapara district. The bore log data of BH1, BH2, BH3, BH4, BH5 and BH6 are processed to obtain various parameters, which are to be used in liquefaction potential evaluation models and the analysis is made by using all the three SPT-based methods: Statistical Method (as per IS: 1893 (Part1):2016), ANN-based Method (as per Juang et al. [4]) and MGGP-based Method (as per Muduli and Das [10]) and the results are shown in the Table 1. From the BH1 data analysis it is found that at the depth 3m, 6.5m and 18.5m below ground level the F_s are 0.55,0.46,0.84; 0.64,0.54,0.98; 0.68,0.60,0.69; by using the above three methods respectively. This indicates the soil in each of the above layers is liquefiable (L) as $F_s \leq 1$ in a future seismic event. All above three methods are at par in predicting the liquefiable (L) soils. It can be noted here that in these layers the soil type is very soft silty clay with liquid limit (LL): 40-53%, plasticity index (PI): 15to 17%, fines content (FC): 45%-60% and natural water content (w_n): 49-62%. These soil properties of the above liquefiable layers are matching with the criteria for liquefaction susceptibility as mentioned in Bray and Sancio [16]. At the depth 9.5m and 21.5m below ground level the soil layers are found to be non-liquefiable (NL) as the F_s are 5.15 and 7.31 respectively, which are greater than 1 in accordance to MGGP-based method. For the above two layers $(N_1)_{60cs}$ are 51.83 and 44.23 respectively. As per the Statistical Method the soil in the above two layers are non- non-liquefiable as $(N_1)_{60cs}$>30. Similarly, as per ANN-based method the soil in the above two layers are also non-liquefiable as $(N_1)_{60cs}$>35.

Similar study is made for the bore log data BH2 the results are shown in Table 1. In this case, it is observed that soil layers below 3.5m and 5m, from ground level are liquefiable in the assumed future seismic event as the calculated F_s are 0.72,0.65; 0.84,0.76; 0.71,0.68 by the Statistical method, ANN-based Method and MGGP based Method respectively. Bray and Sancio [16] observed similar results for the type of soil observed for the above bore log data. At the depth 12.5m below the ground level the F_s of the soil layer is found out to be 2.34, 2.28 and 4.15 by Statistical method, ANN-based Method and MGGP based Method respectively. This indicates that the above layer is non-liquefiable (NL).It is also found that $(N_1)_{60cs}$ value of the soil layer at 23m below ground level is 36.67 which indicates that soil is NL as per both Statistical method and ANN-based Method. Similar result is also found by the MGGP-based Method for the above layer as the Fs is 7.35. All these three methods predict equally well the non-liquefiable (NL) soil layers.

The bore log data, BH3 are analyzed and it is observed that at the depth 9.5m below the ground level soil is liquefiable under the assumed seismic conditions in future as per the results provided in Table 1 in accordance to all the three methods. The soil type in this zone is very soft silty clay with LL: 44%, PI: 17%, FC: 90% and w_n: 47%. At the depth 15.5m and 18.5m below ground level the F_s of the two layers are found to be 1.53,1.79; 1.72, 1.99; 1.04,1.38 by Statistical method, ANN-based Method and MGGP based Method respectively, which confirms that the above soil layers are NL as F_s>1. The soil type in these zones is medium dense silty sand with LL: 39-48%, PI: 22-29%, FC: 68-73% and w_n: 24-30%.

The bore log data of BH4 are analyzed and it is observed from the Table 1 that there is only one liquefiable (L) layer at the depth of 3.5m below ground level as the evaluated F_s is 0.76, 0.90 and 0.77 by Statistical method, ANN-based Method and MGGP-based Method respectively. Soil type in this zone is very soft silty clay with LL: 43%, PI: 18%, FC:90% and w_n: 35%. Further, based on plasticity criteria in accordance to Bray and Sancio [16] the above layer of soil is susceptible to liquefaction.

Similar analysis is made for the bore log data of BH5 and the results are shown in Table 1. It is found that only one layer, 9.5m below ground level is on the verge of margin of liquefaction as the evaluated F_s is found out to be 0.99 by both Statistical Method and MGGP-based Method though the F_s is evaluated to be 1.16 by ANN-based Method, which is close to 1. The two soil layers 15.5m and 21.5m below ground level are found to be NL as per the evaluated F_s (7.63, 15.07) by MGGP-based Method and $(N_1)_{60cs}$ values (52.54,74.41) as shown in the Table 1.

Table 1. Comparison of F_s using different methods for evaluation of liquefaction potential

Depth (m)	SPT $(N_1)_{60}$ value	SPT $(N_1)_{60cs}$ value	Statistical Method (IS 1893 (Part1): 2016) F_s	ANN based Method [4] F_s	MGGP based Method [10] F_s	Remark
			BH1			
3.0	1.69	7.04	0.55	0.64	0.68	L
6.5	1.49	6.79	0.46	0.54	0.60	L
9.5	39.02	51.83	$(N_1)_{60cs}$>30	$(N_1)_{60cs}$>35	5.15	NL
18.5	5.17	11.20	0.84	0.98	0.69	L
21.5	38.03	44.23	$(N_1)_{60cs}$>30	$(N_1)_{60cs}$>35	7.31	NL
			BH2			
3.5	3.27	8.93	0.72	0.84	0.71	L
5	2.83	8.39	0.65	0.76	0.68	L
12.5	27.36	27.71	2.34	2.28	4.15	NL
23	35.92	36.67	$(N_1)_{60,cs}$>30	$(N_1)_{60cs}$>35	7.37	NL
			BH3			
9.5	2.21	7.66	0.58	0.68	0.66	L
15.5	12.74	20.29	1.53	1.72	1.04	NL
18.5	14.12	21.95	1.79	1.99	1.38	NL
			BH4			
3.5	3.15	8.78	0.76	0.90	0.77	L
14.5	18.21	26.85	2.56	2.58	1.93	NL
18	13.70	21.43	2.72	4.46	2.45	NL
			BH5			
9.5	14.54	14.78	0.99	1.16	0.99	L
15.5	39.61	52.54	$(N_1)_{60cs}$>30	$(N_1)_{60cs}$>35	7.63	NL
21.5	57.84	74.41	$(N_1)_{60cs}$>30	$(N_1)_{60cs}$>35	15.07	NL
			BH6			
9.5	32.89	44.46	$(N_1)_{60cs}$>30	$(N_1)_{60cs}$>35	6.15	NL
15.5	55.44	71.53	$(N_1)_{60cs}$>30	$(N_1)_{60cs}$>35	12.52	NL
18.5	83.38	105.06	$(N_1)_{60cs}$>30	$(N_1)_{60cs}$>35	16.05	NL
21.5	108.70	135.44	$(N_1)_{60cs}$>30	$(N_1)_{60cs}$>35	14.95	NL

Bore log data pertaining to BH6 are analysed and liquefaction potential of soil layers at 9.5m, 15.5m, 18.5m and 21.5m below the ground surface are evaluated using all the above methods. It is observed that the F_s at the above depths as evaluated by MGPP-based Method are 6.15, 12.52, 16.05 and 14.95 respectively, which indicates that the soil layers are NL. $(N_1)_{60cs}$ values of the above layers are found out to be 44.46, 71.53, 105.06 and 135.44 respectively. The Statistical Method as well as ANN-based Method confirms that these layers are NL as $(N_1)_{60cs} > 30$.

From the above findings, it is pertinent to mention here that even though MGGP-based CRR model does not include fines content (FC) parameter, it predicts well all the NL soil layers having $(N_1)_{60cs}$>30, which

are considered to be too clay rich to liquefy. This shows the compactness as well as effectiveness of MGGP-based CRR model over the lengthy CRR models of the Statistical and ANN–based methods.
In case of important projects, it is always necessary to verify the result obtained by one method with that of the other available methods to get confirmation and take proper design decisions. All the three deterministic methods: Statistical Method, ANN-based Method and MGGP-based Method are equally efficient in predicting liquefaction potential of soil for a future seismic event.

5. Conclusion

Ratanpur and Satabhaya of Kendrapara district is under seismic zone-III as per the seismic zoning map of Bureau of Indian Standards (IS 1893(Part 1): 2016). Based on composition of soil the liquefaction susceptibility in these areas is high. It is observed that certain layers of soil are liquefiable in five cases except one as studied by the three different deterministic methods. All the three deterministic methods: Statistical Method, ANN-based Method and MGGP-based Method are equally efficient in predicting liquefaction potential of soil for a future seismic event. The coastal alluvial soil is susceptible to liquefaction, which is clear from the present findings. Hence, the National Highway Authority of India (NHAI) has to pay special attention while undertaking the proposed coastal NH project in Odisha under Bharatmala Pariyojana.

References

[1] Seed H B and Idriss I M 1971 Simplified procedure for evaluating soil liquefaction potential *Journal of Soil Mechanics Foundations Div*
[2] Youd T L and Idriss I M 2001 Liquefaction resistance of soils: summary report from the 1996 NCEER and 1998 NCEER/NSF workshops on evaluation of liquefaction resistance of soils *Journal of Geotechnical Geoenvironmental Engineering* **127** 297-313
[3] Cetin K O, Seed R B, Der Kiureghian A, Tokimatsu K, Harder Jr L F, Kayen R E and Moss R E 2004 Standard penetration test-based probabilistic and deterministic assessment of seismic soil liquefaction potential *Journal of Geotechnical Geoenvironmental Engineering* **130** 1314-40
[4] Juang C H, Chen C J, Jiang T and Andrus R D 2000 Risk-based liquefaction potential evaluation using standard penetration tests *Canadian Geotechnical Journal* **37** 1195-208
[5] Samui P, Sitharam T and Contadakis M 2011 Machine learning modelling for predicting soil liquefaction susceptibility *Natural Hazards Earth System Sciences* **11**
[6] Samui P 2007 Seismic liquefaction potential assessment by using relevance vector machine *Earthquake Engineering Engineering Vibration* **6** 331-6
[7] Gandomi A H and Alavi A H 2012 A new multi-gene genetic programming approach to non-linear system modeling. Part II: geotechnical and earthquake engineering problems *Neural Computing Applications* **21** 189-201
[8] Muduli P K, Das M R, Samui P and Kumar Das S 2013 Uplift capacity of suction caisson in clay using artificial intelligence techniques *Marine Georesources Geotechnology* **31** 375-90
[9] Muduli P K and Das S K 2013 SPT-based probabilistic method for evaluation of liquefaction potential of soil using multi-gene genetic programming *International Journal of Geotechnical Earthquake Engineering* **4** 42-60
[10] Muduli P K and Das S K 2014 Evaluation of liquefaction potential of soil based on standard penetration test using multi-gene genetic programming model *Acta Geophysica* **62** 529-43
[11] Muduli P K and Das S K 2015 Model uncertainty of SPT-based method for evaluation of seismic soil liquefaction potential using multi-gene genetic programming *Soils Foundations* **55** 258-75
[12] 1893 Criteria for earthquake resistant design of structures *IS 1893 (Part 1) Bureau of Indian Standards, New Delhi, India* **1**
[13] Muduli P K 2014 Evaluation of liquefaction potential of soil using genetic programming. National Institute of Technology, Rourkela, Odisha, India.

[14] Seed H B, Idriss I and Arango I 1983 Evaluation of liquefaction potential using field performance data *Journal of Geotechnical Engineering* **109** 458-82
[15] Bolton Seed H, Tokimatsu K, Harder L and Chung R M 1985 Influence of SPT procedures in soil liquefaction resistance evaluations *Journal of geotechnical engineering* **111** 1425-45
[16] Bray J D and Sancio R B 2006 Assessment of the liquefaction susceptibility of fine-grained soils *Journal of Geotechnical Geoenvironmental Engineering* **132** 1165-77

Statistical and Time series Analysis of Ground Water Parameters of Kalahandi District, Odisha, India

M. Patnaik, M. Priyadarshini

Department of Civil Engineering, Government College of Engineering, Kalahandi, Bhawanipatna, Odisha, India

E-mail: manaswini@gcekbpatna.ac.in

Abstract. The present research is based on both qualitative and quantitative analysis of ground water of a western state Kalahandi of Odisha, India. 15 physio chemical parameters in addition to ground water depth data has been collected from CGWB, Bhubaneswar, Odisha for all 13 blocks of the district. The main objective of this study is to derive significant relationships between the groundwater parameters and to understand the factors controlling these parameters. Plots of linear regression reveals strong co relationship between magnesium, bicarbonate ions with total alkalinity. Magnesium, chloride and bicarbonate ions are significantly controlling the electrical conductivity of the groundwater. An important implication derived from the time series plot that the pattern of most of the physiochemical and quantitative parameters cannot be assumed linear. Therefore, reliance on formal regression cannot be undertaken for trend analysis. Mann-Kendall test as a non-parametric method of trend testing is taken. Based on S statistics, parameters such as hardness and magnesium show increasing trend whereas pH, sodium and potassium show a significant decreasing trend. Groundwater depth shows surprising results of significantly increasing trend being the S value as high as 103, which leads to ample scope of utilization and reliance on groundwater of the district.

1. Introduction

Water is the most vital natural resource for livings to sustain life on earth. It has a crucial role in shaping and regulating climate. Water is necessity for all forms of life. The domestic utility of water involves drinking, cooking, bathing, cleaning, and for watering garden crops. However, when compared to the worldwide industrial application of water, domestic water utilizations merely form half of its quantity use. Furthermore, water is required to yield crops and to maintain the balance in environment ecology. Worldwide irrigation sector consumes a large share of surface and ground water. As about two – third of the cultivated area is rainfed in India, rest one -third is dependent on irrigation, from which the utilization of groundwater for irrigation consumes nearly 65 % of total water resources of our country. [1-19].

Due to the exponential increase in the population and expansion of industrialization, the globe has seen a tremendous increase in freshwater demand. As the demand is too high overexploitation, compromise with quality and quantity of water in daily life have become so common nowadays. The sharp reduction in surface water quality and groundwater draft in almost all major urban centres in the developing nation risks human health and creates a situation in these places where dwelling becomes almost impossible. Ground water depth in most of the places in the country has fallen to greater depths. Unregulated and unchecked access to groundwater, have resulted in overdraft of groundwater for domestic & irrigation purposes; and to rapidly falling groundwater tables in prime aquifers. India with declining freshwater sources has an immense shortage of potable water of acceptable quality. The socio-economic growth in many parts of country is causally related to the availability of water resources of acceptable quality and is highly constrained by its non-availability. India's demand for water is growing despite its expansion in supplies. In such a scenario, decision makers should come up with challenging policies to cater the water demand of the present population. In this context statistics and probabilistic studies can be viewed as effective decision-making tools to assess the quality and potential of surface and ground water source.

1.1. Groundwater scenario of the district

The area in the present study is no exception to the current scenario. With the growing population there is a marked growth in agriculture, industries, and domestic demands of water. The dependence on unsafe surface water sources to meet domestic needs paves way for unaccountable diseases and major health concern among persons of this region. In such a situation the groundwater is conceived as a major alternative source. But the serious concern associated with its over exploitation and probable contamination due to natural or artificial causes. Chemical contaminants such as fluoride, arsenic and nitrate pose major health issues in India. These Figures are alarmingly high and need to be addressed at least in the regions where dependency on groundwater is large. These chemicals with water from the irrigation fields or channels enter the sub surface to meet ground water table and may potentially hamper the quality of groundwater to a large extent at some downstream location [1-19].

1.2. Groundwater Resources of Kalahandi district

According to the reports of Central Groundwater Board the average depth of groundwater table in meter below the ground level of Kalahandi district during pre-monsoon ranges from 3.44 to 8.00 and during post monsoon it ranges from 1.74 to 5.48. These Figures indicate a fair amount of availability of groundwater resource and a reasonable reliability on groundwater sources.

The theme of the present research is to examine underlying relationship between groundwater potential (in terms of metres of depth) and quality parameters. This research also focusses critically on the variation of groundwater quality parameters over a time scale of about 17 years, to be vigilant of probable contamination of groundwater resulting from anthropogenic or other activities. As discussed in the previous section the overdraft of groundwater being a serious concern, the real time monitoring of groundwater is also worked out to some extent in this research. The main objectives are,

- Performing linear regressions among various physiochemical parameters and groundwater depth to explain the nature and magnitude of relationships.
- Conducting spatially averaged time series test on different concentrations of physio chemical parameters as well as ground water depth parameter to identify trend using Mann-Kendall trend test.

2. Area for the Study

The study area Kalahandi, is a district of Odisha, which span across a geographical area of 7920 sq. KM and lies in 19°03' & 20°45' North latitude and from 82°18' & 83°48' East latitude. The position of Kalahandi district falls in the toposheet nos. 64 L, 64P, 65 I and 65 M of Survey of India. Administratively, Kalahandi comprises of two subdivisions, Bhawanipatna and Dharamgarh and Headquartered at Bhawanipatna. The district has 13 blocks i.e., Bhawanipatna, Jayapatna, Junagarh, Koksara, Madanpur Rampur, Karlamunda, Narla, Dharmagarh, Kesinga, Golamunda, Thuamul Rampur, Lanjigarh, Kalampur. The district is sharing its boundary with Bolangir and Nuapada districts in the North, on the east by Kandhamal, by Koraput on the south and Nabarangpur on the west.

IOP Conf. Series: Materials Science and Engineering **970** (2020) 012034 doi:10.1088/1757-899X/970/1/012034

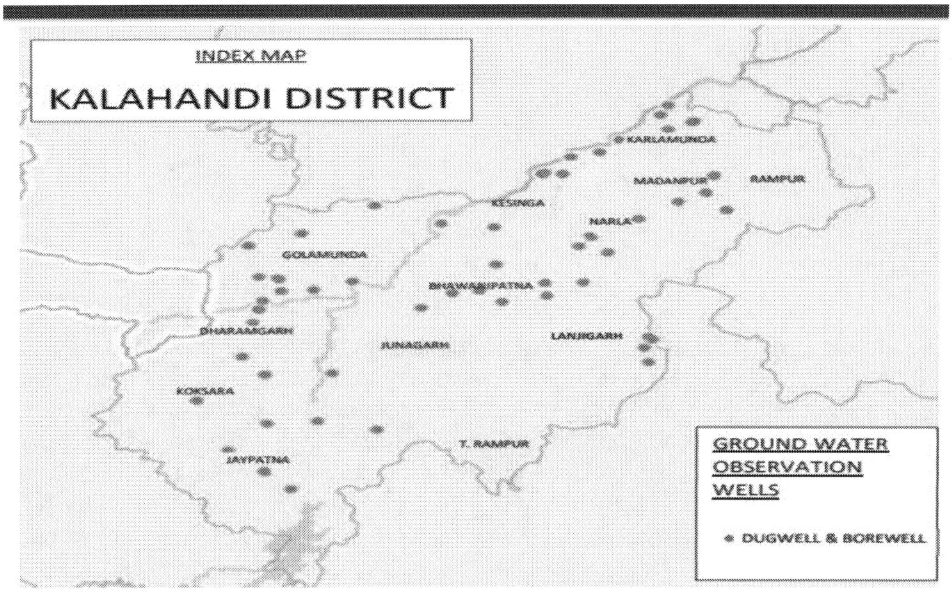

Figure 1. Ground water observation points of the research area.

3. Data Collection

Gound water quality and depth data were collected from *Central Ground Water Board, Bhubaneswar (Odisha)* for Kalahandi district consisting of 13 blocks (Bhawanipatna, Dharmagarh, Golamunda, Jaipatna, Junagarh, Karlamunda, Kesinga, Lanjigarh, Rampur, Narla, Koksara, Madanpur, T.Rampur) and 45 stations. 15 numbers of physiochemical parameters pertaining to 45 stations for pre monsoon seasons were collected from the year 2000 to the year 2017. For groundwater potential study ground water depth (in meters below the ground level) for the same 45 numbers of station of Kalahandi district were collected from the same source.

The ground water data of each station and year are sorted for pre-monsoon period for 15 hydro chemical parameters and one groundwater depth parameter. For the present research, the data has been sorted and arranged basically in two ways:

a) Spatial averaged time values: The data collected for 45 different stations is averaged and arranged over a period of 18 years and 21 years for ground water quality and quantity studies, respectively. This type of data setting is necessary for multivariate analysis like regression studies component analysis studies etc.

b) Temporal averaged space values: Here the data collected for 18 years and 21 years for ground water quality and quantity studies respectively is averaged and placed over 45 data stations. For timeseries analysis or trend studies, such arrangement of data is required.

4. Materials and Method

4.1 Regression Analysis

Linear regression is a method to access the extent of relationships among various parameters where all dependent variables are plotted against independent variables. The process of performing a regression allows to identify the dependent variables that affects the independent variable the most. Regression analysis serves as an indicative study as to ignore or drop the factors that impacts an independent variable the least. The dependence and corelation among the dependent variables can also be reflected from these analyses. Time averaged spatial values of parameters are utilized for these analyses. For the current research, four independent variables such as total alkalinity, electrical conductivity, hardness, and

IOP Conf. Series: Materials Science and Engineering **970** (2020) 012034 doi:10.1088/1757-899X/970/1/012034

groundwater depth are selected, and other variables are plotted against these variables. The regression line fitting the data plot is given by a linear or straight-line equation.

$$Y = \alpha X + \beta \quad (1)$$

The regression line is dependent upon the relationship between independent variable and dependent variable. The aforesaid relationship is represented below by Pearson's correlation co-efficient:

$$r = \frac{n(\sum xy)-(\sum x)(\sum y)}{\sqrt{[n\sum x^2-(\sum x)^2][n\sum y^2-(\sum y)^2]}} \quad (2)$$

4.2 Trend Analysis

A chronological time series analysis of gathering information and attempting to recognize a pattern called Trend Analysis is considering in the study. In statistics, this refers to techniques for understanding an hidden pattern of behaviour in a time series , which would otherwise be partly or completely concealed by noise commonly referred as error. A formal regression analysis is undertaken if the trend can be assumed linear, as described in trend estimation. If the trends assumes a non-linear curve, trend testing can be performed by non- parametric methods, e.g. Mann-Kendall test, which is a type of Kendall rank correlation coefficient. Smoothing techniques aids testing and recognizing of nonlinear trends.

4.2.1 Mann-Kendall trend test (M-K test)

The M-K test is used to analyze data collected over chronological time for consistently increasing or decreasing trends. Moreover Mann-Kendall test is used to determine whether there is a linear monotonic trend in a given time series data. Mann-Kendall trend test is a non-parametric test for trend closely related to the concept of Kendall's correlation coefficient and the data should have no serial correlation. The procedure follows Gilbert (1987). Data $x_1, \dots x_n$ are assumed to be in a sequential order of collection time, or in spatial sequence. The indicator function is defined as

$$Sgn\,x = \begin{Bmatrix} 1, if\ x > 0 \\ 0, if\ x = 0 \\ -1, if\ x < 0 \end{Bmatrix} \quad (3)$$

The S statistic is calculated as below:

$$S = \sum_{i=1}^{n-1} \sum_{j=i+1}^{n} sgn(x_j - x_i) \quad (4)$$

S is taken as negative for a negative trend, zero for no trend, and positive for an increasing trend. For $n \leq 10$, the p value is taken from a table of exact values (Gilbert 1987). For $n>10$, a normal approximation is used, as follows. The total number of groups of ties g and the number of tied values tj within each group, in the sorted sequence is determined. Then the standard deviation of S is estimated by

$$SD = \sqrt{\frac{1}{18}[n(n-1)(2n+5) - \sum_{j=1}^{g} t_j\,(t_j - 1)(2t_j + 5} \quad (5)$$

The Z statistic, is then

$$Z = \frac{|S|-1}{SD} \quad (6)$$

That used to find out p from the cumulative normal distribution. The subtraction of 1 indicates the continuity correction.

5. Results and Discussions

Table 1. Normal statistics parameters of water quality for ground water samples

Parameters	Maximum	Minimum	A_m	Median	Q_1	Q_3	S_d	C_v
pH	8.273	7.43	7.96	7.985	7.86	8.09	0.19	2.43
EC	1336	320	659.14	601.57	517.25	726.13	233.39	35.41
TDS	976	256	479.77	447.83	406	534.07	143.29	29.87
Total Alkalinity	320	68	209.68	208	183	229.65	43.59	20.79
Hardness	722	116	261.69	233.59	204	264	109.27	41.75
Ca^{++}	147.6	24	56.32	50	44.65	59.2	23.80	42.26
Mg^{++}	135.66	3.90	27.64	24.05	21.52	30.26	17.35	62.78
Na^{+}	61.58	5.8	28.87	27.91	17.51	38.99	14.34	49.66
K^{+}	29.36	0.167	3.89	1.96	0.88	3.71	6.11	157.11
Cl^{-}	306.29	11.34	70.34	51.76	36.56	74.49	64.76	92.08
SO_4^{--}	44.34	0	17.21	15.31	9.98	22.02	10.38	60.29
CO_3^{--}	17.89	0	4.08	2.87	0	6.51	4.41	108.17
HCO_3^{-}	390.4	74.59	245.77	251.67	217.16	277.59	61.27	24.93
F^{-}	0.91	0.081	0.49	0.47	0.33	0.63	0.22	45.79
NO_3^{-}	71.73	1.25	28.89	27.61	14.16	36.19	16.97	58.74
GWD	8.286	3.585	5.783	5.812	5.012	6.61	1.05	18.25

5.1 Statistical Results

The central tendency and the measures of variability parameters are determined for the temporal averaged values of spatial ground water quality and depth data. The statistical parameters such as arithmetic mean, median, quartiles, standard deviation and coefficient of variance are calculated and listed in the Table 1. The inferences or the preliminary observations are made from the data of normal statistics is discussed in the following section.

All units except pH and Electrical Conductivity are in mg/l, AM- Arithmetic mean, SD – Standard deviation, CV- Coefficient variation, Q1 – Median of the lower half, Q3 – Median of the upper half

Table 1 reveals a wide range of values in most of the chemical parameters except pH and fluoride. Higher values of standard deviations have been detected for electrical conductivity, total dissolved solids and hardness, which indicate that the values are widespread around the mean. In the district, the nitrate value varies from1.3 to 72 ppm for the pre – monsoon period. The nitrate value for the study area is found to be more than 45 ppm as per WHO (1994) in 6 stations. The highest nitrate concentration is observed in Junagarh block, which is irrigation prone area. The higher concentration in this region owes to over application of fertilizer, which paves its way with water towards ground water reservoir. Chloride is also found to be widely distributed in the present study area, ranging from 11 to 306 ppm. The highest chloride concentration pertains to a location in Golamunda block. Even alkalinity in this location is found to be on higher side, which indicates that there may be interrelation between these two variables.

5.2. Time Series Analysis

The space averaged temporal values of concentrations (in mg/l) are plotted over a time of 18 years. Figure 2 shows the bar chart representation and time series plots of various hydro chemical parameters of groundwater. These graphs present an overall notion of variation of these parameters over a period.

In the Figure 2, it can be generally inferred that the values of physiochemical parameters are well below the permissible limits. But for most of the parameters the values are less than the acceptable limits for most of the time as these parameters are plotted over a time scale of 18 years. These plots revealed some parameters for most of the time less than acceptable limit. Major ions like Ca^{2+} and Mg^{2+} values are found to be 94 % and 67 % of the time respectively, down the acceptable limit. Similarly, values of hardness, TDS and total alkalinity falls below the acceptable limits for 11 %, 56 % and 33 % respectively of the time. Most of the parameters such as Cl^-, SO_4^{2-}, F^- and NO_3^- the values remain less than acceptable limits for all the time. *IS 10500:2012 standard for drinking water* suggests values in excess of those mentioned under 'acceptable' render the water not suitable, but still may be considered in the absence of an alternative source but up to the limits indicated under 'permissible limit in the absence of 'alternate source', above which the sources will have to be rejected.

From the Figure 2 it is also noticeably clear that the underlying pattern of most of the physiochemical and quantitative parameters cannot be assumed to be linear. Therefore, trend analysis cannot be undertaken within formal regression analysis. A non-parametric method of trend testing i.e., Mann-Kendall test is hence adopted for present set of data.

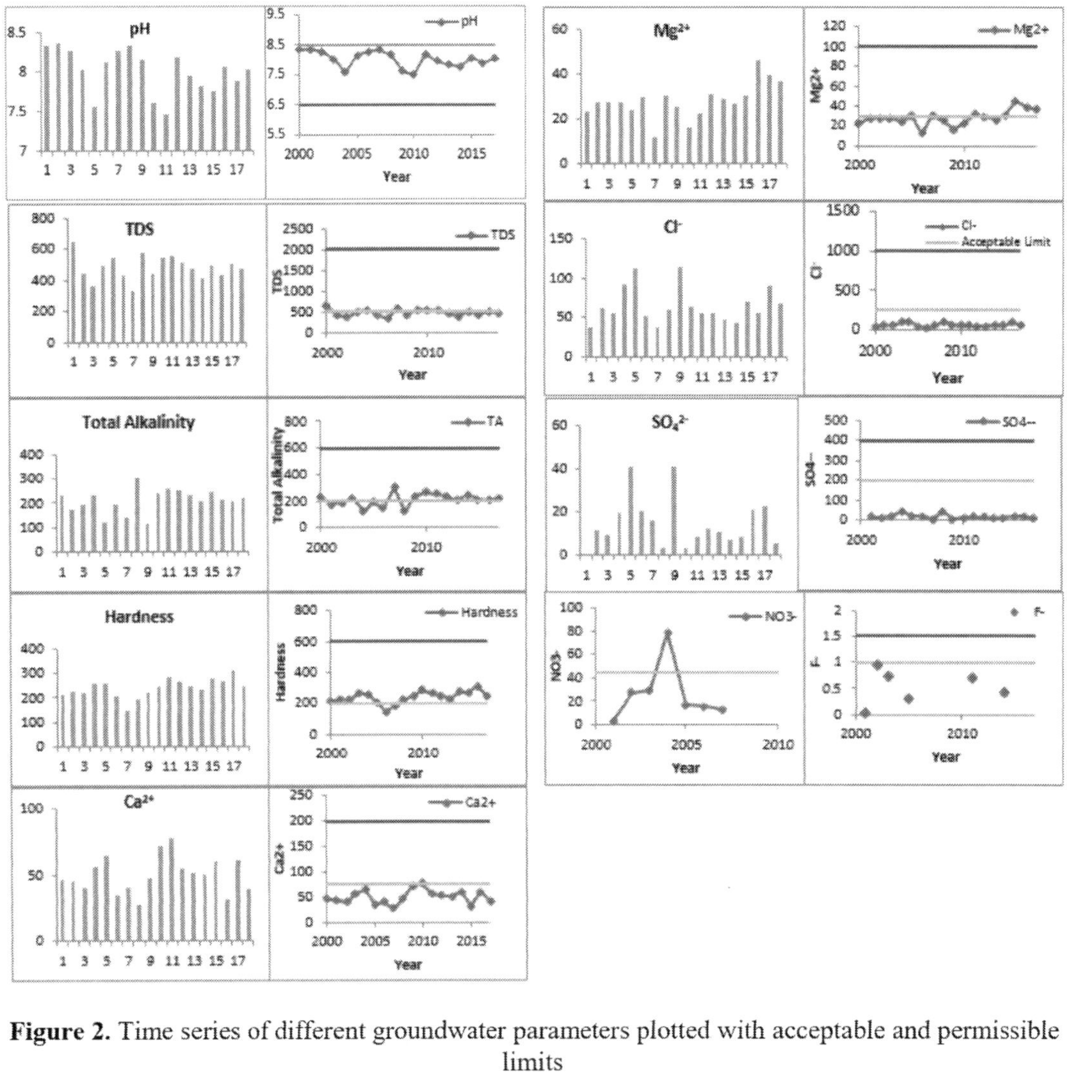

Figure 2. Time series of different groundwater parameters plotted with acceptable and permissible limits

5.3. Mann Kendall Test Result

Mann-Kendall test is performed with the space averaged temporal values of concentration and depth of ground water. Since the present study is focussed on only pre monsoon ground water data, the data set is devoid of any periodicities (or seasonal effects).

Table 2. Mann-Kendall trend test result

Ground Water Parameters	*S*	*Z*	*p*	Remarks
pH	-51	1.8966	0.05788	Statistically significant decreasing trend
Electrical Conductivity	2	0.041193	0.96714	Not statistically significant trend
TDS	-9	0.30302	0.76187	Not statistically significant trend
Total alkalinity	23	0.83331	0.40467	Not statistically significant trend
Hardness	63	2.3484	0.018853	Statistically significant increasing trend
Ca^{++}	13	0.45453	0.64945	Not statistically significant trend
Mg^{++}	65	2.4242	0.015343	Statistically significant increasing trend
Na^{+}	-72	2.9247	0.0034481	Statistically significant decreasing trend
K^{+}	-52	2.1008	0.035656	Statistically significant decreasing trend
Cl^{-}	9	0.30302	0.76187	Not statistically significant trend
SO_4^{2-}	-12	0.45312	0.65046	Not statistically significant trend
CO_3^{2-}	-25	0.90906	0.36332	Not statistically significant trend
HCO_3^{-}	23	0.83331	0.40467	Not statistically significant trend
Groundwater Depth	103	3.0843	0.0020403	Statistically significant increasing trend

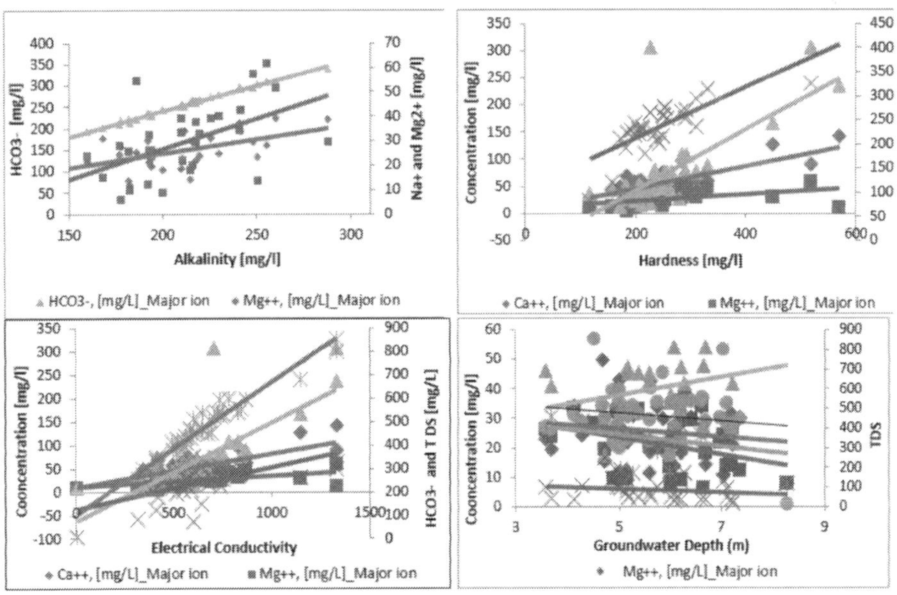

Figure 3. Linear regression plots of independent variables [TA, Hardness, EC and GWD] with tabledependant variables

The Mann-Kendall test results for different qualitative and quantitative ground water parameters has been tabulated in Table 2.Remarkable *S* values has been observed for statistically significant trend either increasing or decreasing. The direction of *S* shows the trend is either increasing or decreasing. The *S* values for hardness and magnesium ions is positive and on higher side resulting in increasing trend. Whereas parameters like pH, sodium and potassium ions for which *S* value is negative shows a significant decreasing trend. Statistically no trend has been found for some parameters like EC, TDS, total alkalinity, calcium ions, chloride ions, sulphate ions, carbonate, and bicarbonate ions. However, groundwater depth shows surprising results of significantly increasing trend being the *S* value as high as 103 To determine the nature and extent of relationships among various qualitative and quantity parameter, the concentrations of all dependent variables are plotted against independent variables. Time averaged spatial values of parameters are utilized for these analyses. In the present study four independent variables has been chosen and plotted against a set of dependent variables. Independent variable as total alkalinity, electrical conductivity, hardness and ground water depth and the choice of it is arbitrary.

Table 3. Regression analysis test results for independent and dependent variables

Independent Variables	Dependent Variables	α	β	R^2	*p value*	*Relationship*
Total Alkalinity	Ca^{++}	0.137	26.99	0.25	0.037	Reasonable
	Mg^{++}	0.135	-2.091	0.68	0	Significant
	Na^{+}	0.132	0.535	0.56	0.0006	Reasonable
	K^{+}	-0.016	7.033	0.02	0.321	Nil
	Cl^{-}	0.460	-24.03	0.366	0.0132	Reasonable
	CO_3^{2-}	0.001	3.491	-	0.909	Nil
	SO_4^{2-}	0.057	3.286	0.29	0.049	Reasonable
	HCO_3^{-}	1.203	-6.168	0.89	0	Significant
	F^{-}	0.046	4.278	0.002	0.783	Nil
	NO_3^{-}	0.116	1.700	0.17	0.054	Nil
Electrical Conductivity	Ca^{++}	0.072	9.216	0.78	0	Significant
	Mg^{++}	0.026	9.216	0.65	0	Significant
	Na^{+}	0.023	13.02	0.58	0.0088	Reasonable
	Cl^{-}	0.212	-64.43	0.79	0	Significant
	CO_3^{2-}	-0.001	4.666	0.008	0.567	Nil
	HCO_3^{-}	0.179	126.0	0.61	0	Significant
	TDS	0.566	102.38	0.94	0	Significant
	GWP	0.000	6.446	0.009	0.52	Nil
Hardness	Ca^{++}	0.202	5.085	0.72	0	Significant
	Mg^{++}	0.063	9.781	0.55	0.00026	Reasonable
	Cl^{-}	0.512	56.34	0.62	0	Significant
	SO_4^{2-}	0.046	3.736	0.64	0	Significant
	CO_3^{2-}	-0.006	5.237	0.02	0.354	Nil
	HCO_3^{-}	0.463	125.3	0.57	0.0045	Reasonable
Ground water Depth	Mg^{++}	-0.915	30.93	0.02	0.312	Nil
	Na^{+}	-0.856	32.72	0.01	0.489	Nil
	Cl^{-}	-0.392	72.25	-	0.946	Nil
	SO_4^{2-}	-0.493	18.03	0.007	0.372	Nil
	CO_3^{2-}	-0.303	5.572	0.019	0.583	Nil
	NO_3^{-}	-1.551	32.91	0.01	0.641	Nil

From the regression analysis the dependency of the observed variables and independent variable are established and it is found to be different and not significant for all parameters. Concentrations of most

variables increased with increasing independent variables (Table 3). The results of the multivariate analysis with the general linear regression model (Figure 3) shows strong positive relationships (*p value*<0.05) of total alkalinity with calcium, magnesium, sodium, chloride, sulphate and bicarbonate ions; electrical conductivity with magnesium, sodium, chloride and bicarbonate ions; hardness with calcium, magnesium, chloride, sulphate and bicarbonate whereas ground water depth is found to have no or meagre co-relation with any other parameters. Some of the dependent variables shows strong relationship with the independent ones. For example, a strong co relationship is revealed between magnesium, bicarbonate ions with total alkalinity, which means these, are the major two underlying factors controlling the alkalinity of the groundwater. All other parameters listed in the table except potassium, carbonate, fluoride and nitrate ions are reasonably related. Similarly, magnesium, chloride and bicarbonate ions are significantly controlling the electrical conductivity of the groundwater, which is obvious. The strong factors affecting hardness are calcium, sulphate, and chloride ions whereas magnesium and bicarbonates are moderately related to hardness.
From these observations, it can be concluded that the hardness is permanent in nature in ground water. Among all independent and dependent variables, the strongest relationship is revealed between EC and TDS with coefficient of correlation $R^2 = 0.94$. Ground water depth fails to establish correlation with the any of the parameters in consideration.

6. Conclusions

The current research is restricted to the preliminary results of multivariate analysis in which linear regression component is only stressed on. Time series and trend analysis forms another part of this research. Nevertheless, prior to this an attempt is to reduce the large data set and search for the inter relations that co-exist between various ground water parameters. The salient features have been concluded in the following points.

- Strong co relationship is established between magnesium, bicarbonate ions with total alkalinity, which means these, are the major two underlying factors controlling the alkalinity of the groundwater. Water alkalinity is primarily a function of geology of the area and dissolution of carbon dioxide from atmosphere. The ions responsible for alkalinity originate from dissolution of geological minerals into rain and ground water. High alkalinity in the groundwater is usually traced in the areas with limestone formations, this is attributed to the dissolution of bicarbonates and carbonates.
- Similarly, magnesium, chloride and bicarbonate ions are significantly controlling the electrical conductivity of the groundwater, which is obvious. Electrical conductivity of water can be immediately related to the concentration of dissolved ions in water. Dissolved solids in water is able to conduct electrical current. Hence, an incredibly significant relationship is observed between EC and TDS.
- The strong factors affecting hardness are calcium, sulphate, and chloride ions whereas magnesium and bicarbonates are moderately related to hardness. From these observations, it can be concluded that the hardness of ground water is permanent in nature.
- The relationship of groundwater depth with other parameters merely can be correlated. For the present study area, any variations in the physiochemical parameter have null effect on the depth parameter.
- Time series plots depicts that the physiochemical parameters of groundwater are well below the permissible limit which gives clear view of contamination free groundwater and is acceptable. Major ions like Ca^{2+} and Mg^{2+} for most of the time remain below the acceptable limit and are the major ions towards unsafe side. Nevertheless, from past few years, Mg^{2+} values are exceeding the acceptable limit, which should be diagnosed rigorously. Since strong co-relationship is depicted between magnesium ions and total alkalinity from linear regression analysis. If such groundwater is withdrawn and utilized for irrigation purpose, it may render salinity problems in culturable area. TDS and total alkalinity moderately exceed acceptable limits. Hardness in the groundwater is an instinct problem of this region. Around 89 % of the time (i.e., over 18 years) hardness exceeds

acceptable limit which is quite high. Most of the parameters such as Cl^-, SO_4^{2-}, F^- and NO_3^- the values remain less than acceptable limits for all the time. These are parameters under safe category. Measures of all major ions do not exceed acceptable limits prescribed in *Indian Standards* for groundwater quality except for Ca^{2+} and Mg^{2+} in scattered areas, indicating the confined water is potable

- Mann-Kendall test reveals significantly increasing trend for hardness and magnesium ions which is a key factor contributing to the total alkalinity. Parameters like sodium and potassium ions shows a significant decreasing trend. Statistically no trend has been found for some parameters like EC, TDS, total alkalinity, calcium ions, chloride ions, sulphate ions, carbonate, and bicarbonate ions. However, groundwater depth shows surprising results of significantly increasing trend being the *S* value as high as 103.

References:

[1] 2012 Drinking Water Specification (Second Revision), Manak Bhavan, 9 Bahadur Shah Zafar Marg *IS 2012 Bureau of Indian Standards, New Delhi, India* **2**

[2] Association A P H, Association A W W, Federation W P C and Federation W E 1915 *Standard methods for the examination of water and wastewater* vol **2**: (American Public Health Association.)

[3] Batabyal A K and Chakraborty S 2015 Hydrogeochemistry and water quality index in the assessment of groundwater quality for drinking uses *Water Environment Research* **87** 607-17

[4] Bathusha M I and Saseetharan M 2006 Statistical study on physio-chemical characteristics of ground water of Coimbatore south zone *Indian journal of environmental protection* **26** 508

[5] Dohare D, Deshpande S and Kotiya A 2014 Analysis of ground water quality parameters: a Review *Research Journal of Engineering Sciences* ***ISSN*** **2278** 9472

[6] Krishan G, Singh S, Kumar C, Gurjar S and Ghosh N 2016 Assessment of water quality index (WQI) of groundwater in Rajkot district, Gujarat, India

[7] Mahananda M, Mohanty B and Behera N 2010 Physico-chemical analysis of surface and ground water of Bargarh District, Orissa, India *International journal of research reviews in applied sciences* **2** 284-95

[8] Mohan U and Singh R 2013 Water quality assessment District Hapur, Uttar Prad *Environment Conservation Journal* **14** 143-9

[9] Nagamani C, Devi C S and Shalini A 2015 Physico-chemical analysis of water samples *International Journal of Scientific Engineering Research* **6** 2149-55

[10] Neerja K, Rajesh K, Yadav S and Singh R 2012 Water quality index assessment of ground water in Koilwar block of Bhojpur (Bihar) *Journal of chemical pharmaceutical research* **4** 1782-6

[11] Organization W H 2017 *Global diffusion of eHealth: making universal health coverage achievable: report of the third global survey on eHealth*: World Health Organization)

[12] Rajankar P N, Tambekar D H and Wate S R 2011 Groundwater quality and water quality index at Bhandara District *Environmental monitoring assessment* **179** 619-25

[13] Ramakrishnaiah C, Sadashivaiah C and Ranganna G 2009 Assessment of water quality index for the groundwater in Tumkur Taluk, Karnataka State, India *E-Journal of chemistry* **6**

[14] Rao G S and Nageswararao G 2013 Assessment of ground water quality using water quality index *Archive of Environmental Sciences* **7**

[15] Selvakumar S, Ramkumar K, Chandrasekar N, Magesh N and Kaliraj S 2017 Groundwater quality and its suitability for drinking and irrigational use in the Southern Tiruchirappalli district, Tamil Nadu, India *Applied Water Science* **7** 411-20

[16] Singh R 1996 Study of Physico-Chemical parameters of groundwaterof Nagaur District: Some Correlations

[17] Sinha D and Saxena R 2006 Statistical Assessment of Underground Drinking Water Contamination and Effect of Monsoon at Hasanpur, J. P. Nagar(Uttar Pradesh, India) *Journal of environmental science & engineering* **48** 157-64

IOP Conf. Series: Materials Science and Engineering **970** (2020) 012034 doi:10.1088/1757-899X/970/1/012034

[18] Sirajudeen J, Manikandan S A and Manivel V 2013 Water quality index of ground water around Ampikapuram area near Uyyakondan channel Tiruchirappalli District, Tamil Nadu, India *Archives of Applied Science Research* **5** 21-6
[19] Sunitha V, Sudarshan V and Reddy B R 2005 Hydrogeochemistry of groundwater, Gooty area, Anantapur district, Andhra Pradesh, India *Pollution Research* **24** 217

Soil Stabilization using Tea waste

P Harish, Ashutosh Mahapatra

Department of Civil Engineering, Government College of Engineering, Kalahandi, Bhawanipatna, Odisha, India

E-mail: harish.perla28@gmail.com

Abstract. Black cotton soils have frequently subjected alternate swelling shrinkage, which gives differential settlement to buildings. In this study Tea Waste (TW) used as partial replacement of black cotton soil with proportions of 2.5 %, 5 %, 7.5 % and 10 %. Soil sample behaviour was observed by Compaction test, California Bearing Ratio (CBR) test, Liquid limit test, Plastic limit and specific gravity Test. The Maximum dry density and Optimum moisture content were decreased when proportion of Tea Waste is increases. When Tea Waste (TW) is mixed with the expansive soil, the Plastic limit and liquid limit decreases by mixing tea waste, which decreases plasticity index. Addition of Tea Waste resulted in appreciable increases in the CBR of the soil and got maximum @ 5 % Tea Waste (TW) with soil. CBR value was increased by 7.68 % at 5 % Tea Waste with soil.

1. Introduction

For any structure, the soil is very important to carry loads from super structures. Black cotton soils are composition of volcanic ash and magma popularly known as "Reguda" in Telugu found in Telangana, Andhra Pradesh, Tamilnadu and Karnataka. Black cotton soils have frequently subjected alternate swelling shrinkage, which gives differential settlement to buildings. In addition to this it has low bearing capacity, highly affected by water content and construction is difficult [1] .These type of soils can be stabilised mechanically and chemically to increase the strength and improve the properties of soil.

Now a day's different wastes are mixed with soil to get soil stabilization that decreases the settlement of structure and increase the bearing capacity of soil. Waste materials rice husk ash, fly ash, lime, banana fibres etc. are enhanced the soil properties like shear strength-engineering properties [2]. Some waste materials such as Fly Ash, rice husk ash, pond ash, Tea Waste may use to make the soil to be stable. Usage of Waste in soil stabilization process reduces handling of waste and decreased the construction cost by improving the soil strength. Wubshet [2013] used bagasse ash mixed with soft soils with proportion of 5 %, 10 %, 15 %, 20 %, 25 % and 30 %.He found that soil stabilised by plasticity index ,swelling index, maximum dry density(MDD) were decreased and Optimum Moisture content and CBR were increased [3]. Kharade et al. (2014) used bagasse as partial replacement in black cotton soil and found that CBR increased by 41.52 %, MDD increased by 5.8 % [4]. Teresa and Joy (2016) investigated banana fibre as soil stabilizer which is a waste material. The proportion of banana fibre varies from 0.25 %, 0.75 %, 1 %and 2 %.

The banana fibres enhance the characteristics of soil. The optimal dosage of banana fibre was obtained at 0.75%for marine clay. It was found that dry density decreased and OMC increased with banana fibre. The shear strength changed from 8.5 kN/m2 to 32.91 kN/m2 with 0.75 % of banana fibre and

CBR changed from 2.79 to 13.2 which make it good for sub-grade [5]. Kamariya (2019) investigated used tealeaves and lime blended with soil with different proportions. The soil is stabilised by increasing CBR values and Maximum Dry Density. The soil texture also changed by decreasing Atterberg's limits of soil sample. In India day-by-day population increases, in the same the way the production of Tea Waste is increased. Tea Waste production is around 5-10 kg from tea shops, restaurants, hotels for every month [1-10].
In this study, Tea Waste is used as partial replacement of black cotton soil with proportions of 2.5 %, 5 %, 7.5 % and 10 %. Soil sample behaviour was observed by Compaction test, California Bearing Ratio (CBR) test, Liquid limit test, Plastic limit and specific gravity Test.

2. Materials and Methodology

2.1. Materials
The materials used in the experimental investigation include:
- Black Cotton soil
- Tea Waste(TW)

2.2 Methodology
Tea Waste is collected from hotels, restaurants near Basra town. Collected Tea Waste kept in oven at 60 ^{0}C to remove moisture to become dry powder. After Tea Waste was stored under room temperature. Black cotton soil collected from construction site Basar campus and it should be air-dried. The soil sample was sieved through BS 425 μm.
Initially, Tests were conducted on natural soil without adding tea waste, such that effect of addition of Tea Waste with soil easily understand by comparison. The samples prepared by weighing for each batch and mixed properly before adding water. Compaction test, Liquid limit, California bearing ratio test and Plastic limit Tests were conducted as per Indian Standard procedure [6-8] to do this project:

3. Results and Discussions

Table 1. Comparison of Max. dry density and Optimum Moisture content in different proportions of soils samples

Soil Type	Max. density MDD, gm/cc	Optimum Moisture Content OMC, %
Soil	1.482	17.4
Soil +2.5 %TW	1.419	18.8
Soil +5.0% TW	1.407	19
Soil +7.5%TW	1.369	20.8
Soil +10% TW	1.364	21.05

3.1. Compaction Test
Table 1 shows the variations of optimum moisture content (OMC) and maximum dry density (MDD) and of different soil types. As the percentage of Tea Waste (TW) is increases in the soil Maximum Dry Density and OMC decreases. Maximum dry density in the soil without TW is 1.482 g/cc decreased to 1.364 g/cc for soil having 10 % TW. Meanwhile Optimum moisture content for soil without TW is 17.05 % decreased to 21.05 % for soil having 10 % TW. Reason for decrease in Maximum Dry Density and increase in OMC can be related to organic nature of material and water retention capability [9].

3.2. California bearing ratio (CBR) Test
Table 2 shows the variation of CBR (unsoaked condition) of the four black cotton soil samples with the addition of Tea Waste (TW) in increasing percentages. From the results, we can say that addition

of 5 % of Tea Waste can be used for the remaining experiments. CBR value was increased by 7.68 % at 5 % Tea Waste with soil [10].

Table 2. Variant of CBR values in different proportions of soils samples

Soil Type	CBR value
Soil	7.33
Soil +2.5 % TW	7.43
Soil +5.0 % TW	7.94
Soil +7.5 %TW	5.02
Soil +10 % TW	4.91

3.3. Liquid limit
Table 3 shows that liquid limit decreases with the addition of 5 % Tea Waste (TW) [9].

Table 3. Camparison of Atterberg's limit for different proportions of soils samples

Soil type	Liquid limit (%)	Plastic limit (%)	Plasticity index (%)
Soil	67	33.33	33.67
Soil +5.0% TW	60	25	28

3.4. Plastic limit
Table 3 shows that addition of Tea Waste (TW) results in a steady decline in the plastic limit of the soils. The addition of Tea Waste (TW) decreases also the plasticity index.

4. Conclusions
Some of the vital conclusions are drawn from the investigation are:

- Maximum dry density and optimum moisture content decreases with the increase in Tea Waste quantity, there is apparent reduction in (MDD) and decrease in.
- The CBR values of all soil samples increased considerably up to the addition of 5 % Tea Waste with soil and then decreased with further addition of Tea Waste (TW). California Bearing Ratio reveals that 5 % Tea Waste (TW) stabilized Black cotton soil gives high strength to the soil
- When Tea Waste (TW) is mixed with extensive soil, liquid limit and plastic limit reduces by mixing tea waste, which also reduces plasticity index.
- Mixing Tea Waste (TW) in precise quantity (5% TW + soil) with the extensive soil will help in reducing swelling, unequal settlement and shrinkage problem.

References
[1] J K 2019 Soil Stabilization using Used Tea Leaves with Lime *International Journal for Research in Applied Science and Engineering Technology* **7** 162–6
[2] Kassa R B, Workie T, Abdela A, Fekade M, Saleh M and Dejene Y 2020 Soil Stabilization Using Waste Plastic Materials *Open Journal of Civil Engineering* **10** 55
[3] Wubshet M 2013 Bagasse ash as a sub-grade soil stabilizing material *Addis Ababa Institute of Technology*
[4] Kharade A S, Suryavanshi V V, Gujar B S and Deshmukh R R 2014 Waste product bagasse ash from sugar industry can be used as stabilizing material for expansive soils *International Journal of Research in Engineering and Technology* **3** 506-12
[5] Sunny T and Joy A 2016 Study on the effects of marine clay stabilized with banana fibre *International Journal of Scientific Engineering and Research* **4** 96-8
[6] 1980 IS: 2720 (Part 8, 1980), Methods of test for soils: Determination of water content-dry density relation using heavy compaction, Bureau of Indian Standards, New Delhi, India

[7] 1980 IS: 2720 (Part 7, 1980), Methods of test for soils: Determination of water content-dry density relation using light compaction, Bureau of Indian Standards, New Delhi, India

[8] 1987 IS: 2720 (Part XVI, 1987), Methods of test for soils: Laboratory determination of CBR, Bureau of Indian Standards, New Delhi, India

[9] James J and Pandian P K 2013 Performance study on soil stabilisation using natural materials *International Journal of Earth Sciences and Engineering* **6** 194-203

[10] Kavyashree M, Renukaprasad M and Maruti R 2016 Black Cotton Soil Stabilization Using Eggshell Powder and Lime *International Journal for Science and Advance Research in Technology (IJSART)* **2** 1-8

AUTHOR INDEX